ULTRASTRUCTURE OF FOSSIL SPORES AND POLLEN

ULTRASTRUCTURE OF FOSSIL SPORES AND POLLEN

Its bearing on relationships among fossil and living groups

Editors

M.H. Kurmann J.A. Doyle

Assistant Editor

M.S. Zavada

Production Editor

S. Dickerson

Published by

The Royal Botanic Gardens
Kew

General Editor: J.M. Lock

Typeset at the Royal Botanic Gardens, Kew, by
Pam Arnold, Christine Beard, Dominica Costello,
Margaret Newman, Pam Rosen
and Helen Ward

Cover design by Media Resources, Royal Botanic Gardens, Kew

ISBN 0 947643 60 5

Printed and bound in Great Britain by
Whitstable Litho Ltd., Whitstable, Kent.

CONTENTS

PREFACE

This volume is based on papers presented at a symposium entitled *Ultrastructure of fossil spores and pollen and its bearing on relationships among fossil and living groups* held at the 8th International Palynological Congress in Aix-en-Provence, France, in September 1992. Studies of the ultrastructure of fossil spores and pollen using transmission electron microscopy (TEM) have added a critical dimension in the search for evolutionary and phylogenetic relationships within and between fossil and living groups of plants. Because of its virtually unlimited magnification range and capability of revealing chemical differences within the exine, TEM has furnished a wide variety of characters beyond those discernible with light, or even with scanning electron microscopy (SEM). This volume attempts to show the breadth of current ultrastructural investigations on fossil spores and pollen, ranging from the earliest palynological evidence of land plants, through Palaeozoic and Mesozoic pteridophyte spores and gymnosperm pollen, to Cretaceous and Tertiary angiosperm pollen. Throughout, the authors have addressed problems in the definition and application of ultrastructural characters and their significance in generating and testing of systematic hypotheses.

Historically, there have been two approaches in the use of ultrastructure in elucidating relationships of fossil spores and pollen. One is the study of *in situ* spores and pollen extracted from fossil reproductive structures, which represents a logical application of TEM to a long-standing focus of Palaeozoic and Mesozoic palaeobotany. Important landmarks in this tradition are Pettitt's (1966) study of *in situ* Devonian and Carboniferous spores and seed plant pollen, and Millay and Taylor's (1976) synthesis of exine structural data on early seed plants that had accumulated over the subsequent decade. The other tradition is a product of ultrastructural investigations on modern spores and pollen, often with a strong developmental and sytematic emphasis, which necessitated a reassessment of many morphological concepts formulated during the initial light microscopic period in the history of palynology. Here an important breakthrough was the recognition by Van Campo (1971) and Van Campo and Lugardon (1971) of the differences among columellar, alveolar, and granular intratectal structures in seed plants, varying patterns of endexine, and their potential value in distinguishing angiosperms from other seed plants in the dispersed pollen record. These distinctions were applied to dispersed Early Cretaceous pollen by Doyle et al. (1975), Walker and Walker (1984) and other workers. Both traditions are well represented in the present volume, while the paper of Pedersen, Friis and Crane illustrates the fusion of these two traditions in their investigations of *in situ* pollen from the reproductive structure of Cretaceous angiosperms.

We should like to thank the Organising Committee of the Congress, especially Mme A. Le Thomas, J.-P. Suc and B. Lugardon, for their efforts. We are grateful to many colleagues (K.L. Alvin, W.R. Anderson, R.M. Bateman, P.R. Bell, W.G. Chaloner, C.P. Daghlian, M.E. Dettman, P.K. Endress, I.K. Ferguson, L.E. Graham, A.R. Hemsley, F.M. Hueber, W.L. Kovach, A.R.H. Martin, M.A. Millay, D.J. Nichols, J. Nowicke, S.K. Srivastava, P.K. Stroher, W.A. Taylor, A. Tryon and G.L. Webster) for kindly reviewing manuscripts, to Ms Ann McNeil for checking the references and to Professor Gren Lucas for his support with the production of this volume. Part of this work was conducted while the first editor was under the tenure of a Board of Trustees Fellowship, which is gratefully acknowledged.

References

Doyle, J.A., Van Campo, M. and Lugardon, B. (1975). Observations on exine structure of *Eucommiidites* and Lower Cretaceous angiosperm pollen. *Pollen & Spores* **17**: 429–486.

Millay, M.A. and Taylor, T.N. (1976). Evolutionary trends in fossil gymnosperm pollen. *Rev. Palaeobot. Palynol.* **21**: 65–91.

Pettitt, J.M. (1966). Exine structure in some fossil and Recent spores and pollen as revealed by light and electron microscopy. *Bull. Brit. Mus. (Nat. Hist.), Geol.* **13**: 223–257.
Van Campo, M. (1971). Précisions nouvelles sur les structures comparées des pollens de Gymnospermes et d'Angiospermes. *Compt. Rend. Hebd. Séances Acad. Sci. Sér. D* **272**: 2071–2074.
Van Campo, M. and Lugardon, B. (1973). Structure grenue infratectale de l'ectexine des pollens de quelques Gymnospermes et Angiospermes. *Pollen & Spores* **15**: 171–187.
Walker, J.W. and Walker, A.G. (1984). Ultrastructure of Lower Cretaceous angiosperm pollen and the origin and early evolution of flowering plants. *Ann. Missouri Bot. Gard.* **71**: 464–521.

Marie H. Kurmann
Royal Botanic Gardens
Kew

James A. Doyle
University of California
Davis

Dedicated to Mme M. Van Campo

In appreciation of her pioneering contributions towards understanding relationships among plant groups based on pollen morphology, and especially her key insights into the value of exine structural characters in clarifying the systematic affinities of fossil pollen.

CONTRIBUTORS

B.E. Balme
The University of Western Australia, Department of Geology, Nedlands 6009, Australia

C. Brousmiche Delcambre
Paléobotanique, URA CNRS 1365, Sciences de la Terre, Université des Sciences et Technologies de Lille, 59655 Villeneuve d'Ascq, France

P.R. Crane
Department of Geology, Field Museum of Natural History, Roosevelt Road at Lake Shore Drive, Chicago, IL 60605, U.S.A.

J.A. Doyle
Division of Biological Sciences, Section of Botany, University of California, Davis, CA 95616, U.S.A.

C.B. Foster
Australian Geological Survey Organisation, GPO Box 378, Canberra 2601, Australia

E.M. Friis
Department of Palaeobotany, Swedish Museum of Natural History, Box 50007, 104 05 Stockholm, Sweden

A.R. Hemsley
Departments of Biology and Geology, Royal Holloway, University of London, Egham, Surrey, TW20 0EX, U.K.

C.L. Hotton
Department of Paleobiology, MRC NHB 121, Smithsonian Institution, Washington, DC 20560, U.S.A.

W.L. Kovach
Palynological Research Centre, Institute of Earth Studies, University of Wales, Aberystwth, Dyfed SY23 3DB, Wales, U.K.

M.H. Kurmann
Palynology Unit, Herbarium, Royal Botanic Gardens, Kew, Richmond, Surrey, TW9 3AE, U.K.

H.A. Leffingwell
Unocal Energy Resources Division, Unocal Corporation, 376 S. Valencia Ave., Brea, CA 92621, U.S.A.

B. Lugardon
Biologie Végétale, Université Paul-Sabatier, 39 allées Jules Guesde, 31000 Toulouse, France

L.A. Milne
Department of Earth Sciences, The University of Queensland, Brisbane 4072, Australia

J.M. Osborn
Division of Science, Northeast Missouri State University, Kirksville, MO 63501, U.S.A.

K.R. Pedersen
Department of Geology, Universitetsparken, University of Aarhus, 8000 Århus C, Denmark

J.J. Skvarla
Department of Botany and Microbiology and Oklahoma Biological Survey, University of Oklahoma, Norman, OK 73109, U.S.A.

T.N. Taylor
Department of Plant Biology and Byrd Polar Research Center, The Ohio State University, Columbus, OH 43210, U.S.A.

W.A. Taylor
Department of Biology, University of Wisconsin Eau-Claire, Eau-Claire, WI 54701, U.S.A.

J.H.A. Van Konijnenburg-Van Cittert
Laboratory of Palaeobotany and Palynology, University of Utrecht, Heidelberglaan 2, 3584 CS Utrecht, The Netherlands

J.V. Ward
Department of Geology, Auburn University, Auburn, AL 36849, U.S.A.

M.S. Zavada
Department of Biology, University of Southwestern Louisiana, Lafayette, LA 70504, U.S.A.

Hemsley, A.R. (1994). Exine ultrastructure in the spores of enigmatic Devonian plants: its bearing on the interpretation of relationships and on the origin of the sporophyte. In: M.H. Kurmann and J.A. Doyle (editors). Ultrastructure of fossil spores and pollen, pp. 1–21, Royal Botanic Gardens, Kew.

1. EXINE ULTRASTRUCTURE IN THE SPORES OF ENIGMATIC DEVONIAN PLANTS: ITS BEARING ON THE INTERPRETATION OF RELATIONSHIPS AND ON THE ORIGIN OF THE SPOROPHYTE

ALAN R. HEMSLEY

Departments of Biology and Geology, Royal Holloway University of London, Egham, Surrey, TW20 0EX, U.K.

Abstract

Recent investigations of the spores of bryophyte-like plants from the Upper Silurian and Lower Devonian have demonstrated a diversity of wall ultrastructure. *Protosalvinia* Dawson and *Parka* Fleming (both probably of algal affinity) exhibit a spore wall composed of laminae, and in this respect, resemble fossil bryophytes (such as *Naiadita* Buckman) and extant members of this group.

The absence of triradiate markings on the spores of *Parka* and other early land plants leads to uncertainty regarding their status as meiotic products. It is possible that these spores may be zygospores like those formed by charophycean algae. Production of numerous coated zygospores could be an important stage in the transition from sporopollenin-coated diploid propagules to coated meiotic products. Fossil permanent tetrads (viewed by some authors as having a bryophytic origin) and the extant thalloid alga *Coleochaete* contribute to a model incorporating this transition by providing evidence for the simple derivation of the multicellular sporophyte generation from a primitive Charophyceae zygospore.

Introduction

At some time in the late Ordovician (around 460 to 475 million years ago, see Gray and Shear, 1992), plants of greater than microscopic size began to colonize the land. Before this momentous event, the vegetational covering of the exposed substrates presumably consisted of microalgae of various types, fungi, and bacterial sheets. All of these would have been largely restricted to littoral zones, river margins and other permanently or frequently moist habitats. Some forms may have been able to survive extended periods of desiccation but the predominantly zooidogamous algae would have required abundant moisture (unless bound in symbiotic association such as lichens).

It is clear that the algal group ancestral to the most successful land plants was the Chlorophyta (green algae). Other groups may have gained a brief 'foothold' on land (e.g. *Protosalvinia* and *Prototaxites*, enigmatic Devonian plants that have both been considered as Phaeophyceae, brown algae), but these are no longer represented as large, multicellular land plants. Many authors have advocated a migration of multicellular aquatic plants directly onto land (e.g. Stewart, 1983), whilst others have argued that larger algal types arose from filamentous algae already present on land (Stebbins and Hill, 1980). Whatever the case, large plants (up to 5 mm in height!), consisting of upright branching axes with sufficient substance to become fossilized, were present by the late Silurian to early Devonian (see Edwards and Fanning, 1985; Fanning et al., 1992, Figs. 1C and 13).

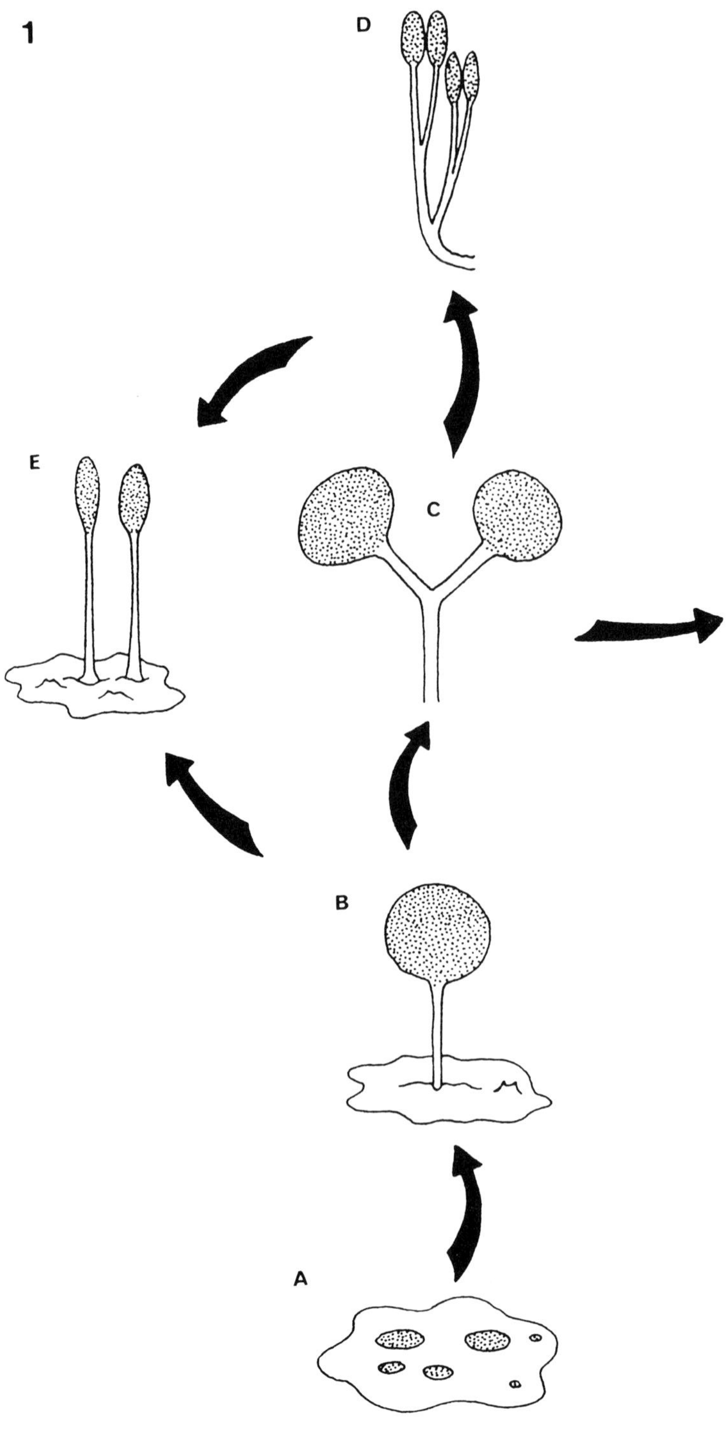
1
D
E
C
F
B
A

Macroalgae, exploiting the drier conditions on land, required a number of features to ensure their survival. Amongst these were: a water-proof covering (to maintain aquatic conditions within the plant body), a method of internal conduction (to transfer water from the ground to the aerial portion (more susceptible to desiccation), and propagules adapted for dispersal out of water. All three of these features leave evidence of their existence in the form of fossil cuticles, tubes (tracheids), and spores. These are known in the fossil record from somewhat before the definitive evidence of the whole organs of probable land plants (e.g. Richardson and Edwards, 1989), a factor which may be viewed as supporting the migratory origin of terrestrial plants.

Of these three components of land plant construction, cuticle and spores (in the form of permanent tetrads) are the first to appear (Gray et al., 1982; Johnson, 1985; Chaloner, 1988; Gensel et al., 1990; Gray, 1991). It would, therefore, seem that the earliest land plants developed protection against desiccation before the development of aerial parts (i.e. they were ground-hugging – thalloid, Fig. 1A). This comes as no great surprise but is important since some of our first large plant fragments are obviously aerial axes with tubular plumbing (Edwards et al., 1992). Some aquatic algae may have demonstrated a 'pre-adaptation' in the sense of having evolved features of use, but perhaps not crucial to their existence, before emergence as fully fledged land plants (advocated by Chapman, 1985).

Because of the remarkable preservation of spore wall material, it is possible, even in these very ancient plants, to observe ultrastructural details. As the number of examples studied increases, the ultrastructure they contain may eventually provide clues regarding the ploidy (haploid or diploid) of their parent plant and is clearly useful in indicating possible relationships with other fossil or extant plant groups. This short review covers the limited ultrastructural work carried out so far on enigmatic, thalloid plant spores and looks at the relationships, particularly with algal and bryophytic groups, that this suggests.

The Palaeobotanical Significance of *Parka* and *Protosalvinia*

A. Thalloid enigmas

Ultrastructural investigation of the spores of *Protosalvinia arnoldii* (Taylor and Taylor, 1987) and *Parka decipiens* (Hemsley, 1989a) has increased our knowledge of these enigmatic plants and fueled further speculation regarding their status as land plants. These two plants of respectively Upper Devonian and Upper Silurian/Lower Devonian age, present morphologies which are difficult to place in relation to other groups of plants. In the cases of *Parka* and *Protosalvinia* the morphological traits seen in the fossils suggest that these plants may have been at least partially adapted to existence out of water, if only periodically, as is the case for many littoral marine algae. That both of these plants produced resistant spores (see Fig. 6 and Taylor and Taylor, 1987) is generally taken as an indication of an adaptation to desiccation.

Protosalvinia arnoldii consisted of dichotomously branching thalli which sometimes bore terminal depressions (conceptacles?) in which tetrads of large, resistant spores were produced (see Fig. 2). As well as the production of spores, *Protosalvinia* shows a distinct

Fig. 1. A simplified illustration of the presumed evolution of early land plants and the possible origins of the major groups (sporangia are stippled). **A**. A simple thalloid plant (possibly similar to *Parka*, described herein) bearing sessile sporangia within or upon the thallus. **B**. The evolution of vascular tissue permits the development of aerial axes, hence better dispersal of spores. **C**. Diversification of simple, erect plants occurs (e.g. *Cooksonia*, see also Fig. 13) which eventually leads to **D**. Relatively large, dichotomously branched plants with a variety of sporangial types. **E**. The origin of the bryophytes remains unclear but may have occurred at a number of stages during the development of the first vascular plants. **F**. Lycopods seem to have diverged quite rapidly from the plants consisting of naked, dichotomous axes and, like the bryophytes, their origin is unclear. Diagram after Mishler and Churchill (1985) and Hemsley, (1992b).

2

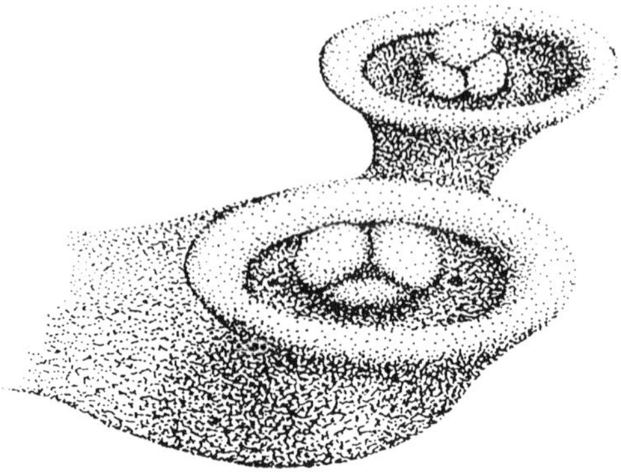

3

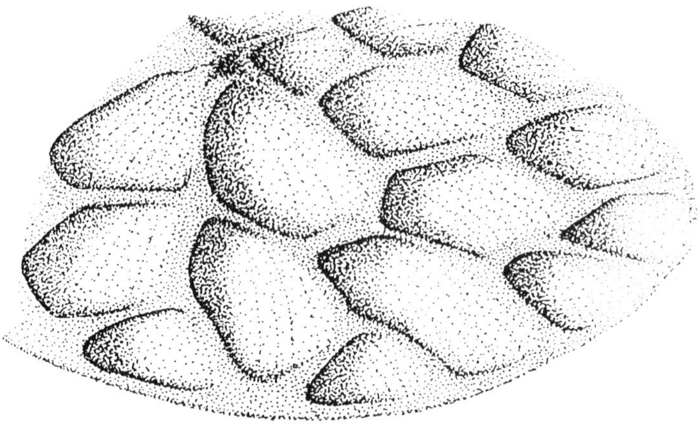

Figs. 2–3. Diagrammatic reconstructions. **2**: *Protosalvinia arnoldii*. Spore tetrads were borne in conceptacle like structures which were produced by dichotomous thalli. Each conceptacle is around 1.5 mm in diameter. **3**: *Parka decipiens*. Spores were in turn borne in relatively large sporangia which were borne either within, or on the surface of the thallus. The sporangia are around 1 mm in diameter.

Figs. 4–8. *Parka decipiens*. Details of the source of the material illustrated are given in Hemsley (1989a). **4**: *Parka* is often found as an impression fossil. A portion of the thallus shows the arrangement of the sporangia and their size. **5**: Some specimens of *Parka* retain sporangial contents. The spore mass is often covered with a thin, coaly layer. **6**: SEM reveals the compressed spores forming the spore mass and their unornamented surface. No haptotypic markings are present. **7**: In TEM section, the ultrastructure of the compressed spore can be seen clearly. This consists of two distinct regions; an outer homogeneous region and an inner, laminated region. In this micrograph, the closed central lumen is bordered by a dark band. **8**: A detailed TEM micrograph of the spore wall showing the two regions illustrated in Fig. 7. The laminations present in the inner region are of variable thickness and appear to interlink. The boundary of the laminate region and the outer homogeneous region is marked by the upper arrow. The lower arrow indicates the darkened lumen. Compare with Fig. 9.

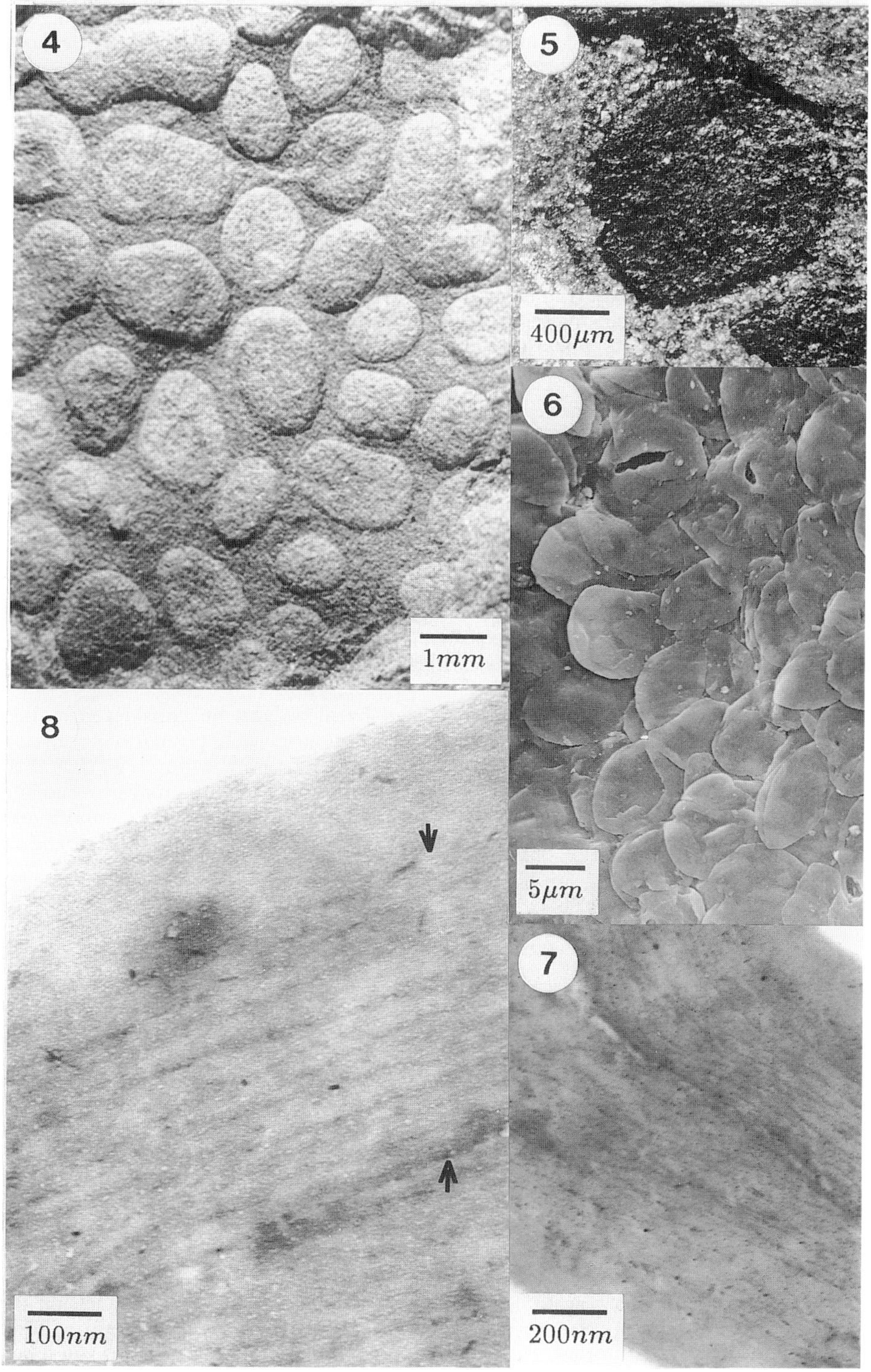
4
1mm
5
400μm
6
5μm
8
100nm
7
200nm

cuticle, another adaptation to a terrestrial existence. *Protosalvinia* appears rather late in the fossil record to be regarded as an ancestor to land plants (vascular or bryophytic) but, like the extant bryophyte *Anthoceros*, it may provide an important analogy for primitive land plants. There is some evidence that *Protosalvinia* may have been a brown alga (Phaeophyceae) (Niklas and Chaloner, 1976), perhaps indicating that algal groups other than the Chlorophyceae may have, at intervals over geological time, attempted to advance onto land.

Parka decipiens was also a thalloid plant (Figs. 3–4) but in this case the thallus was generally more or less circular in outline. It consisted of radiating filaments which bore within the thallus, in a regular manner, discs of spores (spore masses, Fig. 5). The spores are simple, unornamented, alete structures (see Don and Hickling, 1917; Niklas, 1976a). *Parka* shows less of a cuticle, but is important in apparently having been a green alga (Niklas, 1976b), and in showing features akin to those we might expect to find in a land plant ancestor, *Parka* itself appearing rather late to be considered as such.

B. The importance of the spores

The spores of *Parka* and *Protosalvinia* are of great value in terms of the information that they provide; their morphology, mode of production, and distribution on the plant surface. The majority of land plant spores (including pollen in the more recently evolved groups) are coated to some degree with sporopollenin. This substance is highly resistant to degradation under anoxic, acidic conditions, circumstances which often occur during sedimentation, be it clastic or organic. Spores may therefore be extracted from such sedimentary rocks and appear little changed in gross morphology from their time of inclusion within the rock matrix. Often correlated with this is the retention of ultrastructural details (which is abundantly illustrated in this volume). Even spores of considerable age may retain a laminate exine structure as well as indications of differing composition.

The extent to which the material produced by algae can be considered to be sporopollenin is in doubt (see below) and such doubts must also apply to the spore wall material of these transitional land plants. Furthermore, little information regarding the composition of the exine material of bryophytes is available. This too, may prove to be significantly different from that of vascular plants. Despite this, whatever the composition of the material involved, in many cases ultrastructural details are retained and can be a major taxonomic aid.

Spore Ultrastructure in Putative Land Plants

A. *Protosalvinia arnoldii*

Protosalvinia has been the subject of a number of studies (Phillips et al., 1972; Niklas and Phillips, 1976) which have illustrated its morphology and discussed its peculiarities. Taylor and Taylor (1987) undertook a detailed SEM and TEM investigation of the spores with a view to determining the similarities of the spore ultrastructure with that of other algal groups.

The wall of the spores of *Protosalvinia* consists of two distinct regions. The inner region is consistent in structure around the entire spore, whilst the outer region shows differences in structure either where the spore walls are in contact with each other, or where it is exposed on the outside of the tetrad (see Fig. 9).

The inner region occupies the bulk of the exine, around 60 to 85% of the total thickness. Its structure is difficult to discern, but would seem to be amorphous to globular in construction. The outer region consists of laminae where spores are in contact and of globular units elsewhere. A suture is present, around which the laminae forming the outer region become contorted. Further details of exine construction are provided by Taylor and Taylor (1987).

9

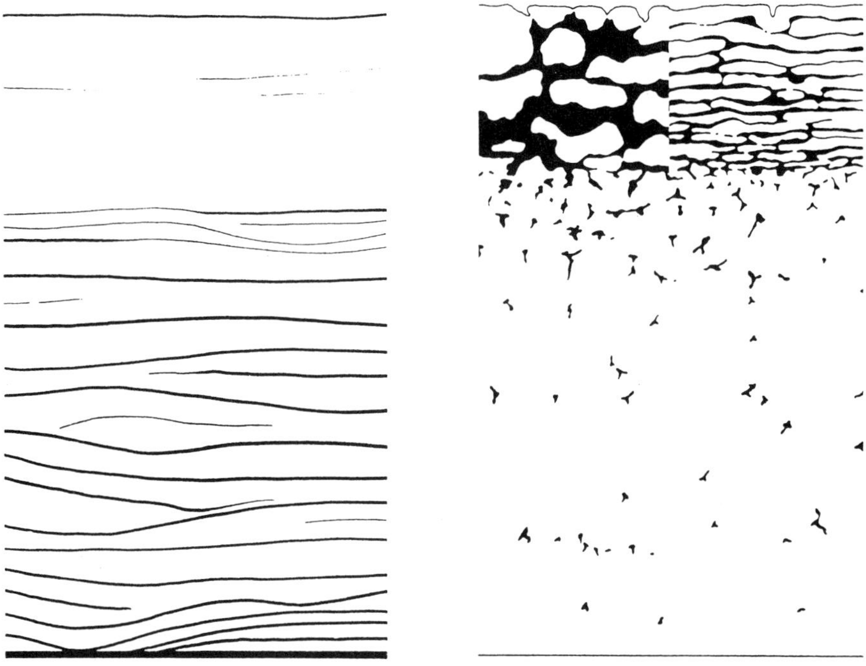

Fig. 9. Diagrammatic interpretations of the spore walls of *Parka* (left) and *Protosalvinia* (right). Both spores have a two-regioned wall although the structure is rather different. Wall ultrastructure differs in *Protosalvinia* between the distal part of the spore (to the left) and that in contact with other spores of the tetrad (to the right). Diagrams after Hemsley (1989a) and Taylor and Taylor (1987). Spore wall thickness are: *Parka*, 0.7 μm; *Protosalvinia*, 4–10 μm.

B. *Parka decipiens*

Like *Protosalvinia, Parka* has been the subject of a number of studies which have served to emphasize the unique nature of this plant. Hemsley (1989a) investigated the spores by SEM and TEM with the following results.

The spores of *Parka* are alete and devoid of external markings beyond those left by the irregular contact of neighbouring spores within the spore mass (Fig. 6). The exine is 0.7 μm in thickness and, like that of *Protosalvinia,* consists of two distinct regions (Figs. 7–9). The outer region (around 30% of total exine thickness) is homogeneous to finely granular in composition. The inner region consists of numerous laminae which interlink. The laminae are irregular in thickness. An inner dark line generally marks the position of the lumen in these compressed spores. Further details are given by Hemsley (1989a).

Comparisons with Fossil and Living Analogues

A. Algal spores

It is important to point out at an early stage that the production of trilete spores in tetrads is not confined to land plants. Many algae, particularly red algae such as *Polysiphonia,* produce spore tetrads. However, amongst the green algae viewed as potential land invaders (Charales, Coleochaetales, Zygnematales), alete zygospores are produced which give rise to motile zoospores.

Atkinson et al. (1972) produced a survey of wall ultrastructure in simple Chlorophyceae (*Scenedesmus*, *Chlorella*) demonstrating the widespread occurrence of the white-line laminae, an apparently basic trilaminate unit forming an important part of the vegetative cell walls. The distribution of this structure within various plant groups is discussed by Blackmore and Barnes (1987). Amongst the potential land invaders mentioned above, ultrastructural investigations of spores and other cell walls have been undertaken by Graham and Taylor (1986) (*Coleochaete*), Buer (1964) (Zygnemataceae), Dyck (1970) and Leitch (1989) (Characeae).

The acetolysis resistant zygospore wall of *Coleochaete* is thick and lamellated (Graham, 1990) and lies beneath a more homogeneous non-resistant region. The constituent laminations are numerous and in some respects, they are similar to those seen in bryophytic spores and in *Parka*. The ultrastructure of vegetative cells in the Zygnemataceae seems to be somewhat more complex consisting of a fibrillar matrix

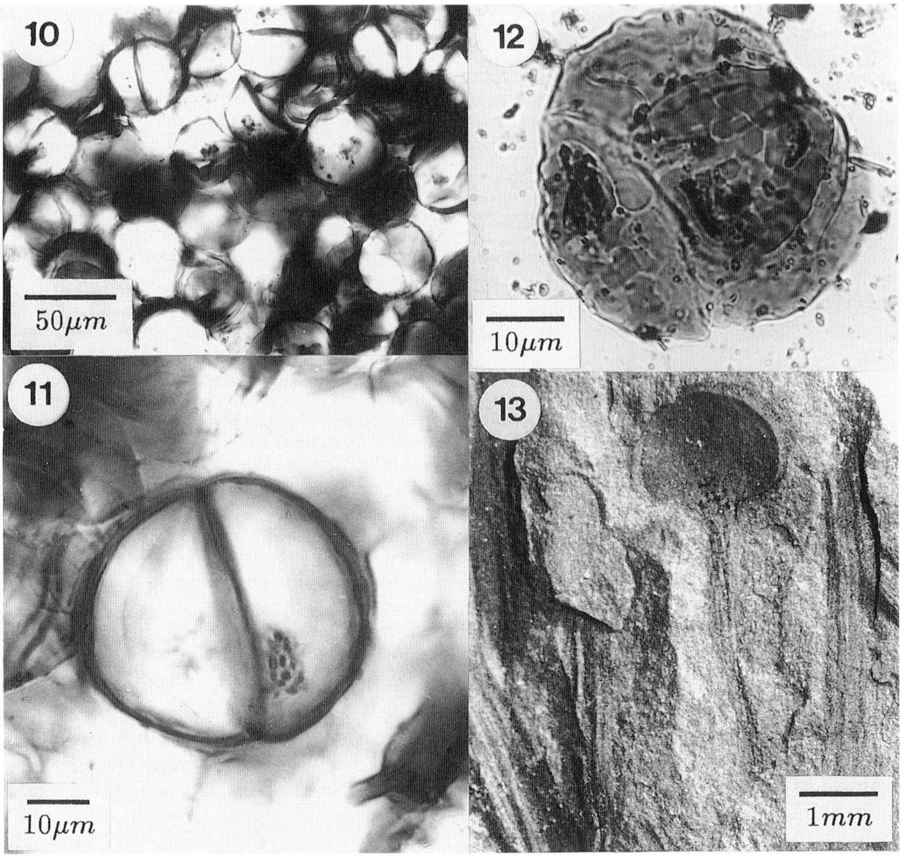

Figs. 10–13. 10: Part of the elongate sporangium of unknown affinity containing numerous dyads illustrated by Lang (1937). **11**: Detail of a single dyad. **12**: *Ambitisporites* as a tetrad. Spores like these are known from some of the earliest land plant sporangia. Mid-Silurian from North Africa (see Chaloner, 1970). **13**: *Cooksonia caladonica* (Edwards, 1970) from Redford, Angus, Scotland (Lower Devonian). The rounded sporangia are apparent.

above a more granular region (Buer, 1964). The ultrastructure of these algal walls is, in some ways, similar to the spore wall structure of *Protosalvinia*, however, they are not acetolysis resistant.

The zygospore wall of the Characeae is a complicated composition of organic and calciferous layers. Laminae are present within the inner layers of the wall, but these differ considerably from the trilaminate structures in possessing a spiralled fine structure (see Leitch, 1989) and being considerably greater in dimensions. Despite their multilaminate walls, the spores of the Characeae cannot be considered similar to bryophytic spores due to the minimal content of sporopollenin (Shaw, 1971).

Taylor and Taylor (1987), to afford a comparison with the spore wall of *Protosalvinia*, examined the mesochite wall ultrastructure of a number of brown algae (Phaeophyceae). In general these were found to have a finely fibrillar to granular ultrastructure, a wall type different from that exhibited by any of the plants discussed so far. *Polysiphonia* (Rhodophyceae) tetraspore walls did not survive acetolysis (see Erdtman, 1960) and when this was not used, spores were found not to take up TEM stains. Certainly none of these spore types are similar to those of either *Protosalvinia* or *Parka*, nor do they resemble those of extant bryophytes.

If it is accepted that multiplication of the trilaminate exine unit around diploid cells occurred at an early stage in the evolution of land plant spores (Blackmore and Barnes, 1987) then we might hope to find similar features in the spores of potential land plant ancestors. Unfortunately, there is no conclusive evidence from any of the wall structures observed so far.

B. Bryophyte spores, extant and extinct

The enigmatic plant *Sporogonites* (somewhat like the hypothetical example in Fig. 1E) demonstrates many of the features that might be expected in a primitive bryophyte, stalked sporangia, a capsule-like sporangium and presumably a basal thallus. However, recent reinvestigation (pers. comm. Shute, 1991) casts some doubts on its possible relationship with extant bryophytic groups.

Although a thalloid mode of growth would seem likely as the primitive type amongst bryophytes, especially in view of the ancestral models described above, it has also been suggested (Schuster, 1979) that the more primitive type may have been upright with leaf-like enations. Much of the fossil evidence, however, suggests that early liverworts were thalloid (Stewart, 1983).

Unfortunately, very few of the Devonian and Carboniferous plants interpreted as bryophytes bear sporangia with well preserved spores. One of the earliest plants fulfilling these criteria is *Naiadita lanceolata* from the early Triassic.

Naiadita has been variously described as a moss and a lycopod. It is now recognised as a liverwort (Hepaticae) with a close affinity to the modern Sphaerocarpales (Schuster, 1979; Hemsley, 1989b). Although its affinity was largely surmised on the basis of morphological characters, the ultrastructure of the spores provided important confirmation of affinity and indicated similarities with the modern sphaerocaralean genus *Geothallus* (exine illustrated in Fig. 18).

The exine of *Naiadita* (Figs. 14–18) consists of two main regions which frequently separate. The outer region is more or less granular and forms the surface ornament. The inner region can be further subdivided into four laminate layers; a thin outer layer of one or two thick laminae, two more or less equal layers of thin laminae, and an inner thick lamina. The thin laminae and innermost lamina show fine white lines at the center of each unit.

Naiadita seems to have been a fresh or brackish water aquatic, somewhat like the extant liverwort *Riella* (exine illustrated in Fig. 18), and may be showing adaptation to this particular environment. However, it is interesting to note that it is an upright, leafy, member of what is now a thalloid group of liverworts.

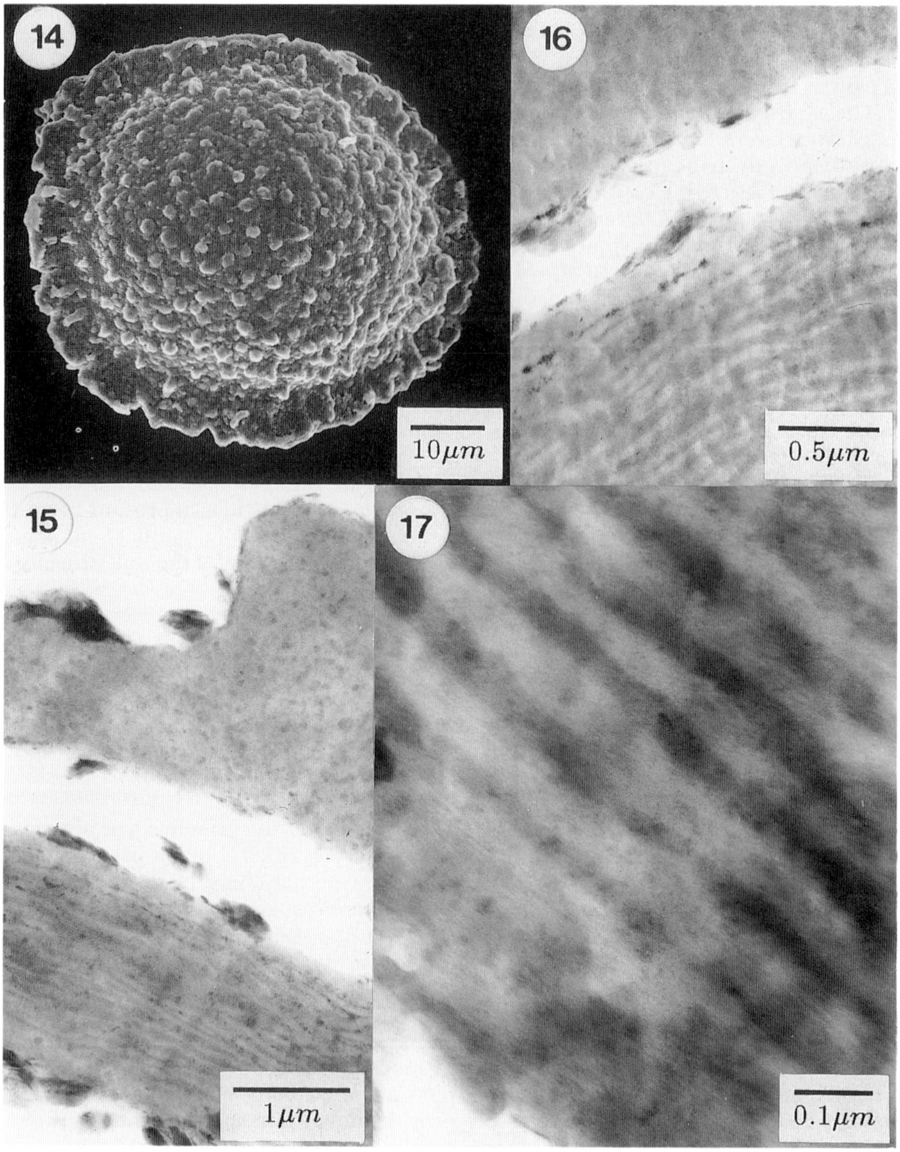

FIGS. **14–17**. *Naiadita lanceolata*. Details of the source of the material illustrated here are given in Hemsley (1989b). **14**: The distal surface and equatorial extension of *Naiaditaspora*. **15**: A TEM section through both of the main regions of the exine. The outer, granular to homogeneous region (top) bears the surface ornament. The inner laminated region (below) contains a number of different layers (see Fig. 18). **16**: Detail of the junction between the two main exine regions. The single, thick lamina adherent to the outside of the laminate region is apparent. **17**: Detail of the inner laminae of the inner region. Fine white lines can clearly be seen at the center of some of the laminae.

Until such time as more ultrastructural work can be undertaken on older material, our understanding of the origins of the multilaminate bryophytic spore wall remains obscure. In common with the development of the sporophyte, the spores of bryophytes seem to

18

FIG. 18. Bryophyte exines. The exine of *Naiaditaspora* is shown (top left) with three comparable extant liverwort exines. All are clearly laminate, at least in part. In both *Riccia* (bottom left) and *Riella* (top right) the entire exine, including the ornament, is laminate. Only in the exine of *Geothallus* (bottom right) is there a close correspondence of features. The similarity of the laminate regions is particularly striking. Diagram from Hemsley (1989b) based on Steinkamp (1973) and Denizot (1971). Exine thickness are: *Naiadita*, 6 μm; *Riella*, 10–20 μm; *Riccia*, 7–15 μm; *Geothallus*, up to 20 μm.

reflect a relatively primitive condition and may, by analogy, be a useful guide to the ultrastructure of primitive land plant spores.

C. Primitive spores from vascular plants

The early vascular land plants such as *Cooksonia* (Fig. 13) produced a variety of spore types (Fanning et al., 1988, 1991). Currently, relatively little is known regarding their ultrastructure but, in common with the spores of *Parka* and *Protosalvinia*, they seem to have had an exine that exhibited two distinct regions (Shute and Edwards, 1989). The significance of the widespread two-regioned exine remains unclear but may well be associated with the dual contributions of gametophytic and sporophytic components to the exine.

The ultrastructure of the spores of the upper Devonian *Psilophyton* is probably too recent to give any important clues regarding the ultrastructure of its predecessors, but it is interesting to note that the exine is not clearly of two regions (Gensel and White, 1983); diagrammatically illustrated in Fig. 19, but consists of separating exine and perispore. The outermost part of the exine of *Psilophyton* exhibits laminations which appear similar to those seen in *Parka* and *Naiadita*. A considerable amount of partially structured tapetal material is retained around the spores and may be indicative of the impending further development of sporophytically derived exine regions. A simple extrapolation of this hypothesis could result in a situation comparable with the exine organization seen in the micro- and megaspores of the upper Devonian progymnosperm *Archaeopteris* (Pettitt, 1966). In these spores, a granular to globular exine region surrounds the inner laminate region.

Speculation on the Evolution of Land Plant Spores

A. Palynology and Morphology

A number of problems occur when equating the morphological concept of the origin of land plants (isomorphy of generations, evidence for which comes from the putative gametophytes described by Remy and Remy, 1980; Remy, 1982) with the palynological evidence. These essentially condense into two main areas of discrepancy.

Among the palynomorphs that are perhaps difficult to include within current concepts are the coated monads (e.g. *Parka*, Fig. 6) and dyads (Figs. 10–11) produced in the sporangia of the early land plants (Lang, 1937; Shute and Edwards, 1989; Fanning et al., 1991). These spore types also predate the earliest known occurrences of land plant macrofossils and occur at the same time as obligate and membrane bound tetrads of *Ambitisporites*-like meiospores (Fig. 12). Both of the former spore types are ambiguously meiotic with a lack of clarity regarding the ploidy (sporophyte or gametophyte) of the spore-producers. Some of the possibilities presented by these enigmatic palynomorphs are reviewed by Johnson (1985), Hemsley (1990), Fanning et al. (1991), Gray (1991) and Strother (1991).

The second problem concerns the apparent transfer of the resistant coating material (sporopollenin?) from the sporophytic zygospore (in charophycean algae) to the meiospores, a transition which was outlined by Blackmore and Barnes (1987). This latter transition is closely linked to the question of the ploidy of the parent plant.

B. Multiplication of the zygospore

Marine or freshwater algae with the potential to exist out of water (either periodically or permanently) may well have reproduced in a similar manner to modern charophycean algae. This involves the production, by a gametophyte, of an oospore which, on fertilization, becomes a diploid zygospore. This structure (the zygote) forms the resting stage. It later divides meiotically to form swimming zoospores (Fig. 20, A–B) which are

19

Fig. 19. *Psilophyton*. The exine (bottom) and tapetal debris (top) of spores of *Psilophyton* based on illustrations by Gensel and White (1983). There is an indication of lamination in the exine toward the outside. These laminations look much like those found in the inner exine of *Parka*. How much this reflects a similar degree of diagenesis is unclear at present. Exine thickness (including debris), 1–1.5 μm.

eventually released. This is the basis from which a multicellular sporophyte (producing numerous haploid spores) is here presumed to have been derived.

The multiplication of the sporophytic stage (zygospore, or spore mother cell?) within a charophycean alga therefore seems an essential step. This may result from mitotic multiplication of the oospore or (after fertilization) the zygospore. The latter seems most probable as this would not be reliant on multiple fertilizations. Such a situation would result in the production of a number, perhaps a large number, of diploid spores (Fig. 20C). Depending on the degree of separation of the resulting meiotic products at the time of deposition of the coating material, the result could be polyads, tetrads, dyads or monads (alete spores). *Parka* may well be exhibiting this grade of land plant evolution,

20

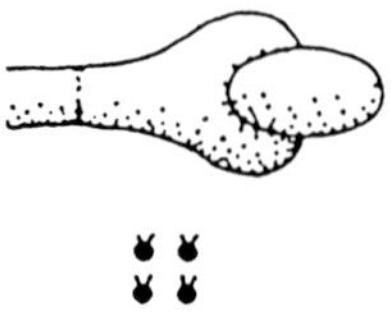

A. Simple algal zygospore gives rise to 4 motile zoospores.

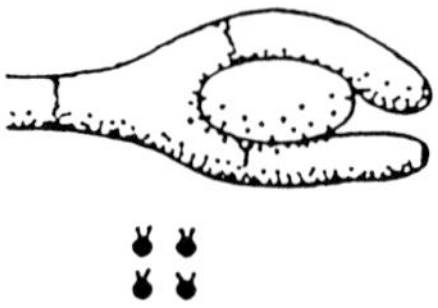

B. Filaments enclosing walled zygospore e.g. *Coleochaete*.

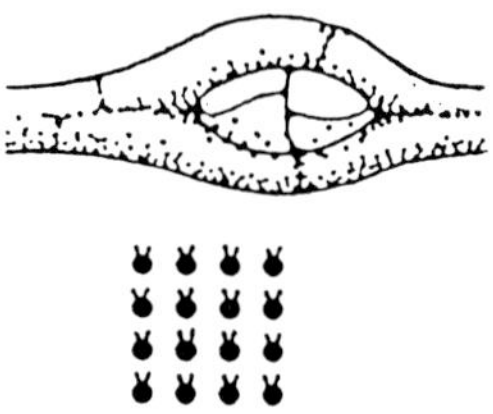

C. Multiple "zygospores". Mitosis of initial zygospore gives rise to numerous walled zygospores (in this example 4). *Parka ?*

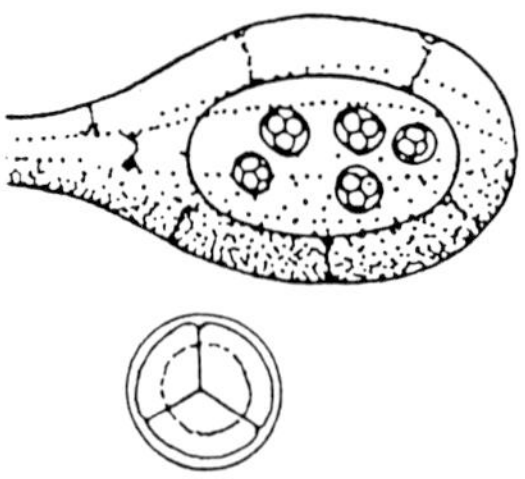

D. Wall material deposited on both zygospores and meiotic products. Membrane bound tetrads.

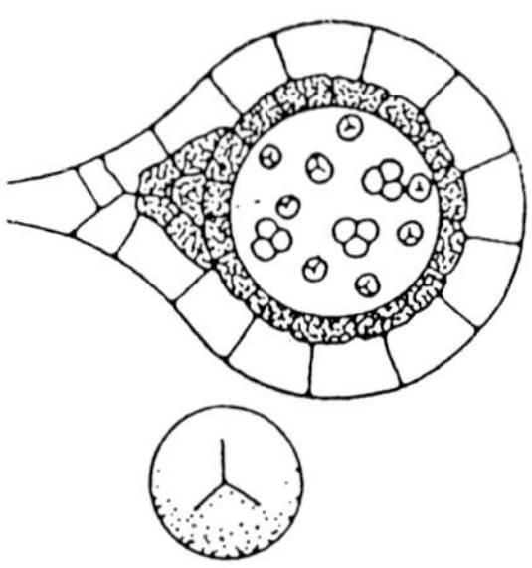

E. Outer zygospores attain sterility (sporophyte). Fertile zygospores no longer with a wall. Meiotic products with wall ... trilete spores.

producing multiple, alete, diploid spores. Supporting this brief is evidence from extant *Coleochaete* which, although not producing multiple zygospores, does produce numerous zoospores (16, 32 etc) that result from multiplication of the DNA content of the zygospore (Hopkins and McBride, 1976). This is indicative that *Coleochaete* once had, or has the potential to produce numerous mitotic zygospores.

C. Membrane bound tetrads

The transfer of coating material from zygospore to meiotic products suggests the loss of water as a medium for the dispersal of swimming zoospores. There are two aspects involved: the protection of the now vulnerable gametophytes and, perhaps of greater importance, their dissemination without water currents/tides. It would seem that the first stage in this transition involved the production of a coating on the meiotic products in addition to the coating around the spore mother cell (zygospore) (Figs. 20D, 21). This would give rise to the membrane bound tetrads (see Gray, 1985) known from the early Silurian and, in a decreasing proportion, to the early Devonian. The coating of the meiotic contents of the zygospore most probably required the production of sporopollenin by the gametophytic generation, thus it might be expected that the 'exine' of these enclosed spores might resemble the inner layers of a typical land plant spore exine (e.g. that of *Psilophyton*) whilst the zygospore wall may be more akin to the wall layers produced by the tapetum, generally the outer layers of the exine. Johnson (1985) explores similar ideas.

The impetus behind this shift of wall material deposition is therefore one of protection of the gametophytic spores and a move toward their individual dissemination. However, the consequences of this shift result in obligate and membrane bound tetrads and in the interaction of both the sporophytic and gametophytic generations in the production of the land plant spore exine.

In view of the known ultrastructure of the charophycean zygospore wall and that of the earliest land plant spores, the ultrastructure of these membrane bound tetrads would provide considerable information regarding their actual status.

The improved genetic variability obtained by the separate dissemination of the individual members of the tetrad would lead to the rapid establishment of forms without a surrounding membrane, although, as a pioneering strategy, membrane bound and obligate tetrads may well have retained some advantage.

D. A simple sporophyte

In most extant land plants, both the gametophytic stage (the spores) and the sporophytic stage (the tapetum) produce sporopollenin. Generally though, the tapetum is responsible for the bulk of sporopollenin production. It might seem reasonable to suggest therefore, that in order to produce substantial exines, early land plant spores required a tapetum as an additional source of sporopollenin. A 'division of labour' amongst potential spore mother cells could result in some cells that would undergo meiosis to produce spores, some cells that would not divide but would produce sporopollenin and finally, some cells which acted only in a protective role. If arranged concentrically from the inside, the result would be akin to a sporangium, the sterile protective and 'tapetal' cells forming a sporophyte. Such a sporophyte (Fig. 20E) would quite possibly remain largely enclosed within the gametophytic plant and would be similar to the sporophytes produced by the simple liverworts such as *Riccia*. Simple sporangia such as this, may have exhibited a gradual transition between the particular

FIG. **20**. A simple derivation of the sporangium (a minimal sporophyte) from a charophycean alga as described in the text. Stages **A** to **D** may be considered as various stages of algal specialization. Dark stippled cells in **E** represent the minimal sporophyte and a bryophytic condition.

21

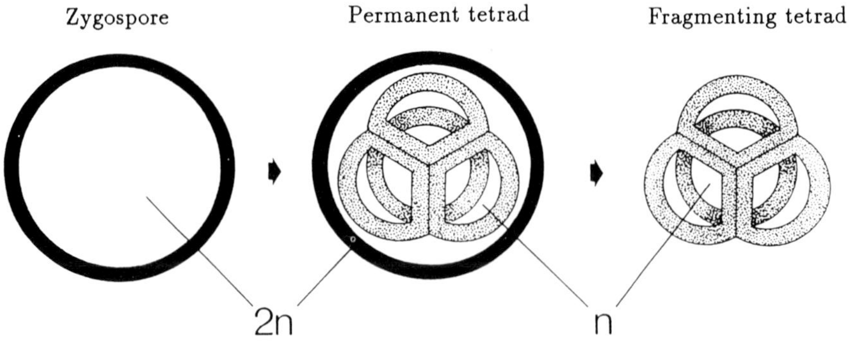

Fig. 21. A simplistic diagram illustrating the origin (in terms of ploidy) of the components of the wall of three stages in the development of the tetrahedral tetrad.

regions rather than the sharp distinctions seen in extant land plant sporangia.

This sequence of hypothetical events, although highly speculative, does at least fit much of the available evidence, both from the fossil material known and from the characteristics of living analogues. At all times it should be borne in mind that early land plants must have had a reasonably flexible physiology in order to make the transition from water to land. It seems likely that the developmental processes (which are complex and varied even within green algae) may also have exhibited considerable flexibility. None of the sequence envisaged here discounts the significance of the gametophytic plants described by Remy (1982). The flexibility possessed by the early land plants may well have been retained in plants of the Lower Devonian causing a 'blurring' of the distinction between bryophytes and pteridophytes.

Conclusions

In this brief review of fossil enigmatic and primitive plant spore ultrastructure, two general points have arisen. A partly laminate exine ultrastructure is exhibited by all of the spore types described above and the two-regioned exine also seems to predominate. It is interesting that although *Protosalvinia* seems to have been derived from a different group of plants, it too exhibits a two-regioned exine. Both a laminate ultrastructure and a fundamentally two-regioned construction seem to be consistent within all of these emergent plants.

The previous interpretation of Silurian and Devonian membrane-bound and permanent tetrads as bryophytic (Gray, 1985) in origin is unlikely. Bryophytes, at least as we see them today, have evolved considerably from their inception. It is not clear whether the 'bryophytic' features attributed to the fossil spore tetrads have not been a more recent aquisition by the extant bryophytes. It seems more probable that these structures were produced by plants which were transitional between the zygospore-bearing algae and the simplest land plant sporophytes (independent or sessile upon a dominant gametophyte), these containing a variety of spore types. Such plants may have had many, if not all, of the essential characteristics of modern bryophytes but to call them such implies far too much about possible relationships. As explained by Hemsley (1990) and Fanning et al. (1991), the variety of fossil spore types leads to a corresponding variety of possible developmental scenarios and inferences of ploidy, most of which will be difficult to resolve using material from the fossil record.

22

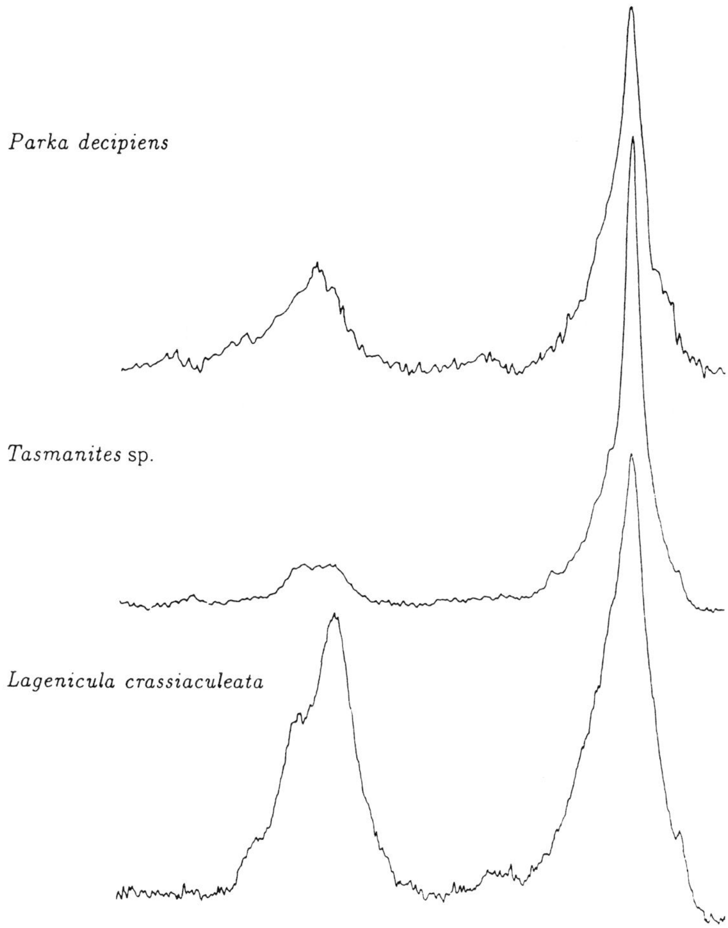

FIG. 22. [13]Carbon solid state Nuclear Magnetic Resonance spectra of fossil spore wall materials. This technique is outlined by Hemsley et al. (1992). The three spectra shown here are presented for comparison. *Parka* spores (top) have a composition (spectrum) transitional between the fossil algal cysts of *Tasmanites* (middle) and the sporopollenin wall of the lycopod spore *Lagenicula* (bottom).

To some degree, comparative spore ultrastructure is becoming a less stable basis for inferring affinity for enigmatic fossil plants. In the past, spore wall material (sporopollenin) has been assumed to be moderately consistent in composition in a wide range of plant groups, including algae (see Brooks, 1971; Delwiche et al., 1989). This now seems to be less certain and if, as seems to be the case, algal wall material is diverse and

23

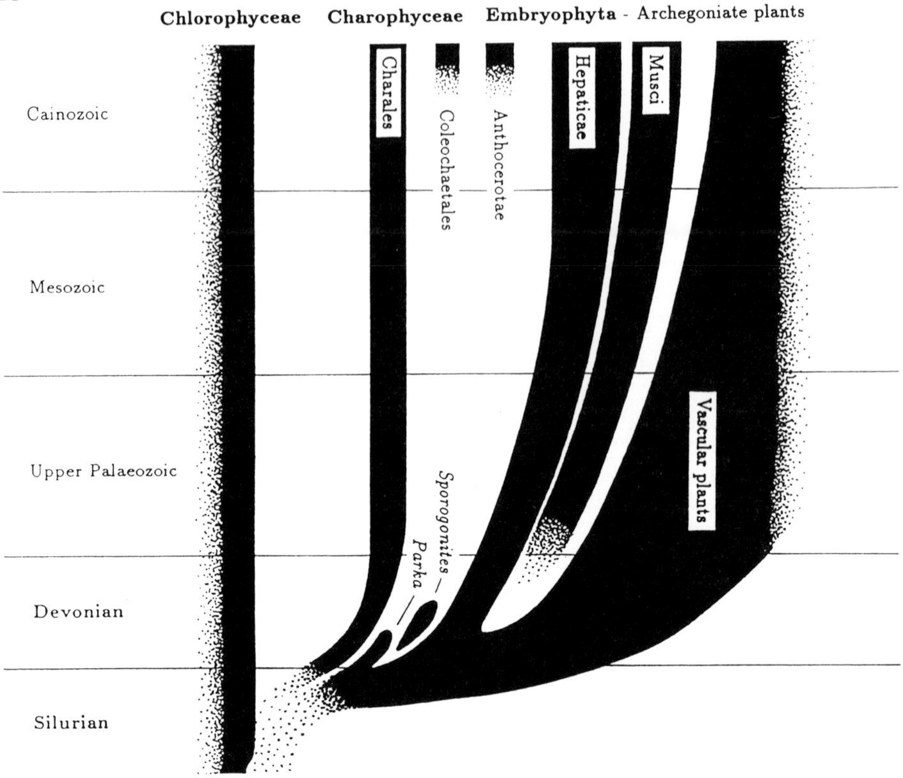

FIG. 23. A phylogeny of the main groups discussed here. Plants such as *Parka* and *Sporogonites* may have occupied a transitional position between algae and archegoniate plants. At the present time, *Coleochaete* and *Anthoceros* might be considered to occupy similar positions. Based on Grambast (1964), Schuster (1979), Stewart (1983), and Lee (1989).

different in basic composition to spore wall material in vascular plants (Zelibor et al., 1988; de Leeuw et al., 1991; Hemsley et al., in press), is a possible assumption of homology of laminate structures valid? It is interesting that the wall material of *Parka* is intermediate in composition between that of a fossil algal cyst (*Tasmanites*, see Wall, 1962) and that of a fossil lycopod spore (*Lagenicula*, see Hemsley, 1992a) (see Fig. 22). The essential differences between the alga and the spore are that the algal wall lacks a major aromatic component (and is thus like a fossil cuticle in composition) whereas the sporopollenin of the spore wall has approximately equal amounts of aromatic and long chain carbon compounds.

Despite the above concerns, spore ultrastructure provides key information from plants with a very simple morphology, lacking in characteristics and often poorly preserved. This review has outlined the little that is known about the ultrastructure of early land plant spores, and has shown how this may be used to raise questions of affinity. In addition, however, spore ultrastructure in these early plants is interlinked with the broader question of the origin of the sporopollenin-coated structures so prevalent in the fossil record.

Gradually a concept of the phylogeny of land plants is unfolding (Fig. 23), but there remains, however, a considerable gap between algae and the relative complexity of a simple bryophyte or vascular plant. Despite this, both at a time contemporaneous with the initial crossing of the gap and, apparently, at subsequent intervals, a number of plants seem to have occupied an intermediate position (*Parka, Sporogonites, Anthoceros*?). These plants indicate perhaps, that although the gap is wide, it may be less difficult to cross than is generally believed.

Acknowledgements

I thank Cedric Shute of the Natural History Museum, London, for permission to include Figs. 10 and 11, and Margaret Collinson (RHBNC) for helpful comments on the manuscript.

References

Atkinson, A.W., Gunning, B.E.S. and John, P.C.L. (1972). Sporopollenin in the cell wall of *Chlorella* and other algae: ultrastructure, chemistry and incorporation of ^{14}C-acetate studied in synchronous culture. *Planta* **107**: 1–32.

Blackmore, S. and Barnes, S.H. (1987). Embryophyte spore walls: origin, development, and homologies. *Cladistics* **3**: 185–195.

Brooks, J. (1971). Some chemical and geochemical studies on sporopollenin. In: J. Brooks, P.R. Grant, M. Muir, P. van Gizel and G. Shaw (editors). Sporopollenin, pp. 351–407, Academic Press, London.

Buer, F.v. (1964). Licht- und elektronenmikroskopische Untersuchungen an Zellwänden und Gallerten einiger Zygnemataceen. *Flora* **154**: 349–375.

Chaloner, W.G. (1970). The rise of the first land plants. *Biol. Rev. Cambridge Philos. Soc.* **45**: 353–377.

Chaloner, W.G. (1988). Early land plants - the saga of a great conquest. In: W. Greuter and B. Zimmer (editors). Proceedings of the XIV International Botanical Congress, pp. 301–316, Königstein, Koeltz.

Chapman, D.J. (1985). Geological factors and biochemical aspects of the origin of land plants. In: B.H. Tiffney (editor). Geological factors and the evolution of plants, pp. 23–45, Yale University Press, New Haven, Conn.

de Leeuw, J.W., van Bergen, P.F., van Aarssen, B.G.K., Gatellier, J-P.L.A., Sinninghe Damsté, J.S. and Collinson, M.E. (1991). Resistant biomacromolecules as major contributors to kerogen. *Philos. Trans., Ser. B* **333**: 329–337.

Delwiche, C.F., Graham, L.E. and Thompson, N. (1989). Lignin-like compounds and sporopollenin in *Coleochaete*, an algal model for land plant ancestry. *Science* **245**: 399–401.

Denizot, J. (1971). Sur la presence d'ensembles exiniques élémentaires dans le sporoderme de quelques Marchantiales et Sphaerocarpales. *Compt. Rend. Hebd. Séances Acad. Sci., Sér. D* **272**: 2166–2169.

Don, A.W.R. and Hickling, G. (1917). On *Parka decipiens*. *Quart. J. Geol. Soc. London* **71**: 648–666.

Dyck, L.A. (1970). Morphological, chemical and developmental studies of *Chara* oosporangial walls. 145 pp. Unpublished Ph.D. Thesis, Washington University, St Louis, USA.

Edwards, D. (1970). Fertile Rhyniophyta from the Lower Devonian of Britain. *Palaeontology* **13**: 451–461.

Edwards, D. and Fanning, U. (1985). Evolution and environment in the late Silurian-early Devonian: the rise of the pteridophytes. *Philos. Trans., Ser. B* **309**: 147–165.

Edwards, D., Davies, K.L. and Axe, L. (1992). A vascular conducting strand in the early land plant *Cooksonia*. *Nature* **357**: 683–685.
Erdtman, G. (1960). The acetolysis method: a revised description. *Svensk Bot. Tidskr.* **54**: 561–564.
Fanning, U., Richardson, J.B. and Edwards, D. (1988). Cryptic evolution in early land plants. *Evol. Trends Pl.* **2**: 13–24.
Fanning, U., Richardson, J.B. and Edwards, D. (1991). A review of *in situ* spores in Silurian land plants. In: S. Blackmore and S.H. Barnes (editors). Pollen and Spores: Patterns of Diversification, pp. 25–47, Clarendon Press, Oxford.
Fanning, U., Edwards, D. and Richardson, J.B. (1992). A diverse assemblage of early land plants from the Lower Devonian of the Welsh Borderland. *Bot. J. Linn. Soc.* **109**: 161–188.
Gensel, P.G. and White, A.R. (1983). The morphology and ultrastructure of spores of the early Devonian trimerophyte *Psilophyton* (Dawson) Hueber et Banks. *Palynology* **7**: 221–233.
Gensel, P.G., Johnson, N.G. and Strother, P.K. (1990). Early land plant debris (Hooker's "waifs and strays"). *Palaios* **5**: 520–547.
Graham, L.E. (1990). Meiospore formation in charophycean algae. In: S. Blackmore and R.B. Knox (editors). Microspores: Evolution and Ontogeny, pp. 43–54, Academic Press, London.
Graham, L.E. and Taylor, C. (1986). Occurrence and phylogenetic significance of 'special walls' at meiosporogenesis in *Coleochaete*. *Amer. J. Bot.* **73**: 178–186.
Grambast, L. (1964). Précisions nouvelles sur la phylogénie des charophytes. *Naturalia Monspel., Sér. Bot.* **16**: 71–77.
Gray, J. (1985). The microfossil record of early land plants: advances in understanding of early terrestrialization, 1970–1984. *Philos. Trans., Ser. B* **309**: 167–195.
Gray, J. (1991). *Tetrahedraletes, Nodospora* and the 'cross' tetrad: an accretion of myth. In: S. Blackmore and S.H. Barnes (editors). Pollen and Spores: Patterns of Diversification, pp. 49–87, Clarendon Press, Oxford.
Gray, J. and Shear, W. (1992). Early life on land. *Amer. Sci.* **80**: 444–456.
Gray, J., Massa, D. and Boucot, A.J. (1982). Caradocian land plant microfossils from Libya. *Geology* **10**: 197–201.
Hemsley, A.R. (1989a). Ultrastructure of the spores of the Devonian plant *Parka decipiens*. *Ann. Bot.* **64**: 359–367.
Hemsley, A.R. (1989b). The ultrastructure of the spores of the Triassic bryophyte *Naiadita lanceolata*. *Rev. Palaeobot. Palynol.* **61**: 89–99.
Hemsley, A.R. (1990). *Parka decipiens* and land plant spore evolution. *Historical Biol.* **4**: 39–50.
Hemsley, A.R. (1992a). Evolution of exine ultrastructure in Palaeozoic lycopod megaspores. *Courier Forschungsinst. Senckenberg* **147**: 93–107.
Hemsley, A.R. (1992b). Vascular pipe dreams. *Nature* **357**: 641–642.
Hemsley, A.R., Chaloner, W.G., Scott, A.C. and Groombridge, C.J. (1992). Carbon-13 solid-state nuclear magnetic resonance of sporopollenins from modern and fossil plants. *Ann. Bot.* **69**: 545–549.
Hemsley, A.R., Barrie, P.J., Chaloner, W.G. and Scott, A.C. (in press). The composition of sporopollenin and its use in living and fossil plant systematics. *Grana*.
Hopkins, A.W. and McBride, G.E. (1976). The life history of *Coleochaete scutata* (Chlorophyceae) studied by a Feulgen microspectrophotometric analysis of the DNA cycle. *J. Phycol.* **12**: 29–35.
Johnson, N.G. (1985). Early Silurian palynomorphs from the Tuscarora Formation in central Pennsylvania and their paleobotanical and geologic significance. *Rev. Palaeobot. Palynol.* **45**: 307–360.
Lang, W.H. (1937). On the plant remains from the Downtonian of England and Wales. *Philos. Trans., Ser. B* **227**: 245–291.

Lee, R.E. (1989). Phycology. (2nd edn.) 645 pp. Cambridge University Press, Cambridge.

Leitch, A.R. (1989). Formation and ultrastructure of a complex multilayered wall around the oospore of *Chara* and *Lamprothamnium* (Characeae). *Brit. Phycol. J.* **24**: 229–236.

Mishler, B.D. and Churchill, S.P. (1985). Transition to a land flora: phylogenetic relationships of the green algae and bryophytes. *Cladistics* **1**: 305–328.

Niklas, K.J. (1976a). Morphological and ontogenetic reconstructions of *Parka decipiens* Fleming and *Pachytheca* Hooker from the Lower Old Red Sandstone, Scotland. *Trans. Roy. Soc. Edinburgh* **69**: 483–499.

Niklas, K.J. (1976b). The chemotaxonomy of *Parka decipiens* from the Lower Old Red Sandstone, Scotland (U.K.). *Rev. Palaeobot. Palynol.* **21**: 205–217.

Niklas, K.J. and Chaloner, W.G. (1976). Chemotaxonomy of some problematic plants. *Rev. Palaeobot. Palynol.* **22**: 81–104.

Niklas, K.J. and Phillips, T.L. (1976). Morphology of *Protosalvinia* from the Upper Devonian of Ohio and Kentucky. *Amer. J. Bot.* **63**: 9–29.

Pettitt, J.M. (1966). Exine structure in some fossil and recent spores and pollen as revealed by light and electron microscopy. *Bull. Brit. Mus. (Nat. Hist.) Geol.* **13**: 221-257.

Phillips, T.L., Niklas, K.J. and Andrews, H.N. (1972). Morphology and vertical distribution of *Protosalvinia* (*Foerstia*) from the New Albany Shale (Upper Devonian). *Rev. Palaeobot. Palynol.* **14**: 171–196.

Remy, W. (1982). Lower Devonian gametophytes: relation to the phylogeny of land plants. *Science* **215**: 1625–1627.

Remy, W. and Remy, R. (1980). Devonian gametophytes with anatomically preserved gametangia. *Science* **208**: 295–296.

Richardson, J.B. and Edwards, D. (1989). Sporomorphs and plant megafossils. In: C.H. Holland and M.G. Basset (editors). A Global Standard for the Silurian System, pp. 216–226, National Museum of Wales, Cardiff.

Shaw, G. (1971). Chemistry of Sporopollenin. In: J. Brooks, P.R. Grant, M. Muir, P. van Gizel and G. Shaw (editors). Sporopollenin, pp. 305–348, Academic Press, London.

Schuster, R.M. (1979). The phylogeny of the Hepaticae. In: G.C.S. Clarke and J.G. Duckett (editors). Bryophyte Systematics, pp. 41–82, Academic Press, London.

Shute, C.H. and Edwards, D. (1989). A new rhyniopsid with novel sporangium organization from the Lower Devonian of South Wales. *Bot. J. Linn. Soc.* **100**: 111–137.

Stebbins, G.L. and Hill, G.J.C. (1980). Did multicellular plants invade the land? *Amer. Naturalist* **115**: 342–353.

Steinkamp, M.P. (1973). Spore wall ultrastructure in the Hepaticae with special reference to the Sphaerocarpales. 301 pp. Unpublished Ph.D. Thesis, University of California, Santa Cruz, USA.

Stewart, W.N. (1983). Paleobotany and the evolution of plants. 405 pp. Cambridge University Press, Cambridge.

Strother, P.K. (1991). A classification schema for the cryptospores. *Palynology* **15**: 219–236.

Taylor, W.A. and Taylor, T.N. (1987). Spore wall ultrastructure of *Protosalvinia*. *Amer. J. Bot.* **74**: 437–443.

Wall, D. (1962). Evidence from Recent plankton regarding the biological affinities of *Tasmanites* Newton, 1875 and *Leiosphaeridia* Eisenack, 1958. *Geol. Mag.* **99**: 353–362.

Zelibor Jr, J.L., Romankiw, L., Hatcher, P.G. and Colwell, R.R. (1988). Comparative analysis of the chemical composition of mixed and pure cultures of green algae and their decomposed residues by ^{13}C nuclear magnetic resonance spectroscopy. *Appl. Environm. Microbiol.* **54**: 1051–1060.

Kovach, W. (1994). A review of Mesozoic megaspore ultrastructure. In: M.H. Kurmann and J.A. Doyle (editors). Ultrastructure of fossil spores and pollen, pp. 23–37, Royal Botanic Gardens, Kew.

2. A REVIEW OF MESOZOIC MEGASPORE ULTRASTRUCTURE

WARREN L. KOVACH

Palynological Research Centre, Institute of Earth Studies, University of Wales, Aberystwyth, Dyfed SY23 3DB, Wales, U.K.

Current address: **Kovach Computing Services, 85 Nant-y-Felin, Pentraeth, Anglesey LL75 8UY, Wales, U.K.**

Abstract

Although many megaspores have been described from the Mesozoic, few have been analysed ultrastructurally. Those that have show many variations of the typical "spongy" lycopsid or "granular" aquatic fern wall structure. Studies of modern lycopsids, particularly *Selaginella*, have demonstrated that megaspore ultrastructure shows promise as a systematic tool. Many ultrastructural types seen in Mesozoic spores are very similar to modern ones, suggesting systematic affinities and reflecting the conservative nature of these groups. There are also some wall types that had not fully developed before the Cenozoic, indicating that some modern groups had not yet arisen. Other types are not seen in the modern flora, indicating now-extinct groups.

Introduction

This symposium volume demonstrates that the study of spore and pollen ultrastructure is widely used to infer systematic relationships. However, in the study of Mesozoic lycopsid and aquatic fern megaspores comparatively little work has been done. Batten and Kovach (1990) published a comprehensive catalogue of Mesozoic and Tertiary megaspores. It contains over 3000 records of 774 species of megaspores, gleaned from 507 publications. Most of these (640 species, 343 publications) are from the Mesozoic. This body of literature was surveyed again for the present study, looking for papers that describe the ultrastructure of Mesozoic megaspores. Only 26 papers describe the wall structure of 78 species (see Appendix) . Many of these have only brief descriptions, and some are based on either SEMs of hand-sectioned or broken specimens or light micrographs, not TEM. There has been no review or synthesis of the existing data. However, there is still enough to give an overview of the types of ultrastructure seen in Mesozoic megaspores and to evaluate the potential usefulness of wall characters in systematic studies.

There are two basic questions to be kept in mind in this review of Mesozoic megaspore ultrastructure: 1) Can ultrastructure help in determining the relationships between fossil megaspore species and modern plant groups, and 2) is it useful for defining natural relationships between fossil taxa?

Aquatic Ferns vs. Lycopsida

There are two main groups of modern plants that produce freely dispersed megaspores, the aquatic ferns and the lycopsids. These two groups are unrelated and very different in the morphology of the whole plant (Tryon and Tryon, 1982). Their spores are also very different and easily distinguishable. Aquatic ferns have spores that are often very elaborate. *Azolla* exemplifies the height of complexity, with its system of floats, columella, and collar surmounting the main megaspore body (Fowler and Stennett-Willson, 1978). In general, the aquatic ferns have megaspores with a small, inconspicuous trilete mark that is often hidden by leafy proximal appendages. The exine is in whole or in part granular, with a "prismatic layer" forming the outer portion of the exine in the Marsileales (Pettitt, 1966). In the Salviniaceae the megaspores are covered with a fibrous perine (Hall, 1974; Fowler and Stennett-Willson, 1978). Lycopsid megaspores, on the other hand, have prominent trilete marks and a spongy exinal structure (Pettitt, 1966; Tryon and Tryon, 1982). They do not have a fibrous perine, although most modern *Isoetes* have a siliceous outer layer that is considered by some to be the perine (Pettitt, 1971; Robert et al., 1973).

There are many fossil megaspores that are clearly members of the aquatic ferns. Hall (1974, 1975) reviewed the Cretaceous members of the Salviniaceae, including comparisons of their ultrastructure, and Fowler (1975) and Collinson (1980) reviewed the genus *Azolla* through time. Batten and Kovach (1990) list numerous Cenozoic species that have been assigned to *Azolla* and *Salvinia*. The Cretaceous genera *Arcellites* and *Molaspora* are considered marsileaceous (Ellis and Tschudy, 1964 ; Kovach and Dilcher, 1988), but little work has been done on their ultrastructure. The appendix lists Mesozoic species of aquatic ferns that have had their ultrastructure illustrated and described.

Modern Lycopsids

The rest of this paper will focus on lycopsid megaspores, which are far more common in the Mesozoic than the megaspores of aquatic ferns (Kovach and Batten, 1993). There are two groups of modern heterosporous lycopsids, the Isoetales and the Selaginellales. These differ in the morphology of the whole plant and the microspores (Tryon and Tryon, 1982), but their megaspores are less easy to distinguish morphologically. However, some ultrastructural differences have been noted that may be useful in distinguishing between these two groups (Kovach and Dilcher, 1985; Kovach, 1989).

Lycopsid megaspore exines are composed of anastomosing sporopollenin elements. These range from rod-like to granular, forming a porous network. *Isoetes* spores tend to have their rod-like elements oriented parallel to the surface of the spore and have a high porosity (Fig. 1). *Selaginella*, on the other hand, has sporopollenin elements that are more randomly oriented and tend to have less open space between them (Fig. 1). Methods have been developed to quantify these characteristics (Kovach, 1989). Statistical analyses of quantitative data on various modern and fossil megaspores have shown that, although there is some overlap, these two characteristics together can help to distinguish between the two groups (Kovach, 1989; Taylor, 1994).

A. *Selaginella*

Megaspore ultrastructure might also be systematically useful within these two lycopsid orders. *Selaginella* has over 700 species divided variously into two to five subgenera (Walton and Alston, 1938; Tryon and Tryon, 1982; Jermy, 1986). Ultrastructural reviews of *Selaginella* have been published by Morbelli (1977), Minaki (1984), and Taylor (1989). Morbelli, who sectioned eight Argentinian species, distinguishes two basic types of *Selaginella* ultrastructure: granular, in which the wall is composed of closely packed and regularly distributed spherical units, and spongy, formed from anastomosing rod-like elements. Minaki sectioned 38 species and described four basic types (Fig. 2), two spongy or "labyrinthine" (one with an inner layer of large granules) and two granular (one irregular and one ordered). These types are further divided into 19 subtypes, based on the number of layers in the wall and the background pattern of the surface of the spore.

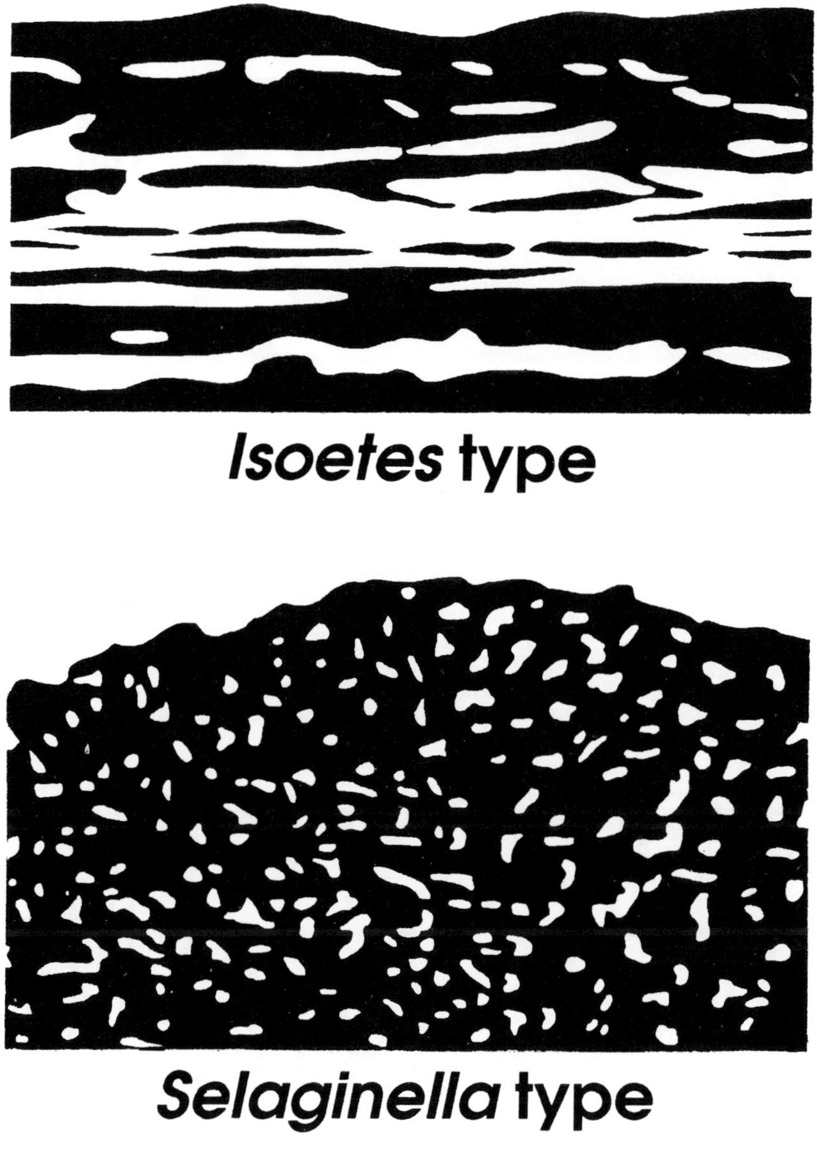

Fig. 1. Diagrams showing the typical megaspore ultrastructure seen in the two modern genera of heterosporous lycopsids.

Taylor (1989), after studying 19 species, divides the spores into three types: ordered (= granular of Morbelli, 1977, and ordered granular [GMT-D] of Minaki, 1984), and two distinct spongy types (Fig. 3), laminar and laterally fused. The laterally fused type is composed of anastomosing rod-like to spherical units, whereas the laminar type is formed from wider sheet-like elements, often forming closed vesicles. Some species studied by Taylor were intermediate between these two spongy types, but most clearly belonged to one or the other. No direct correlation can be made between the Minaki and Taylor schemes. The laminar specimens of Taylor would be placed in one of the two

GMT-D
GMT-C
GMT-B
GMT-A

labyrinthine groups (GMT-A and B) defined by Minaki. However, various members of the laterally fused groups could be assigned to either GMT-A or B if the units are rod-like, or the unordered granular type (GMT-C) if they are composed of more spherical units.

The ordered granular type reported by Minaki (1984) and Taylor (1989) is an unusual and distinct ultrastructural pattern that has received much attention in recent years (Tryon and Lugardon, 1978; Hueber, 1982; Taylor and Taylor, 1987 ; Taylor, 1991 ; Hemsley et al., 1992 ; Collinson et al., 1992 ; Morbelli, 1992). The precise nature of the structure of this wall type has been debated, but it seems to be formed from spherical units of similar size organized into a close packing structure. The size and arrangement of the spherical units is the same as that seen in opaline silica, and the spores often exhibit an opal-like iridescence (Hueber, 1982). Hemsley et al. (1992) and Collinson et al (1992) suggest that these exines were formed from a colloidal solution through self-assembly.

Minaki (1984) compared his 19 megaspore types to Walton and Alston's (1938) subgeneric classification of *Selaginella*. Some megaspore types were restricted to certain subgenera, but others were not so clearly related. In particular, the several species that comprise his megaspore types MT-4 and MT-5 correlate with the subgenera *Homostachys* and *Heterostachys* respectively. The distinct ordered granular type is confined to the subgenus *Stachygynandrum*, but not to any one series within the subgenus. Also, several other types occur within this, the largest subgenus. Therefore this ultrastructural type does not directly correlate with the existing subclassification of *Stachygynandrum*, which is based on life form. However, this wall structure is correlated with two features of sporophyte morphology: strobilus organization and stelar anatomy (Minaki, 1984). Thus it seems that these characters together may define a systematically useful group, and that *Stachygynandrum* might be better subdivided on the basis of these spore and sporophyte characters.

B. *Isoetes*

Isoetes has fewer species (approximately 150). There have been various attempts to divide the taxon into subgenera. The earlier schemes were based on habitat preference or megaspore surface ornamentation (Baker, 1880; Pfeiffer, 1922) but these have not been very successful systematically (Tryon and Tryon, 1982, Hickey, 1986). Hickey (1990) recently erected two subgenera based on leaf characters, which may be more successful. Some ultrastructural work has been done on various species of *Isoetes* (Pettitt, 1966, 1971; Robert et al., 1973; Taylor, 1992, 1993), but no large scale survey has been done as yet. Taylor's (1993) study of six species of *Isoetes* shows that they are fairly uniform in their ultrastructure. There are as yet not enough data to attempt any wall classification within the genus.

Mesozoic Lycopsid Types

The above discussion has reviewed the various ultrastructural types found in modern lycopsid megaspores and their systematic importance. We can now compare these types with those found in Mesozoic megaspores to assess their similarity to the modern groups. We can also consider whether ultrastructure could be useful in determining the validity of some of the Mesozoic form genera.

A. Selaginellalean types

The distinctive ordered granular type of wall ultrastructure (Fig. 4) is seen commonly in the Mesozoic record. Most species of *Erlansonisporites* have this type of ultrastructural

FIG. 2. Four groups of megaspore wall types (GMTs) in *Selaginella*, redrawn from Minaki (1984). GMT-A: labyrinthine with inner layer of large granules. GMT-B: labyrinthine, one layer. GMT-C: irregular granular. GMT-D: ordered granular.

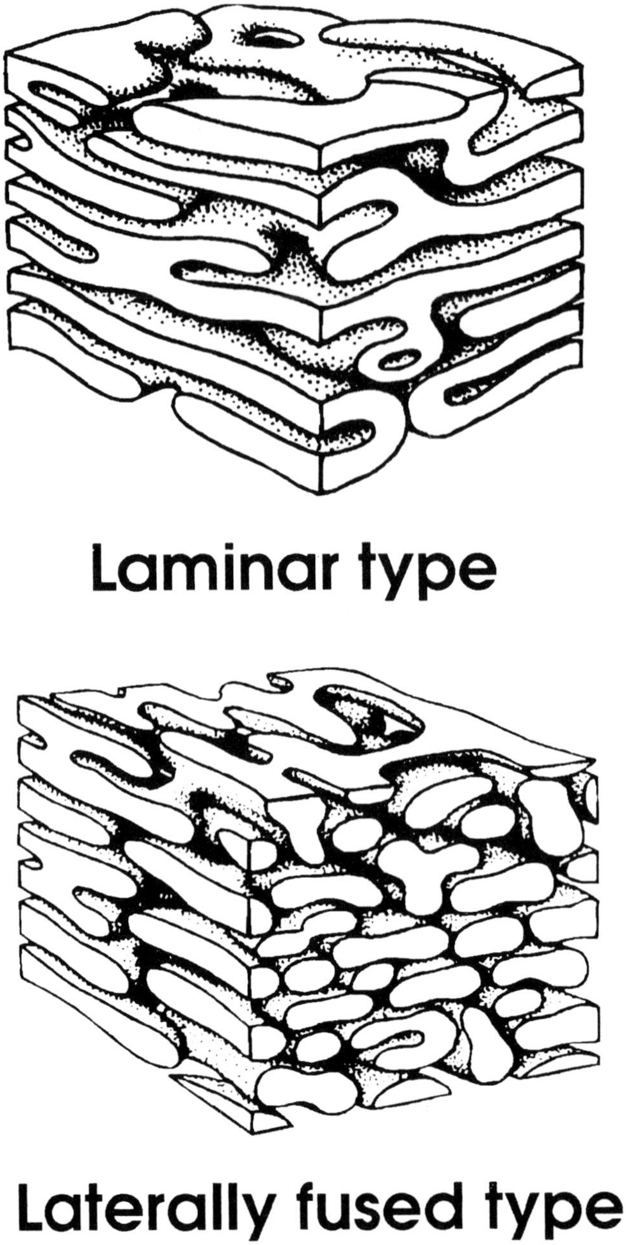

Fig. 3. Two types of spongy megaspore wall ultrastructure, redrawn from Taylor (1989).

organization in at least part of the wall (Bergad, 1978; Kovach and Dilcher, 1988; Koppelhus and Batten, 1989; Hemsley et al., 1992; Collinson et al., 1992). It is also present in *Thylakosporites retiarius* (Hueber, 1982; Koppelhus and Batten, 1989), *Ricinospora cryptoreticulata* (Bergad, 1978), *Rugotriletes diktyotus* and *R. costatus* (Batten, 1988), and

Trileites persimilis (Koppelhus and Batten, 1989). Taylor and Taylor (1988) also recorded this type from a new species they placed in *Horstisporites* (*H. iridodea*). However, the ultrastructural similarities and the rounded muri indicate that this species may be better placed in *Thylakosporites*.

Most of the selaginellalean megaspores that have been sectioned could be assigned to the laterally fused type of ultrastructure (Figs. 5 and 6). Some particularly good examples are *Banksisporites pinguis* (Kempf, 1971a), *Bacutriletes majorinus* (Koppelhus and Batten, 1989), *Horstisporites semireticulatus* (Kempf, 1971b), *H. harrisii* (Bergad, 1978), and *Hughesisporites patagonicus* (Gamerro, 1975; Taylor and Taylor, 1988; Archangelsky and Villar de Seoane, 1990). *Banksisporites pinguis* has been found *in situ* in *Selaginella* megafossils by Lundblad (1950a,b), thus lending support to the suggested selaginellalean affinities.

During this review no spores were found that displayed a well-developed laminar type of ultrastructure, despite it being common in modern *Selaginella*. *Horstisporites* sp., figured by Morbelli (1990) from the Jurassic of Argentina has an outer layer that consists of laminar units, but the bulk of the wall is of the laterally fused construction. The dearth of this type may indicate that it had only begun developing in the Mesozoic and that it did not fully develop until the Cenozoic. However, few lycopsid megaspores have been reported from the Cenozoic (Kovach and Batten, 1993), and even fewer have been sectioned (only *Erlansonisporites*; Hemsley et al., 1992 and Collinson et al., 1992), so it is not known whether this pattern occurs during that time. It also may be that it does occur but that not enough Mesozoic megaspores have been sectioned.

A distinctive variation on the laterally fused type can be seen in the genus *Cabochonicus* (Fig. 6). In this type the units have a distinct angular character, rather than being rounded as in most of the laterally fused patterns. Also, the units are evenly distributed throughout the wall. *Cabochonicus* was erected by Batten and Ferguson (1987) for megaspores with rounded sculptural elements scattered on the spore surface. Several species were transferred from *Verrutriletes* and *Trileites* to this genus based on their morphology alone (Batten and Ferguson, 1987; Batten, pers. comm.). Besides the type species, *C. carbunculus*, *C. gamerroi* has also been sectioned (Baldoni and Taylor, 1985). This species exhibits the same distinctive ultrastructure, thus supporting the view that this genus is a natural grouping rather than an artificial classification. Similar wall structure is seen in the modern *Selaginella selaginoides* (Kempf, 1970), indicating a possible systematic relation between it and *Cabochonicus*.

B. Isoetalean types

Several Mesozoic species of megaspores clearly demonstrate the isoetalean ultrastructural pattern described above. Some authors have specifically discussed possible relationships between these fossils and modern *Isoetes* (Kovach and Dilcher, 1985, 1988; Batten, 1988; Koppelhus and Batten, 1989), and in some cases the presumed isoetalean affinities have been supported by other evidence, such as that from associated microspores, palaeoecology, or association with megafossils (Batten, 1969, 1974; Melchior, 1977; Hickey, 1977; Ash, 1979; Kovach and Dilcher, 1985, 1988; Koppelhus and Batten, 1989).

Many fossil species have exines with very high porosity and thread-like sporopollenin elements that are oriented strongly parallel to the surface (Fig. 7), similar to the modern *Isoetes* type shown in Fig. 1. This type is most commonly seen in *Minerisporites* and *Auriculozonosporites* (Kovach and Dilcher, 1988; Archangelsky and Villar de Seoane, 1989, 1990), although it also can be seen in *Herbosisporites*, *Sepisporites* (Li et al., 1987), and *Scabritriletes* (Batten, 1988). Most of these have thin exines, some of them being only 6-8 μm thick, and almost all have a very thin inner layer, often separated from the much thicker outer layer. The sporopollenin elements in these megaspores are usually evenly distributed throughout the exine.

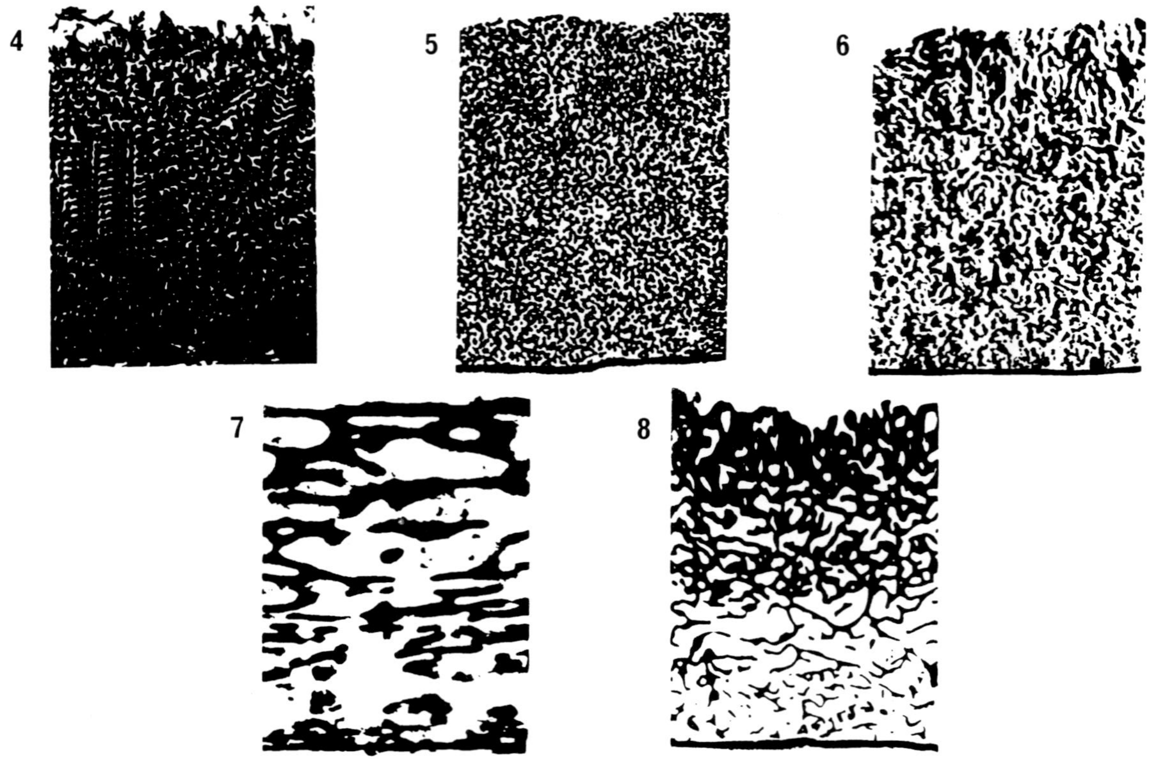

FIGS. 4–8. Five types of megaspore wall ultrastructure seen in the Mesozoic fossil record. Types in Figs. 4–6 are probably selaginellalean, while those in 7–8 are most likely isoetalean. **4**: ordered granular type. **5**: laterally fused type. **6**: *Cabochonicus* type. **7**: *Minerisporites* type. **8**: *Nathorstisporites/Paxillitriletes* type. Note that these five types are meant to be illustrative, not all-encompassing. Also, these are not necessarily drawn to the same scale; in particular, megaspore walls of the type in Fig. 7 are usually thinner than others.

The second isoetalean type (Fig. 8) is exemplified by *Nathorstisporites* (Dettmann, 1961; Kempf, 1971a; Hemsley and Scott, 1989). This type also has a high porosity and thread-like elements arranged parallel to the surface. However, the parallel nature of the elements is not as strongly developed in this type as in the *Minerisporites* type. The sporopollenin elements are often much thicker, particularly in the outer portion of the wall, resulting in a lower porosity. The thickness of the walls is also much greater, up to 25μm. This type is also present in the morphologically similar *Paxillitriletes* (Kovach and Dilcher, 1985, 1988; Baldoni, 1987). In this genus, the outer elements are not thickened to the same extent as in *Nathorstisporites* and the wall is thinner (10–20μm). The species *Minerisporites dissimilis* (Kovach and Dilcher, 1988) and *Tenellisporites* cf. *T. aachenensis* (Koppelhus and Batten, 1989) have similar exines, but appear to be intermediate between the *Nathorstisportes/Paxillitriletes* type and the *Minerisporites* type.

C. Variation within Mesozoic genera

Many of the Mesozoic genera are artificial, erected on the basis of certain types of sculpturing and ornamentation or gross morphology. If megaspore ultrastructure has systematic significance, as is suggested above, then variation of the ultrastructural patterns within these genera can indicate the degree of artificiality. Also, ultrastructure may help in identifying species that have been misclassified.

The genus *Horstisporites* is quite variable in its ultrastructure, containing almost the entire range of known wall types. Most species fit the laterally fused *Selaginella* pattern of Taylor (1989). These include *H. semiretculatus* (Kempf, 1971b) and *H. harrisii* (Bergad, 1978). As mentioned above, *H. iridodea*, described by Taylor and Taylor (1988), exhibits the ordered granular pattern. The presence of an ultrastructural type otherwise not present in this genus, as well as the different morphology, suggests that *H. iridodea* has been misclassified.

On the other hand, *H. kendalli* (Kempf, 1971a) has an exine structure more similar to that seen in isoetalean spores. The wall has a high porosity, and in the inner two-thirds the rod-shaped units are arranged parallel to the surface. The outer third consists of small granular units that intermingle with the outer portion of the rod-shaped layer. Granular exine is usually only seen in *Selaginella* and its possible fossil relatives, not the isoetalean megaspores. Thus this species seems to exhibit characteristics of both groups. However, this granular outer layer is also reminiscent of the siliceous perine of modern *Isoetes*, shown in section by Pettitt (1971). There is no indication as to whether this layer in *H. kendalli* is composed of silica, and there is no mention in the literature of siliceous perines in dispersed fossil megaspores (although W.A. Taylor, pers. observ., has noted silica coatings on megaspores found *in situ* in Jurassic *Isoetites* megafossils). It may be that the siliceous perines of isoetalean megaspores are usually lost during transport, diagenesis or preparation, and that *H. kendelli* is an unusually well-preserved specimen. Although the silica coats of modern *Isoetes* are initially very hard, examination of old herbarium sheets shows that they become very fragile and degrade with age (Kovach, pers. ovserv.).

Most species of *Erlansonisporites* exhibit the distinctive ordered granular wall type (Fig. 4). However, a specimen of *E. sparassis* sectioned by Taylor and Taylor (1988) does not have this pattern; instead it has a more open structure composed of fused granular to rod-shaped units. Likewise, sections of a specimen of *E. erlansonii* from the Dakota Formation of Kansas show that it also lacks the highly ordered structure seen in many spores from this genus (M.E. Collinson, pers. comm.). Minaki (1984) noted that this ordered type is confined to one subgenus, but that other ultrastructural patterns also occur within the same group. Given this finding, it is not surprising that some *Erlansonisporites* species lack the ordered pattern; this should not be taken as indicating misclassification of these species.

All *Minerisporites* spores have the isoetalean wall structure (Fig. 7), but there is some variation within the genus. *Minerisporites succrassulus* (Bergad, 1978) has an unusually thick exine (20–30 μm), and the tendency towards parallel orientation of the wall elements is weaker than in other species. *Minerisporites marginatus*, on the other hand, has a very thin wall (6–8 μm) and is composed of three layers.

Conclusions

Work on the megaspore ultrastructure of modern aquatic ferns and lycopsids has indicated that characters of the wall structure can have systematic significance. The two major groups can be easily distinguished, and *Isoetes* and *Selaginella* can usually be separated. Ultrastructure also has promise of systematic utility at the subgeneric level. However, more work needs to be done on surveying ultrastructural patterns within the genera and correlating these with other vegetative and reproductive characters.

Many of the basic patterns seen in the modern spores also can be distinguished in the Mesozoic fossil record. Some of the more distinct types, such as the ordered granular type seen in *Erlansonisporites* and *Selaginella* subgenus *Stachygynandrum* and the angular type of *Cabochonicus/S. selaginoides*, have changed little in the intervening time. This demonstrates the conservative nature of the lycopsids, particularly in the Selaginellales, which have representatives very similar to modern ones extending back to the Carboniferous (Thomas and Quansah, 1991; Thomas, 1992). Some modern types, such as the laminar *Selaginella* ultrastructure, are poorly developed in the Mesozoic, indicating that although the group may be conservative, there was still much development through the Cenozoic, with new ultrastructural types (and possibly new subgenera) arising.

The type of ultrastructure seen in modern *Isoetes* also occurs in the Mesozoic in genera like *Minerisporites*, some of which have been found in megafossils very similar to modern *Isoetes* (Melchior, 1977; Hickey, 1977). However, the wall structure seen in *Paxillitriletes* and *Nathorstisporites* has not yet been reported in modern megaspores. This indicates that these and other similar genera might represent an isoetalean group that has now become extinct. These spores also have very distinct appendages of various sorts, which are absent in Recent plants, supporting the suggestion that they are an extinct group.

Acknowledgements

I would like to thank Willy Taylor, Margaret Collinson and Alan Hemsley for many stimulating discussions about megaspore ultrastructure, David Batten for collaboration in compiling the literature for our megaspore catalogue, and Catherine Duigan and two anonymous reviewers for providing comments on the manuscript.

References

Archangelsky, S. and Villar de Seoane, L. (1989). Ultraestructura de dos nuevas megasporas cretácicas de Santa Cruz, Argentina. *Bol. Asoc. Latinoamer. Paleobot. Palin.* **12**: 13–25.

Archangelsky, S. and Villar de Seoane, L. (1990). Morfología y estructura de megasporas cretácicas de Patagonia, Republica Argentina. *Revista Esp. Micropaleontol.* **22**: 419–450.

Archangelsky, S. and Villar de Seoane, L. (1991). Notas sobre la flora fosil de la Zona de Ticó, provincia de Santa Cruz: XI. Morfología y estructura de tres megasporas. *Ameghiniana* **28**: 353–364.

Ash, S.R. (1979). *Skilliostrobus* gen.nov., a new lycopsid cone from the Early Triassic of Australia. *Alcheringa* **3**: 73–89.

Baker, J.G. (1880). A synopsis of the species of *Isoetes*. *J. Bot.* **18**: 65–70, 105–110.

Baldoni, A.M. (1987). Dos nuevas especies de megasporas de la Formación Kachaike, Cretácico inferior de Santa Cruz, Argentina. Anais Congr. Brasil. Paleontol., 10th, Rio de Janeiro, 1987, pp. 669–689.

Baldoni, A.M. and Taylor, T.N. (1985). Megasporas cretácicas de la Formación Springhill en el subsuelo de Argentina y Chile austral. *Ameghiniana* **21**: 151–167.

Baldoni, A.M. and Taylor, T.N. (1987). Ultraestructura de la pared de *Paxillitriletes kachaikense*, una nueva especie en el Cretácico inferior de la Provincia de Santa Cruz, Argentina. Simp. Argent. Paleobot. Palin., 7th, Buenos Aires 1987, Actas pp. 89–92.

Baldoni, A.M. and Taylor, T.N. (1988). Ultraestructura de una nueva especie de *Arcellites* en el Cretácico inferior de la Provincia de Santa Cruz, Argentina y sus vinculaciones con la familia Marsileaceae. Congr. Argent. Paleontol. Biostratigr., 4th, Mendoza, 1986, Actas 3, pp. 15–22.

Batten, D.J. (1969). Some British Wealden megaspores and their facies distribution. *Palaeontology* **12**: 333–350.

Batten, D.J. (1974). Wealden palaeoecology from the distribution of plant fossils. *Proc. Geol. Assoc. (London)* **85**: 433–458 (1974 publ. 1975).

Batten, D.J. (1988). Revision of S.J. Dijkstra's Late Cretaceous megaspores and other plant microfossils from Limburg, The Netherlands. *Meded. Rijks Geol. Dienst* **41(3)**: 1–55.

Batten, D.J. and Ferguson, D.J.P. (1987). *Cabochonicus*, a new genus for species of gemmate megaspores previously referred to *Verrutriletes*. *J. Micropalaeontol.* **6**: 65–75.

Batten, D.J. and Kovach, W.L. (1990). Catalog of Mesozoic and Tertiary megaspores. *Contr. Ser. Amer. Assoc. Stratigr. Palynologists* **24**: 1–227.

Bergad, R.D. (1978). Ultrastructural studies of selected North American Cretaceous megaspores of *Minerisporites, Erlansonisporites, Horstisporites* and *Ricinospora*, n. gen. *Palynology* **2**: 39–51.

Collinson, M.E. (1980). A new multiple floated *Azolla* from the Eocene of Britain with a brief review of the genus. *Palaeontology* **23**: 213–229.

Collinson, M.E., Hemsley, A.R. and Taylor, W.A. (1992). Sporopollenin exhibiting a colloidal crystal organization in spore walls. Abstr. Int. Palynol. Congr., 8th, Aix-en-Provence, p.27.

Dettmann, M.E. (1961). Lower Mesozoic megaspores from Tasmania and South Australia. *Micropalaeontology* **7**: 71–86.

Ellis, C.H. and Tschudy, R.H. (1964). The Cretaceous megaspore genus *Arcellites* Miner. *Micropalaeontology* **10**: 73–79.

Fowler, K. (1975). Megaspores and massulae of *Azolla prisca* from the Oligocene of the Isle of Wight. *Palaeontology* **18**: 483–507.

Fowler, K. and Stennett-Willson, J. (1978). Sporoderm architecture in modern *Azolla*. *Fern Gaz.* **11**: 421–428.

Gamerro, J.C. (1975). Megasporas del Cretácico de Patagonia: 1. Ultraarquitectura de la pared megasporal en *Hughesisporites patagonicus* Archang. y *Horstisporites feruglioi* Archang. *Ameghiniana* **12**: 97–108.

Hall, J.W. (1974). Cretaceous Salviniaceae. *Ann. Missouri Bot. Gard.* **61**: 354–367.

Hall, J.W. (1975). *Ariadnaesporites* and *Glomerisporites* in the Late Cretaceous: ancestral Salviniaceae. *Amer. J. Bot.* **62**: 359–369.

Hemsley, A.R. and Scott, A.C. (1989). The ultrastructure of four Australian Triassic megaspores. *Pollen & Spores* **31**: 133–154.

Hemsley, A.R., Collinson, M.E. and Brain, A.P.R. (1992). Colloidal crystal-like structure of sporopollenin in the megaspore walls of Recent *Selaginella* and similar fossil spores. *Bot. J. Linn. Soc.* **108**: 307–320.

Hickey, L.J. (1977). Stratigraphy and paleobotany of the Golden Valley Formation (early Tertiary) of western North Dakota. *Mem. Geol. Soc. Amer.* **150**: 1–183.

Hickey, R.J. (1986). The early evolutionary and morphological diversity of *Isoetes*, with descriptions of two new neotropical species. *Syst. Bot.* **11**: 309–321.

Hickey, R.J. (1990). Studies of neotropical *Isoetes* L. I. *Euphyllum* , a new subgenus. *Ann. Missouri Bot. Gard.* **77**: 239–245.

Hueber, F.M. (1982). Megaspores and a palynomorph from the Lower Potomac Group in Virginia. *Smithsonian Contr. Paleobiol.* **49**: 1–69.
Jermy, A.C. (1986). Subgeneric names in *Selaginella. Fern Gaz.* **13**: 117–118.
Jux, U. and Kempf, E.K. (1971). Microstructures of the Mesozoic megaspore *Tasmanitriletes* n.g. *Grana* **11**: 95–100.
Kempf, E.K. (1970). Elektronenmikroskopie der Sporodermis von Megasporen der Gattung *Selaginella* (Pteridophyta). *Rev. Palaeobot. Palynol.* **10**: 99–116.
Kempf, E.K. (1971a). Electron microscopy of Mesozoic megaspores from Denmark. *Grana* **11**: 151–163.
Kempf, E.K. (1971b). Electron microscopy of the megaspore *Horstisporites semireticulatus* from Liassic strata of Germany. *Grana* **11**: 18–22.
Koppelhus, E.B. and Batten, D.J. (1989). Late Cretaceous megaspores from southern Sweden: morphology and paleoenvironmental significance. *Palynology* **13**: 91–120.
Kovach, W.L. (1989). Quantitative methods for the study of lycopod megaspore ultrastructure. *Rev. Palaeobot. Palynol.* **57**: 233–246.
Kovach, W.L. and Batten, D.J. (1993). Diversity changes in lycopsid and aquatic fern megaspores through geological time. *Paleobiology* **19**: 28–42.
Kovach, W.L. and Dilcher, D.L. (1985). Morphology, ultrastructure, and paleoecology of *Paxillitriletes vittatus* sp. nov. from the mid-Cretaceous (Cenomanian) of Kansas. *Palynology* **9**: 85–94.
Kovach, W.L. and Dilcher, D.L. (1988). Megaspores and other dispersed plant remains from the Dakota Formation (Cenomanian) of Kansas, U.S.A. *Palynology* **12**: 89–119.
Li, W.B. and Batten, D.J. (1986). The Early Cretaceous megaspore *Arcellites* and closely associated *Crybelosporites* microspores from northeast Inner Mongolia, P.R. China. *Rev. Palaeobot. Palynol.* **46**: 189–208.
Li, W.B., Batten, D.J., Zhang, D.H. and Zhang, L.D. (1987). Early Cretaceous megaspores from the Jalainor Group of northeast Inner Mongolia, P.R. China. *Palaeontographica, Abt. B, Paläophytol.* **206**: 117–135.
Lundblad, A.B. (1950a). Studies in the Rhaeto-Liassic floras of Sweden, I. *Kungl. Svenska Vetenskapsakad. Handl., Ser. 4.* **1(8)**: 1–82.
Lundblad, A.B. (1950b). On a fossil *Selaginella* from the Rhaetic of Hyllinge, Scania. *Svensk Bot. Tidskr.* **44**: 477–487.
Melchior, R.C. (1977). On the occurrence of *Minerisporites mirabilis* in situ. *Sci. Publ. Sci. Mus. Minnesota, New Ser.* **3**: 1–11.
Minaki, M. (1984). Macrospore morphology and taxonomy of *Selaginella. Pollen & Spores* **26**: 421–480.
Morbelli, M.A. (1977). Esporas de las especies Argentinas de *Selaginella* (Selaginellaceae-Pteridophyta). *Obra Centen. Mus. La Plata* **3**: 121–150.
Morbelli, M.A. (1990). Austral South American Jurassic megaspores. *Rev. Palaeobot. Palynol.* **65**: 209–216.
Morbelli, M.A. (1992). Megaspore wall in Lycophyta. Ultrastructure and function. Abstr. Int. Palynol. Congr., 8th, Aix-en-Provence, p.104.
Pettitt, J.M. (1966). Exine structure in some fossil and recent spores and pollen as revealed by light and electron microscopy. *Bull. Brit. Mus.* (*Nat. Hist.*), *Geol.* **13**: 223–257.
Pettitt, J.M. (1971). Developmental mechanisms in heterospory I. Megasporocyte degeneration in *Selaginella. Bot. J. Linn. Soc.* **64**: 237–246.
Pfeiffer, N.F. (1922). Monograph of the Isoetaceae. *Ann. Missouri Bot. Gard.* **9**: 79–232.
Robert, D., Roland-Heydacker, F., Denizot, J., Laroche, J., Fougeroux, P. and Davignon, L. (1973). Étude de la paroi siliceuse chez la mégaspore d' *Isoetes setacea. Compt. Rend. Hebd. Séances Acad. Sci., Sér. D.* **276**: 2521–2524.
Taylor, W.A. (1989). Megaspore wall ultrastructure in *Selaginella. Pollen & Spores* **31**: 251–288. (1989 publ. 1991).
Taylor, W.A. (1991). Ultrastructural analysis of sporoderm development in megaspores of *Selaginella galeotii* (Lycophyta). *Pl. Syst. Evol.* **174**: 171–182.

Taylor, W.A. (1992). Megaspore wall development in *Isoetes melanopoda* : morphogenetic post-initiation changes accompanying spore enlargement. *Rev. Palaeobot. Palynol.* **72**: 61–72.

Taylor, W.A. (1993). Megaspore wall ultrastructure in *Isoetes*. *Amer. J. Bot.* **80**: 165–171.

Taylor, W.A. (1994). Tests and applications of a method of quantitative analysis of fossil and extant lycopsid megaspore walls. In: M.H.Kurmann and J.A.Doyle (editors). Ultratructure of fossil spores and pollen, pp. 39–52, Royal Botanic Gardens, Kew.

Taylor, W.A. and Taylor, T.N. (1987). Subunit construction of the spore wall in fossil and living lycopods. *Pollen & Spores* **29**: 241–247.

Taylor, W.A. and Taylor, T.N. (1988). Ultrastructural analysis of selected Cretaceous megaspores from Argentina. *J. Micropalaeontol.* **7**: 73–87.

Thomas, B.A. (1992). Paleozoic herbaceous lycopsids and the beginning of extant *Lycopodium* sens. lat. and *Selaginella* sens. lat. *Ann. Missouri Bot. Gard.* **79**: 623–631.

Thomas, B.A. and Quansah, N. (1991). The palaeobotanical case for dividing *Selaginella* (Selaginellaceae: Pteridophyta). *Fern Gaz.* **14**: 59–64.

Tryon, A.F. and Lugardon, B. (1978). Wall structure and mineral content in *Selaginella* spores. *Pollen & Spores* **20**: 315–340.

Tryon, R. and Tryon, A.F. (1982). Ferns and allied plants. 857 pp. Springer Verlag, New York.

Walton, J. and Alston, A.H.G. (1938). Lycopodiinae. In: F. Verdoorne (editor). Manual of Pteridology, pp. 500–506, M. Nijoff, The Hague.

APPENDIX: Mesozoic megaspores that have been examined ultrastructurally and the method used (TEM — transmission electron microscopy, SEM — scanning electron microscopy, Light — light microscopy).

SPECIES	AUTHOR(S)	METHOD USED
Aquatic ferns		
Arcellites disciformis	Hueber, 1982	SEM
Arcellites disciformis	Kovach and Dilcher, 1988	SEM
Arcellites plicatus	Li and Batten, 1986	TEM
Arcellites pyriformis	Hueber, 1982	SEM
Arcellites santacrucensis	Baldoni, 1987	TEM
Arcellites santacrucensis	Baldoni and Taylor, 1988	TEM
Arcellites yiminensis	Li and Batten, 1986	TEM
Ariadnaesporites ariadnae	Kovach and Dilcher, 1988	SEM
Ariadnaesporites varius	Hall, 1975	TEM
Glomerisporites pupus	Hall, 1974	TEM
Glomerisporites pupus	Hall, 1975	TEM
Molaspora lobata	Batten, 1988	SEM
Molaspora salinum	Kovach and Dilcher, 1988	SEM
Lycopsids		
Auriculozonospora spongiosus	Archangelsky and Villar de Seoane, 1990	TEM
Bacutriletes ferulus	Koppelhus and Batten, 1989	SEM
Bacutriletes guttula	Archangelsky and Villar de Seoane, 1991	TEM
Bacutriletes majorinus	Koppelhus and Batten, 1989	SEM
Bacutriletes sp.	Taylor and Taylor, 1988	TEM
Bacutriletes triangulatus	Taylor and Taylor, 1988	TEM
Banksisporites dejerseyi	Hemsley and Scott, 1989	TEM
Banksisporites pinguis	Kempf, 1971a	TEM
Banksisporites viriosus	Hemsley and Scott, 1989	TEM
Bowenispora tumidosa	Hemsley and Scott, 1989	TEM
Cabochonicus carbunculus	Batten and Ferguson, 1987	TEM
Cabochonicus carbunculus	Koppelhus and Batten, 1989	SEM
Cabochonicus carbunculus	Archangelsky and Villar de Seoane, 1991	TEM
Cabochonicus gamerroi	Baldoni and Taylor, 1985	TEM
Erlansonisporites cf. *E. erlansonii*	Li et al., 1987	TEM
Erlansonisporites erlansonii	Kovach and Dilcher, 1988	SEM
Erlansonisporites scanicus	Koppelhus and Batten, 1989	TEM
Erlansonisporites septus	Koppelhus and Batten, 1989	SEM
Erlansonisporites sparassis	Taylor and Taylor, 1988	TEM
Erlansonisporites spinosus	Bergad, 1978	TEM
Favososporites feruglioi	Gamerro, 1975	SEM
Herbosisporites pilosus	Li et al., 1987	TEM
Horstisporites harrisii	Bergad, 1978	TEM
Horstisporites iridodea	Taylor and Taylor, 1988	TEM
Horstisporites kendalli	Kempf, 1971a	TEM
Horstisporites semireticulatus	Kempf, 1971b	TEM

SPECIES	AUTHOR(S)	METHOD USED
Horstisporites sp.	Morbelli, 1990	TEM
Hughesisporites patagonicus	Gamerro, 1975	SEM
Hughesisporites patagonicus	Taylor and Taylor, 1988	TEM
Hughesisporites patagonicus	Archangelsky and Villar de Seoane, 1990	TEM
Hughesisporites rugulatus	Archangelsky and Villar de Seoane, 1991	TEM
Hughesisporites sp.	Morbelli, 1990	TEM
Minerisporites dissimilis	Kovach and Dilcher, 1988	SEM
Minerisporites elegans	Archangelsky and Villar de Seoane, 1989	TEM
Minerisporites laceratus	Archangelsky and Villar de Seoane, 1990	TEM
Minerisporites marginatus	Kovach and Dilcher, 1988	SEM
Minerisporites patagonicus	Archangelsky and Villar de Seoane, 1989	TEM
Minerisporites succrassulus	Bergad, 1978	TEM
Nathorstisporites flagellulatus	Dettmann, 1961	Light
Nathorstisporites hopliticus	Dettmann, 1961	Light
Nathorstisporites hopliticus	Kempf, 1971a	TEM
Nathorstisporites muricatus	Hemsley and Scott, 1989	TEM
Paxillitriletes dakotaensis	Kovach and Dilcher, 1988	SEM
Paxillitriletes kachaikense	Baldoni and Taylor, 1987	TEM
Paxillitriletes menendezii	Baldoni and Taylor, 1985	TEM
Paxillitriletes vittatus	Kovach and Dilcher, 1985	TEM
Paxillitriletes vittatus	Kovach and Dilcher, 1988	SEM
Racinospora cryptoreticulata	Bergad, 1978	TEM
Ricinospora laevigata	Li et al., 1987	TEM
Rugotriletes costatus	Batten, 1988	SEM
Rugotriletes diktyotus	Batten, 1988	SEM
Scabratriletes cristatus	Batten, 1988	SEM
Sepisporites annulatus	Li et al., 1987	TEM
Striatriletes minimus	Batten, 1988	SEM
Tasmanitriletes pedinacron	Jux and Kempf, 1971	TEM
Tenellisporites aachenensis	Batten, 1988	SEM
Tenellisporites cf. *T. aachenensis*	Koppelhus and Batten, 1989	SEM
Tenellisporites pubescens	Batten, 1988	SEM
Thylakosporites retiarius	Hueber, 1982	SEM
Thylakosporites retiarius	Koppelhus and Batten, 1989	TEM
Trileites murrayi	Kempf, 1971a	TEM
Trileites persimilis	Koppelhus and Batten, 1989	TEM
Trileites turbanaeformis	Kempf, 1971a	TEM
Verrutriletes dubius	Batten, 1988	SEM
Verrutriletes tomtaylori	Archangelsky and Villar de Seoane, 1990	TEM

Taylor, W.A. (1994). Tests and applications of a method of quantitative analysis of fossil and extant lycopsid megaspore walls. In: M.H. Kurmann and J.A. Doyle (editors). Ultrastructure of fossil spores and pollen, pp. 39–52, Royal Botanic Gardens, Kew.

3. TESTS AND APPLICATIONS OF A METHOD OF QUANTITATIVE ANALYSIS OF FOSSIL AND EXTANT LYCOPSID MEGASPORE WALLS

WILSON A. TAYLOR

Department of Biology, University of Wisconsin-Eau Claire, Eau Claire, WI 54701, U.S.A.

Abstract

Quantitative methods of analysis show promise in determining the affinities of dispersed megaspores in the fossil record. This contribution tests some assumptions of these methods, including 1) intrageneric structural differences in modern *Selaginella* and *Isoetes* are not as great as intergeneric differences, 2) structural differences within a single megaspore are not as great as intergeneric differences, and 3) developmental stage of a megaspore should not have a substantial effect on quantitative analysis. Trials conducted for this investigation in combination with previous developmental studies support the validity of these assumptions. Despite slight variation in precise numerical results compared to previous studies using modern *Isoetes* and *Selaginella*, general relationships seem to hold true and results appear to be repeatable. Application of quantitative analyses to fossil megaspores of known taxonomic position results in placement of isoetalean megaspores far from their modern isoetaceous counterparts, based on megaspore wall porosity and wall unit orientation. This discrepancy may be affected by mode of preservation (primarily compression) and/or structural changes as a result of different preparatory techniques (SEM vs. TEM). With regard to taxa of "intermediate" taxonomic status, the range of *Oxroadia* overlaps both *Selaginella* and *Isoetes*, but the mean is significantly different from the former but not the latter, while there is no significant difference between the megaspores of *Selaginellites crassicinctus* and those of modern *Selaginella*.

Introduction

Dispersed megaspores comprise a significant percentage of plant remains in the fossil record. Their utility as stratigraphic indicators has been demonstrated (e.g., Sweet, 1979; Hughes, 1958), but their potential as indicators in biological investigations has been hampered by the difficulty in conclusively determining the taxonomic affinities of the plants that produced them. This limitation has been addressed using ultrastructural characteristics of the megaspore wall, and the resulting determinations were incorporated into a palaeoecological investigation (Kovach and Dilcher, 1985). More recently, an attempt has been made to quantify megaspore wall ultrastructure, resulting in a test for differentiating modern isoetalean and selaginellalean megaspores, as well as fossil representatives of both these groups (Kovach, 1989).

The present investigation was undertaken as both a test and an application of this quantitative method. The tests involved: 1) assessing more thoroughly the structural variability of megaspore walls found in modern lycopsids, 2) assessing the structural variability of different regions in a single modern megaspore wall, and 3) assessing different developmental stages of modern megaspore walls. The applications involved first applying the quantitative methods to the megaspores of two fossil taxa of known taxonomic affinity – *Chaloneria* and *Cylostrobus*. *Chaloneria* is a genus of Late Carboniferous, cormose heterosporous lycopsids included in the Isoetales (Pigg and Rothwell, 1983), and produced megaspores assignable to the dispersed spore genus *Valvisisporites*. The genus *Cylostrobus* was established by Helby and Martin (1965) for small bisporangiate cones from the Triassic of Australia. These cones are thought to have been produced by the cormose lycopsid *Pleuromeia longicaulis* (Retallack, 1975). The examination of these taxa also represents a test of the discriminating power of the quantitative methods on specimens older than the Cretaceous.

For the second application of the quantitative methods, two lycopsid taxa were selected that have received some attention particularly with regard to the characteristics they share with both modern *Isoetes* and *Selaginella*. *Selaginellites crassicinctus* is a Late Carboniferous cone that was described by Hoskins and Abbott (1956). Based upon vegetative similarities to certain members of the genus *Selaginella* – most notably *S. selaginoides* – and organic attachment to the rooting organ *Paurodendron*, this plant was transferred to *Selaginella* as *S. fraipontii* (Schlanker and Leisman, 1969). This assignment has been challenged in the literature (Rothwell and Erwin, 1985). Microspores of *S. crassicinctus* have certain ultrastructural characteristics that exist in microspores of *Isoetes* and some species of *Selaginella* (Taylor and Taylor, 1990). While debate continues as to the precise taxonomic position of this organism, it is hoped that spore wall ultrastructure may play a role in firmly establishing the affinities of this plant. *Oxroadia* is a low trailing lycopsid described by Alvin (1965) from the Lower Carboniferous of Scotland. Recent whole plant reconstructions of this plant place it close to *Paurodendron* in cladistic analyses (Bateman et al., 1992).

Materials and Methods

Living and preserved specimens of several species of *Isoetes* and *Selaginella* were used as megaspore sources for this investigation (see Table 1). Megaspores of *Chaloneria* (*Valvisisporites*) were removed from isolated sporangia contained in calcium carbonate petrifactions (coal balls) from Pennsylvanian (Upper Carboniferous) sediments of the mid-continent United States. Cellulose acetate peels of the type material of *Selaginellites crassicinctus* (Upper Carboniferous) from the Cleveland Museum of Natural History were dissolved in acetone for recovery of megaspore wall fragments. The block that contains the type specimen has been lost. Dispersed megaspores assignable to *Setispora subpalaeocristata* identical to those found in megasporangia of *Oxroadia gracilis* from Oxroad Bay, East Lothian, Scotland, and megaspores from *Cylostrobus sydneyensis* from the Triassic Turimetta Head locality of Australia were also obtained. All materials were prepared for transmission electron microscopy (TEM) using standard techniques (fixed in glutaraldehyde [2.5%; 1 hour], post-fixed in OsO_4 [1 hour], dehydrated in a graded ethanol series, embedded in Spurr resin [acetone intermediate], sectioned with a diamond knife, sections collected on 1 × 2 mm formvar coated slot-grids and stained). Enlarged cross-sections of megaspore walls were analyzed for porosity and wall unit orientation. Porosity estimates are based on an average count of 300 points/micrograph on a random dot overlay. Data on wall unit orientation (r-value) are based on measurements of 30 angles/section. These methods were developed and are described in detail by Kovach (1989). The total number of spores/species, sections/spore, and average r-value and porosity are given in Table 1.

Results

Of the four megaspores described here, only one – *Valvisisporites* (Gastaldo, 1981; Taylor, 1990) – has been previously studied ultrastructurally. Preliminary results on three other taxa are presented here, with a more thorough description to follow in a subsequent publication.

A. *Cylostrobus sydneyensis*

Megaspores removed from *Cylostrobus sydneyensis* are assignable to the dispersed spore genus *Banksisporites*. The megaspores (Fig. 1) are highly compressed and have an approximate equatorial diameter of 600 μm and a smooth to slightly pitted surface. The slightly elevated laesurae are sinuous in places and extend nearly to the spore equator. At the ultrastructural level (Fig. 2), the sporoderm consists of tightly appressed, flattened wall units, which are thickest near the outer surface (900 nm) and thin gradually to the inner surface (70 nm). The overall average unit thickness is approximately 200 nm. Fifteen to 20 layered units are present on any one radius. Units of adjacent layers (outside and inside) are often partially fused. Single units are laterally traceable (parallel to the circumference of the megaspore) for a considerable distance (6 μm or more). Some units have limited lateral extent (as low as 100 nm). These smaller units are more common within 1 μm of the inner surface.

B. *Chaloneria*

Megaspores produced by this plant are assignable to the dispersed spore genus *Valvisisporites*. They are 630–970 μm in proximal diameter and are highly variable with respect to two prominent surface features. The elaboration of the junction of the three laesurae at the proximal pole – the gula – may be prominent, absent, or any condition in between. The bladder-like swellings at the ends of the laesurae – the auriculae – also may be prominent, absent or intermediate in size (Fig. 3). Ultrastructurally, the exospore is 8–9 μm in thickness, consisting of units whose precise thickness is difficult to discern given the degree of compression, particularly adjacent to the spore lumen (Fig. 4).

C. *Oxroadia gracilis*

Megaspores from this Early Carboniferous cone are assignable to the dispersed spore taxon *Setispora subpalaeocristata*. These megaspores are approximately 2 mm in proximal diameter. A dense collection of branched, flattened processes is attached to the megaspore surface directly over the proximal suture (Fig. 5). The laesurae extend to the spore equator. Interradial areas close to the proximal pole are sparsely ornamented to varying degrees with similar although reduced processes like those over the laesurae (Fig. 5). Scanning electron microscopy (SEM) of the inner sporoderm reveals interconnected flattened units, which occasionally coalesce to form radiating structures (Fig. 6, arrow). Under TEM, these various degrees of coalescence provide all manner of cross-sectional shapes, from round "rods" as small as 0.2 μm in diameter, grading continuously up to flattened "plates" 1.7 μm in thickness and 14 μm in lateral extent (Fig. 7). A poorly defined but distinct layering of the units is present in cross-section, with only 10–20% of the units showing clear fusion between layers. The central region of each of the wall units, including the ornamental surface processes, stains darker than the surface. An inner separable layer is present. Overall sporoderm thickness is approximately 30 μm.

D. *Selaginellites crassicinctus* (cone of *Paurodendron*)

Megaspores of this cone are approximately 400 μm in proximal diameter. The sporoderm is approximately 13 μm in thickness and is constructed of a highly interconnected network of wall units, which are thickest and least compressed in the center 50% (Fig. 8). Units become thinner and more compressed toward the spore surface and toward the spore lumen.

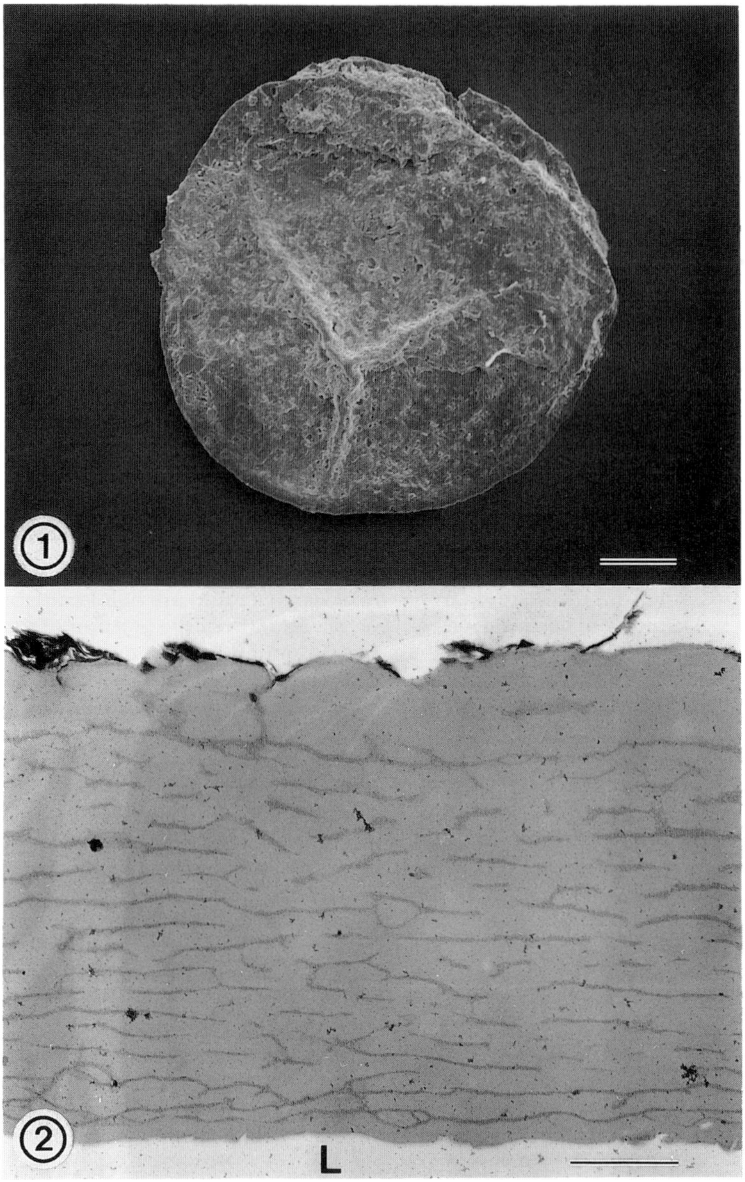

FIGS. 1–2. Megaspores of *Cylostrobus syndeyensis* from the Triassic of Australia. **1**: Proximal view. Size bar = 100 μm. **2**: Exospore cross-section. Size bar = 1 μm.

FIGS. 3–4. Megaspores of *Chaloneria* from the Upper Carboniferous of mid-continent North America. **3**: Six megaspores from a single megasporangium. Note the variability of gula and auriculae structure. Size bar = 500 μm. **4**: Exospore cross-section of megaspore. Size bar = 5 μm.

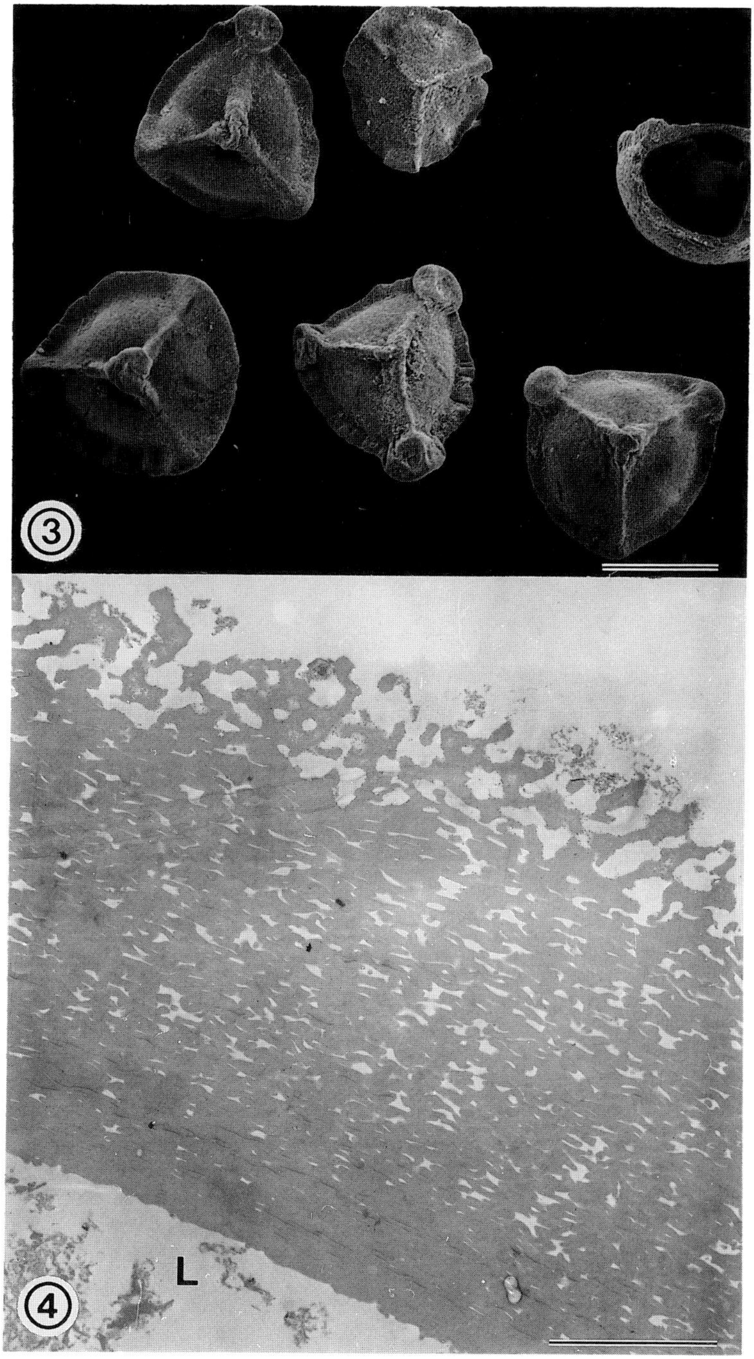
3
4
L

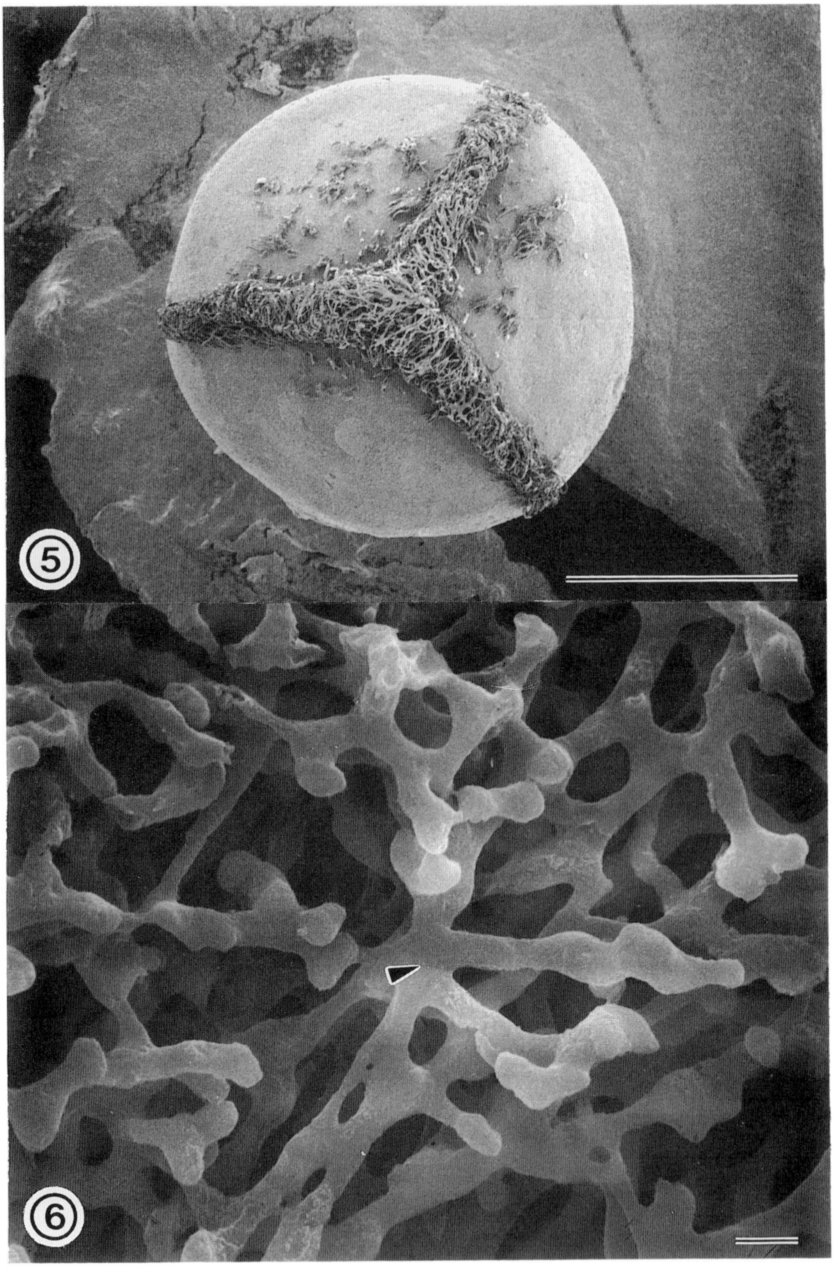

FIGS. 5–6. Megaspores of *Oxroadia gracilis* from the Lower Carboniferous of Scotland. **5**: Proximal view of a megaspore. Size bar = 1mm. **6**: High magnification view of inner exospore units. Arrow indicates a coalescence point of several of the flattened wall units. Size bar = 2 μm.

FIGS. 7–8. *Oxroadia* (Fig. 7) and *Selaginellites crassicinctus* from the Upper Carboniferous of mid-continent North America (Fig. 8). **7**: Exospore cross-section of megaspore. The inner separable layer is not visible in this micrograph. Size bar = 5 μm. **8**: Exospore cross-section of a megaspore from the type specimen. Size bar = 2 μm.

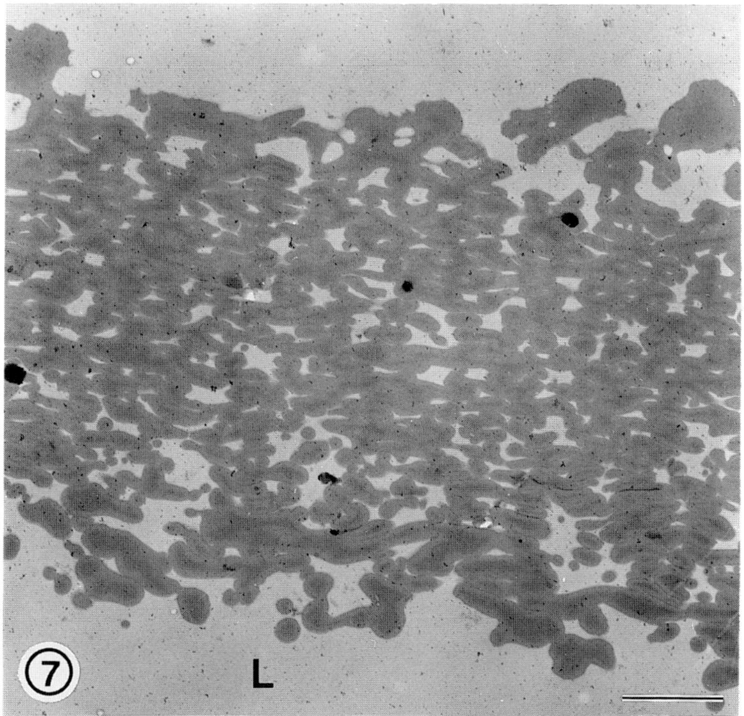
7
L

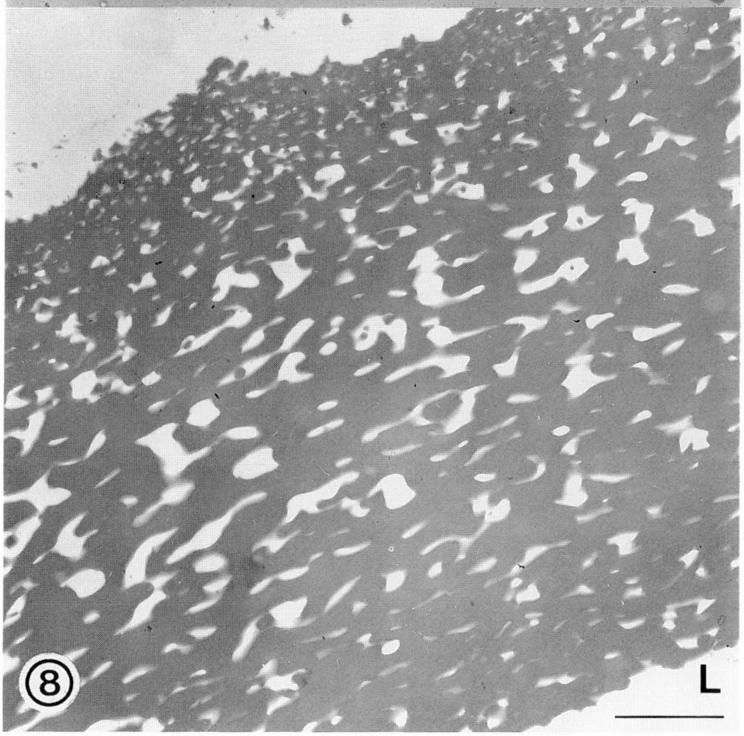
8
L

TABLE 1. Details of specimens examined including source of material, number of spores, number of sections, average porosity and r-values.

Taxon	Source of material	Number of spores	Number of sections	Porosity	r-value
SELAGINELLA					
S. fissidentoides	Royal Botanic Gardens, Kew Herbarium	1	2	30%	.94
S. elmeri	Royal Botanic Gardens, Kew Living Collections Division, no. 1971–2023	1	1	15%	.98
S. viridangula	Royal Botanic Gardens, Kew Living Collections Division, no. 1971–2760	1	1	16%	.94
S. pilifera	Royal Botanic Gardens, Kew Herbarium	1	2	33%	.92
S. usta	Royal Botanic Gardens, Kew Living Collections Division, no. 1963–59908	1	5	16%	.92
ISOETES					
I. flaccida	Ohio State University Herbarium, no. 3339	1	1	51%	.92
I. echinospora	Ohio State University Herbarium, no. 2271	1	1	37%	.93
I. velata	Ohio State University Herbarium, no. N79	1	1	28%	.96

TABLE 1 cont.

Taxon	Source of material	Number of spores	Number of sections	Porosity	r-value
ISOETES contd.					
I. durieui	Ohio State University. Herbarium, no. N104	1	1	38%	.97
I. lacustris	Ohio State University Herbarium, no. 2770	1	1	40%	.98
I. melanopoda	Fairbanks, Louisiana Coll. R.D. Thomas	3	3	46%	.93
CHALONERIA (*Valvisisporites*)	Ohio State University Palaeobotanical Collections, Coal Ball no. 7021 G top	3	11	8%	.92
CYLOSTROBUS (*Banksisporites*)	N.S.W. Australia Turimetta Head	1	6	15%	.96
SELAGINELLITES					
S. crassicinctus	Cleveland Museum of Natural History Type material	1	8	18%	.95
OXROADIA	Oxroad Bay, Scotland R. Bateman	1	5	30%	.95

E. Quantitative tests

The results of quantitative analysis carried out on the specimens described above as well as *Isoetes* and *Selaginella* are shown in Fig. 9. The horizontal axis represents the product of a measure of wall unit orientation with respect to the megaspore surface (r-value) and porosity (Kovach, 1989). Average values for each of these parameters are included in Table 1.

Discussion

In his initial investigation, Kovach (1989) achieved the best quantitative separation of modern *Isoetes* and *Selaginella* megaspores based on a product of r-value and porosity. As a test of repeatability, several species of *Selaginella*, representing two distinctly different structural types – the "laminar" type (Taylor, 1989), e.g., *Selaginella viridangula*, and the "laterally fused" type (Taylor, 1989), e.g., *Selaginella elmeri* – were subjected to quantitative analysis. In addition, proximal, lateral and distal sporoderm cross-sections of *Selaginella usta* were compared to see if different regions of the same spore could give variant results. Neither the different structural types nor the different wall regions showed any significant differences at the 0.05 level using pairwise Watson's F-tests of two circular means. These statistics were pooled to provide data on the genus as a whole. The results are shown in Fig. 9 and are generally consistent with, but slightly higher than those of Kovach (1989).

A hypothetical source of error in this type of analysis involves megaspores at different developmental stages. In both *Selaginella* and *Isoetes*, the arrangement and orientation of megaspore wall units are established early in development, and subsequent changes involve only the enlargement of the units (Taylor, 1991, 1992). Analysis of two developmental stages of *Isoetes melanopoda* megaspores produced values consistent with those of mature megaspores of *Isoetes*. This suggests that the increases in wall unit thickness and porosity are proportional.

Less ultrastructural variation exists in megaspore walls of *Isoetes* (Taylor, 1993) than in those of *Selaginella*. Mature megaspores of six species and two early developmental stages showed no significant differences in wall unit orientation at the 0.05 level using pairwise Watson's F-tests of two circular means. All of these species and developmental stages were combined to serve as data on the genus as a whole.

A. A test of quantitative analysis using megaspores of known affinity

How do fossil isoetaleans compare with modern forms with regard to quantifiable characteristics?

Both *Cylostrobus* and *Chaloneria* megaspores fall solidly on the selaginellalean side of the graph. The range of *Chaloneria* is lower and the mean significantly different (Table 2) than either *Cylostrobus* or *Selaginella* (Fig. 9). *Cylostrobus* megaspores are not successfully separated from *Selaginella* using these techniques. Both *Cylostrobus* and *Chaloneria* are significantly different from *Isoetes*. Several factors may contribute to this discrepancy. First, *Cylostrobus* is preserved as a compression fossil and its megaspores have lost much of the porosity that may once have been present. In the case of *Chaloneria*, there may be a difference in sporoderm thickness between specimens viewed with SEM and those viewed with TEM. TEM preparation may result in a comparable loss of porosity and decrease in wall thickness to that seen in compressed fossils. However, specimens of *Valvisisporites* from authigenically cemented nodules sectioned by Gastaldo (1981) showed no such compression. Furthermore, due to the large depth of field of the SEM, caution must be exercised when attempting to measure cross-sectional thicknesses. Investigations are continuing to confirm or refute these preparatory alterations. Ultimately, preservational and preparatory changes resulting in compression constitute the most serious barriers to meaningful use of these quantitative techniques.

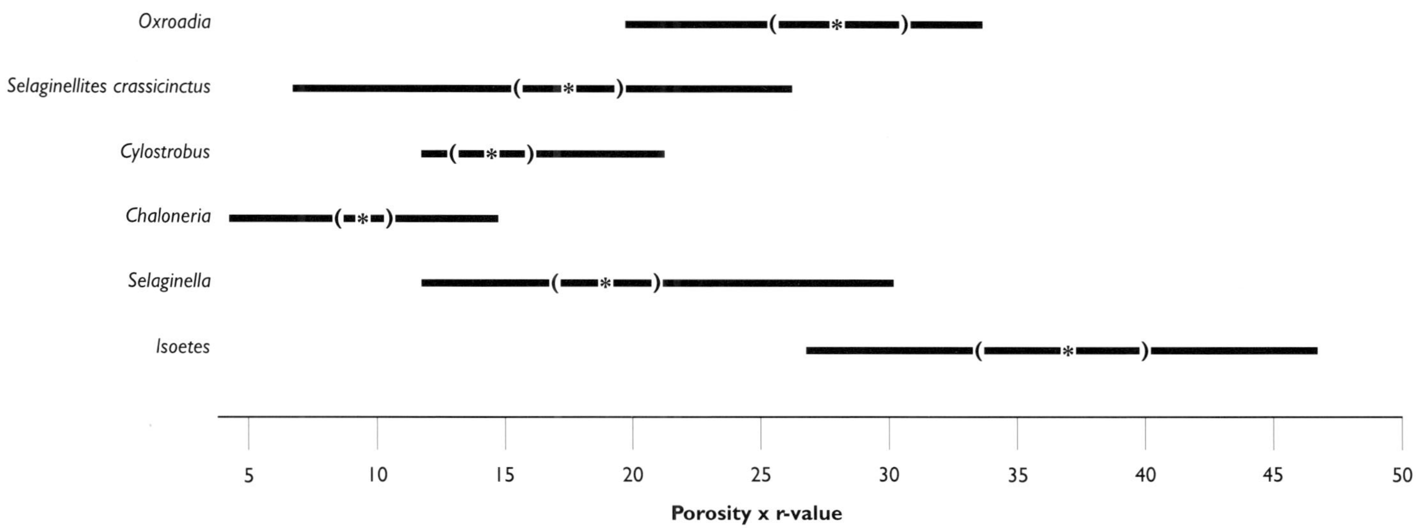

Fig. 9. Graph summarizing quantitative analyses of fossil and extant lycopsid megaspore walls. Horizontal line indicates the range, asterisk represents the mean and the parentheses indicate the standard error.

B. A test using plants of "intermediate" taxonomic status

How do the quantifiable characteristics of "intermediate" fossil forms compare with modern forms?

There is no significant difference between the statistics of *Selaginellites crassicinctus* megaspores and those of modern *Selaginella* (Table 2). The range of *Oxroadia* overlaps both *Selaginella* and *Isoetes*, but the mean is significantly different from the former but not the latter (Fig. 9; Table 2). *Selaginellites crassicinctus* and *Oxroadia* are probably neither selaginellalean nor isoetalean, but share several plesiomorphic characteristics with modern Selaginellales (R. Bateman, pers. comm.).

TABLE 2. T-tests comparing megaspores from a variety of fossil and living lycopsids based on combined measure of wall porosity and r-value. Upper half of matrix contains t values and degrees of freedom (in parentheses), lower half contains the corresponding probabilities. Asterisks indicate means demonstrated to be significantly different.

	Selaginella	*Isoetes*	*Chaloneria*	*Cylostrobus*	*S. crass*	*Oxroadia*
Selaginella	–	4.64(14)	4.59(20)	1.72(15)	.69(17)	2.40(14)
Isoetes	0.00*	–	10.91(14)	6.73(9)	5.07(11)	2.10(8)
Chaloneria	0.00*	0.00*	–	3.17(15)	3.74(17)	8.48(14)
Cylostrobus	>0.05	0.00*	<0.01*	–	1.05(12)	4.76(9)
S. crass	>0.05	0.00*	<0.01*	>0.05	–	2.99(11)
Oxroadia	<0.05*	>0.05	0.00*	<0.01*	<0.05*	–

C. General discussion

Kovach (1989) achieved preliminary success at differentiating dispersed Cretaceous megaspores of the Selaginellales and Isoetales. In this investigation, isoetalean megaspores from lower in the geologic column were not successfully identified, due perhaps to compression, preparatory techniques, or taxonomic differences at the familial level (*Valvisisporites* – Chaloneriaceae; *Cylostrobus* – Pleuromeiaceae; *Isoetes* – Isoetaceae). If compression is the issue, the question arises, how much is too much? This is particularly problematical when the only measure of too much compression of an isoetalean megaspore is its "erroneous" placement on the selaginellalean side of the graph.

The possible utility of these techniques depends upon whether the effects of structural changes resulting from compression/preparation mask the real differences in the various megaspores. Additional analyses of uncompressed and/or variously prepared specimens will be needed to support or refute the effects of structural changes. To these ends, statistical comparison of porosity alone, r-value alone, or some other parameters may prove useful in separating structurally altered specimens.

Even if these preservational/preparatory changes can be compensated for (perhaps by computer image processing?), the issue of the real significance of the wall structure remains. The structural differences may reflect real evolutionary changes within the isoetalean clade, or ecological factors affecting megaspore size. In the latter case, one would not expect to see a consistent trend of change through time, but changes related to life habit and reproductive strategy specific to each organism. There is some support for the view that wall structure is correlated with dispersal strategy in Pennsylvanian lycopsids (Taylor, 1990). Once again, additional analyses are required. Ultimately, the repeatability and rigorous manner in which these structural organizations can be treated using these quantitative techniques provide an important additional tool that can be used to augment descriptive methods.

Acknowledgements

The author thanks Drs. Shya Chitaley, Kathleen Pigg, and Richard Bateman for fossil material used in this investigation, Dr. Warren Kovach for the statistical package used for the quantitative analyses, and especially Ms. Kelly Ohlsson and Mr. Jeffrey Lauritzen for many painstaking hours drawing lines and counting dots. Funds for travel to Aix-en-Provence were provided by the University of Wisconsin-Eau Claire University Research and Creative Activities Grants Program.

References

Alvin, K.L. (1965). A new fertile lycopod from the Lower Carboniferous of Scotland. *Palaeontology* **8**: 281–293.

Bateman, R.M., DiMichele, W.A. and Willard, D.A. (1992). Experimental cladistic analysis of anatomically preserved arborescent lycopsids from the Carboniferous of Euramerica: an essay on paleobotanical phylogenetics. *Ann. Missouri Bot. Gard.* **79**: 500–559.

Gastaldo, R.A. (1981). An ultrastructural and taxonomic study of *Valvisporites auritus* (Zerndt) Bhardwaj, a lycopsid megaspore from the middle Pennsylvanian of southern Illinois. *Micropaleontology* **27**: 84–93.

Helby, R. and Martin, A.R.H. (1965). *Cylostrobus* gen. nov., cones of lycopsidean plants from the Narrabeen Group (Triassic) of New South Wales. *Austral. J. Bot.* **13**: 389–404.

Hoskins, J.H. and Abbott, M.L. (1954). *Selaginellites crassicinctus*, a new species from the Desmoinesian Series of Kansas. *Amer. J. Bot.* **43**: 36–46.

Hughes, N.F. (1958). Palaeontological evidence for the age of the English Wealden. *Geol. Mag.* **95**: 41–49.

Kovach, W.L. (1989). Quantitative methods for the study of lycopod megaspore ultrastructure. *Rev. Palaeobot. Palynol.* **57**: 233–246.

Kovach, W.L. and Dilcher, D.L. (1985). Morphology, ultrastructure, and paleoecology of *Paxillitriletes vittatus* sp. nov. from the mid-Cretaceous (Cenomanian) of Kansas. *Palynology* **9**: 85–94.

Pigg, K.B. and Rothwell, G.W. (1983). *Chaloneria* gen. nov.: heterosporous lycophytes from the Pennsylvanian of North America. *Bot. Gaz.* **144**: 132–147.

Retallack, G. (1975). The life and times of a Triassic lycopod. *Alcheringa* **1**: 3–29.

Rothwell, G.W. and Erwin, D.M. (1985). The rhizomorph apex of *Paurodendron*; implications for homologies among the rooting organs of Lycopsida. *Amer. J. Bot.* **72**: 86–98.

Schlanker, C.M. and Leisman, G.A. (1969). The herbaceous Carboniferous lycopod *Selaginella fraiponti* comb. nov. *Bot. Gaz.* **130**: 35–41.

Sweet, A.R. (1979). Jurassic and Cretaceous megaspores. *Contr. Ser. Amer. Assoc. Stratigr. Palynologists* **5B**: 1–29.

Taylor, W.A. (1989) Sporoderm ultrastructure in selected members of the genus *Selaginella. Pollen & Spores* **31**: 251–288.
Taylor, W.A. (1990). Comparative analysis of megaspore ultrastructure in Pennsylvanian lycophytes. *Rev. Palaeobot. Palynol.* **62**: 65–78.
Taylor, W.A. (1991). Ultrastructural analysis of sporoderm development in megaspores of *Selaginella galeottii. Pl. Syst. Evol.* **174**: 171–182.
Taylor, W.A. (1992). Megaspore wall development in *Isoetes melanopoda*: morphogenetic post-initiation changes accompanying spore enlargement. *Rev. Palaeobot. Palynol.* **72**: 61–72.
Taylor, W.A. (1993). Megaspore wall ultrastructure in *Isoetes. Amer. J. Bot.* **80**: 165–171.
Taylor, W.A. and Taylor, T.N. (1990). Persistent ultrastructural features in microspores of heterosporous lycophytes. *Grana* **29**: 219–228.

Lugardon, B. and Brousmiche Delcambre, C. (1994). Exospore ultrastructure in Carboniferous sphenopsids. In: M.H. Kurmann and J.A. Doyle (editors). Ultrastructure of fossil spores and pollen, pp. 53–66, Royal Botanic Gardens, Kew.

4. EXOSPORE ULTRASTRUCTURE IN CARBONIFEROUS SPHENOPSIDS

BERNARD LUGARDON[1] AND CLAUDINE BROUSMICHE DELCAMBRE[2]

[1]Biologie Végétale, Université Paul-Sabatier, 39, allées Jules Guesde, 31000 Toulouse, France
[2]Paléobotanique, URA CNRS 1365, Sciences de la Terre, Université des Sciences et Technologies de Lille, 59655 Villeneuve d'Ascq, France

Abstract

Calamospora, *Laevigatosporites* and *Vestispora* spores extracted from Upper Carboniferous sphenopsid fructifications share several important exospore features: (1) the aperture is characterized by a protruding fold that, in well-preserved specimens, delimits a narrow slit opening inwards and closed at the fold apex by the greatly reduced but unbroken wall; (2) ultrastructurally, the wall consists of a thin inner part (foliated substructure) that lines the spore cavity and can include outwardly separated sheets, and a thick outer layer of amorphous sporopollenin. Well-marked differences in foliated substructure organization distinguish the three genera of *sporae dispersae*, which are not different ontogenetic stages of the same spore type. Structural features and substructure variations of the exospore in these fossil sphenopsids are remarkably similar to those that characterize the exospore of the filicopsids. These similarities suggest that both pteridophyte groups have a common origin. Furthermore, the exospore of living Equisetaceae has a special structure that seems wholly different from the structure in Carboniferous sphenopsids, implying that the exospore of the sphenopsids has undergone extensive evolutionary modifications through geological time.

Introduction

Spores extracted from fructifications of ancient sphenopsids have been investigated by many workers with the light microscope (e.g. Hartung, 1933; Remy, 1959, 1960, 1961; Smith and Butterworth, 1967; Good, 1977; Storch, 1980) and, in recent years, with the scanning electron microscope (e.g. Serret and Brousmiche, 1987; Zodrow and Gao, 1991). These studies have proved that the fossil plants produced spores corresponding to several genera of *sporae dispersae* (called 'sporomorphs' in the text), such as *Calamospora*, *Elaterites*, *Laevigatosporites* and *Vestispora*. They also have raised several problems, two of which are especially puzzling. The diverse sporomorphs are all the more different because some of them show only an exospore, whereas others have in addition a more or less complex extra-exosporal wall. In addition, several authors (Good, 1975; Serret and Brousmiche, 1987) have reported the presence of two or three distinct sporomorphs in the same strobilus. For these reasons, it has been suggested that some of these sporomorphs may correspond to immature forms of other spores.

Ultrastructural studies with the transmission electron microscope, which are especially appropriate for solving such problems, have been few. Kurmann and Taylor (1984) carried out a fine-structure study of *Elaterites triferens*, showing the structure of the exospore, extra-exosporal wall and elaters of this complex spore, and emphasizing several differences between *Elaterites* and the spores of extant *Equisetum*. Also, the ultrastructure of *Calamospora* spores has been described by Brousmiche and Lugardon (1990).

The present study continues this last work and includes investigations of the exospore of *Vestispora* and *Laevigatosporites*.

Material and Methods

All the figured spores were taken from cones collected from borehole cores through the Upper Carboniferous of the Sarre-Lorraine Coalfield (see details in figure captions).

Spores of *Calamospora* were extracted from fructifications attached to vegetative organs belonging to three species of Calamitaceae: *Calamostachys germanica* Weiss (Fig. 1), *C. tuberculata* (Sternberg) Jongmans (Fig. 5) and *Palaeostachya feistmantelii* Nemejc (Fig. 10). The spore morphology is quite homogeneous, although a large size range was observed (Figs. 2, 6, 11). Difficulties inherent in their specific identification were discussed by Serret and Brousmiche (1987). The sporoderm consists of a single wall, the exospore, which is smooth or slightly ornamented, and shows a trilete aperture with rather short (less than 40% of the spore radius) apertural arms.

The typically operculate spores of *Vestispora* were taken from two isolated, rather poorly preserved fructifications. The first, found near a very small sterile fragment of *Sphenophyllum emarginatum* (Brongniart) Brongniart, produced spores that are ornamented with numerous foveolae (Fig. 14) and are attributable to *V. fenestrata* (Kosanke and Brokaw) Wilson and Venkatachala. The second fructification, which looks like *Cingularia* Weiss, contained ovoid *Vestispora* spores with an irregular extrareticulum and a secondary finer reticulum (Fig. 16) that are characteristic of *V. pseudoreticulata* Spode. In this spore type, the smooth exospore is covered with a thick outer wall that bears the specific ornamentation and includes the circular operculum.

The spores of *Laevigatosporites* came from isolated sporangia associated with whorls of sterile foliage so fragmented that it is impossible to determine whether they belong to *Sphenophyllum oblongifolium* (Germar and Kaulfuss) Unger or to *Leeites oblongifolis* Zodrow and Gao. Since species distinctions in the genus *Laevigatosporites* are based only on size and shape characters, the figured monolete, laevigate spores (Figs. 20, 21) are assigned to *L. minimus* (Wilson and Coe) Schopf et al. The sporoderm of this spore type consists only of the exospore.

The sporangia were mechanically removed from fructifications, then macerated in Schultze's solution. Spores were prepared for light microscopy according to the usual protocols (Smith and Butterworth, 1967). Specimens prepared for examination within the scanning electron microscope were sputter coated with 100% gold. The macerated, usually flattened specimens selected for TEM were positioned in parallel in small blocks of agar-agar. The blocks were then treated with 4% paraformaldehyde and 2% osmium tetroxide, dehydrated in a graded series of ethanol, transferred into propylene oxide and embedded in Epon. Before epoxy resin polymerization, each block was arranged in a flat mould in a position suitable for obtaining transverse sections of the flattened spores. The ultra-thin sections were stained with an aqueous solution of uranyl acetate and Reynold's lead citrate.

Observations

A. *Calamospora*

In the non-apertural regions of the spores, the exospore is uniformly thick and consists of two unbroken, compact layers. Both layers are devoid of noteworthy ultrastructural features, and are distinguishable from each other only when they differ in electron density. The inner, usually darker (more electron dense) layer ranges from 30 to 50 nm in thickness depending on the species. The outer layer is about ten times as thick as the inner layer and forms the slight ornamentation of the spore surface (Figs. 4, 13).

The apertural area is more elaborate. The exospore thickens noticeably in the vicinity of the aperture, and forms a protruding apertural fold that is clearly visible in transverse sections of the apertural arms. In well-preserved spores, this fold delimits a narrow slit opening inwards and closed outwards by the greatly reduced but unbroken wall (Figs. 3, 4, 12, 25A). Near the proximal pole, the fold apex forms a narrow ridge that differs in height among specimens and consists of a regular, very thin exospore region bordering the outermost part of the slit (this ridge is especially obvious, though broken, in Fig. 9). In deformed spores, the slit may widen inwards (Fig. 13) and in the most damaged spores (or spores that germinated prior to fossilization) each side of the exospore fold is separated from the other owing to tearing of the wall above the slit top (Figs. 7–9).

Ultra-thin sections reveal significant modifications of the exospore fine structure in this apertural area, particularly because of the interpolation of a third layer with very special characteristics between the two layers that are joined together in the other regions of the wall (Figs. 3, 4, 7, 8, 12, 13). The inner layer borders the apertural slit without any discontinuity. It gradually thickens from the lateral limit of the wall thickening, reaches its maximum thickness (up to ten times its thickness in non-apertural regions) at the base of the slit, and becomes very thin in the upper part of the apertural fold. The median layer is made up of a very complex system of interconnected septa of nearly constant thickness, delimiting numerous small, tangentially elongated cavities that are either empty or, more often, filled with opaque material. It stretches from the limit of the exospore thickening to the basal and middle parts of the inner layer fold. Wall sections with sufficiently marked contrast variations (Figs. 4, 8, 9) show that each septum includes a darker thin sheet, uniformly covered on both sides with a coat of less electron dense material. The coating material consists of sporopollenin similar to that of the outer exospore layer. The various median sheets show the same electron density as the inner wall layer and are connected to this layer. They are foliations partly detached from, but belonging to, the inner structure of the exospore. The outer layer covers the median zone with an unchanged thickness and is traversed by radial channels connecting cavities of this zone with the wall surface (Figs. 8, 9, 13). It is markedly reduced at the aperture, where it joins the apical part of the inner layer fold.

B. *Vestispora*

As noted earlier, this sporomorph is distinguished by the presence of the extra-exosporal wall, which consists of a thin lamellar layer more or less adhering to the exospore, and a very thick, solid layer forming the specific ornamentation of the spores (Figs. 15, 19; this wall, with its curious operculum, merits a further study). The investigated specimens have not shown any exospore with an intact aperture, the two sides of the sectioned apertures sometimes being severely distorted (Fig. 19), more often quite separated and wide apart (Figs. 15, 17, 18). Although the exospore thickness differs considerably in the two studied species, the fine details of the wall are similar.

In distal and lateral areas (Figs. 18, 19), the exospore consists of a thin inner layer and a much thicker outer layer, which are homogeneous, compact, and usually differ slightly in contrast. At the aperture, both layers form a protruding fold and are greatly reduced at the fold apex (where the wall is broken in most studied spores). In the peri-apertural area (Figs. 15, 17–19, 25B), these layers are separated by a median zone showing several tangential cavities, some of them connected to the outer surface of the exospore by radial channels (Fig. 18). Although the details of the fine structure are not entirely clear in the examined ultra-thin sections, it is certain that this zone consists of septa including sheets that are linked to the inner layer. The septa and the intercalated cavities are infrequent and constitute a layer that is thinner and less complex than the corresponding formation of *Calamospora*, but extends appreciably further from the aperture (Fig. 18).

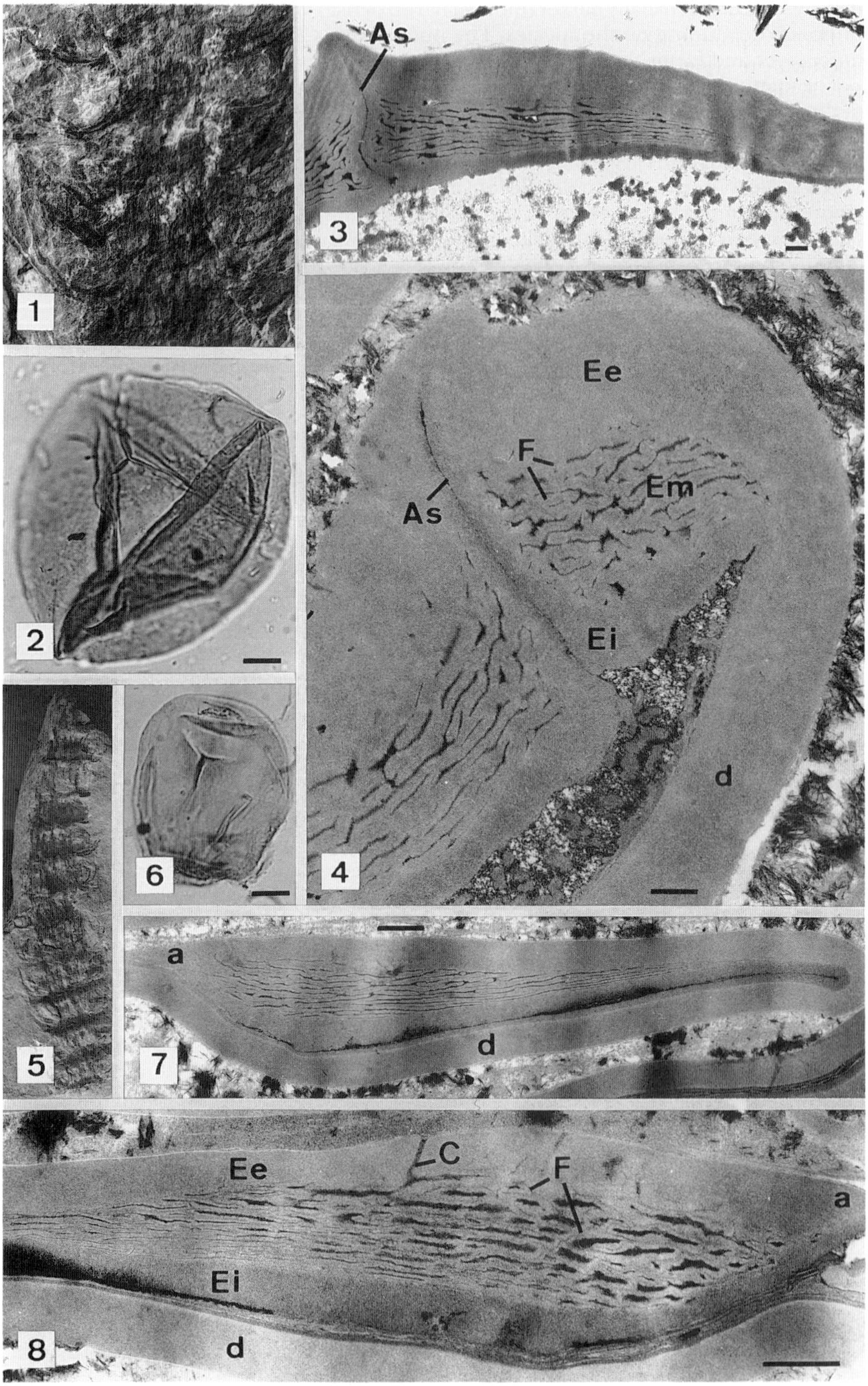
1
2
3
As
4
Ee
F
Em
As
Ei
d
5
6
7
a
d
8
Ee
C
F
a
Ei
d

C. *Laevigatosporites*

The exospore is quite uniform in thickness, with a nearly plain surface, outside the apertural area (Fig. 22). At the aperture, it forms a moderately prominent but well-marked fold, which, in undeformed spores, delimits a very narrow slit slightly widening inwards (Figs. 22–24). The fold apex forms a small, but clearly visible, narrow ridge near the proximal pole (Figs. 22, 23). Sections with adequate variation in contrast unambiguously reveal the bilayered structure of the exospore (Figs. 22–24, 25C). The inner layer is very thin, with a maximum thickness of about 30–40 nm at the apertural fold base. The rest of the wall is composed of the homogeneous outer layer. Minute, punctiform cavities with opaque contents are located in the innermost part of the outer layer, very close to the inner layer. They are relatively abundant in the middle and basal parts of the apertural fold (Figs. 23, 24), but more scattered in the non-apertural exospore regions; some are connected to the wall surface by radial channels.

Unlike *Calamospora* and *Vestispora* exospores, the wall of *Laevigatosporites* appears to be devoid of a periapertural median zone made up of interconnected septa delimiting tangentially elongated cavities. Thin sheets detached from the wall inner layer, similar to those present in the middle of the septa in the two other sporomorphs, seem to be missing in most investigated exospores; only a few specimens show highly localized separation of one sheet from the inner layer. The absence of a complex median zone in these spores is probably correlated with the lack, or the extreme reduction, of the sheets.

Discussion

A. Pteridophyte exospore ultrastructure

The fine structure of spore walls, and especially that of the exospore, is known mainly for spores of extant pteridophytes. Besides their greater availability, extant spores have a considerable advantage over fossil spores in allowing integration of detailed studies of wall ontogeny with the standard ultrastructural description of mature sporoderms. Such combined studies have proved necessary to establish the structural features that characterize the exospores of the different pteridophyte groups (see references in Lugardon, 1990; Tryon and Lugardon, 1991). These features can be divided into two sets: the aperture features, which concern the configuration of the exospore in the apertural area, and the stratification features, which relate to the characteristics of the layers distinguishable in the wall.

The aperture features are usually clearly visible in suitable sections of mature exospores, and are quite different in spores of the filicopsids, homosporous and

Common abbreviations: As, apertural slit; C, channel; d, exospore outside the apertural area; Ee, outer exospore layer; Ei, inner exospore layer; Em, exospore median zone; F, separate sheets of foliated exospore substructure.
Scale bars equal 10 μm for LMs, 5 μm for SEMs, 0.5 μm for TEMs.

FIGS. **1–8**. *Calamostachys germanica*, Hargarten 2, 498.40 m, first levels of Lorraine Stephanien (Figs. **1–4**) and *C. tuberculata*, La Houve terril, upper Westphalien D or first levels of Lorraine Stephanien (Figs. **5–8**). Strobilus and spores. **1**: part of strobilus, × 3. **2**: *Calamospora* spore, the trilete aperture near the top, LM × 500. **3**: exospore section, well-preserved apertural area showing the three-layered wall structure on both sides of the narrow apertural slit, TEM, × 6000. **4**: detail of apertural area ultrastructure, sheets discernible within septa separating cavities of the peri-apertural median zone, TEM, × 12,000. **5**: strobilus, × 1. **6**: *Calamospora*, proximal face, LM, × 500. **7**: one side of open aperture (a, apex of apertural fold) and exospore outside the apertural area; the fine structure is quite similar to that of the previous species, despite the great difference in wall thickness, TEM, × 12,000. **8**: three-layered exospore in apertural area (a, apex of aperture fold, the inner layer corroded in places, sheets distinct inside septa of the median zone, canals connecting cavities of this zone to the exospore outer surface, TEM, × 20,000.

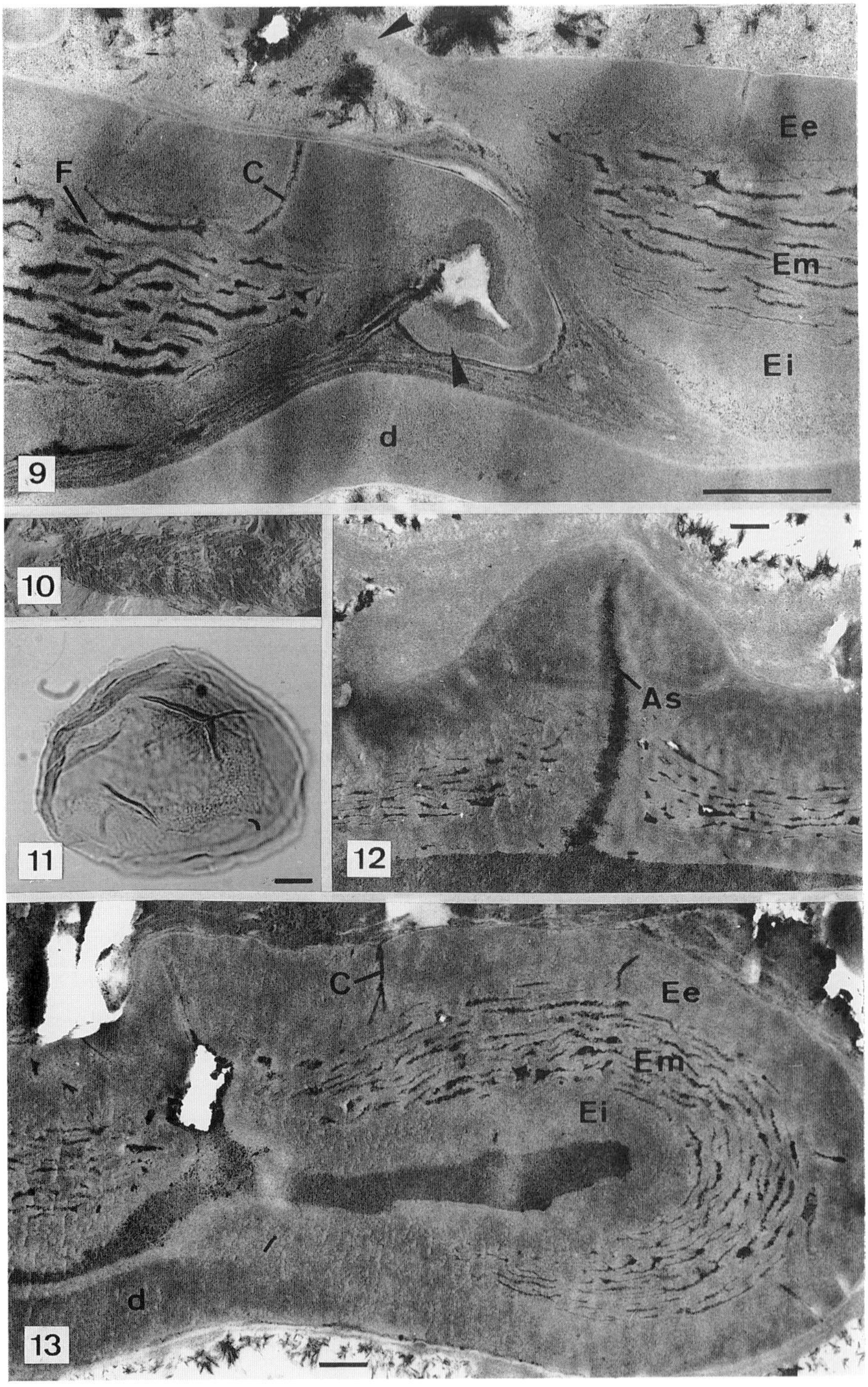
Ee
F
C
Em
Ei
d
9
10
11
As
12
C
Ee
Em
Ei
d
13

heterosporous lycopsids, and living sphenopsids (i.e. Equisetaceae). They supply valuable information relating spores to one of these major groups, without the necessity for ontogenetic studies (yet highly beneficial in that they elucidate the special mode of formation of each aperture type).

The features of exospore stratification are more informative than the aperture features, and can distinguish spores of closely related plant groups, but often they are largely obscured in the mature wall. Observations of successive stages of sporogenesis reveal that the exospore includes, usually in the innermost part, thin structured elements that are either short, overlapping lamellae, or long, interconnected sheets, depending on the pteridophyte groups. These elements develop early, before formation of the outer wall layer consisting of amorphous sporopollenin. They are arranged characteristically in each group of closely related plants; frequently they are combined with small cavities, fissures and channels. They are quite well-defined in early stages of exospore formation, but they often become largely or wholly indistinguishable from the amorphous sporopollenin in the course of wall maturation. Therefore, in ultra-thin sections of mature exospores produced by many living species, very small details like cavities or channels and, in the most favourable cases, more or less marked differences in contrast, are the only signs of the presence and the specific arrangement of the structured elements included in the wall. Consequently, ontogenetic studies in some species belonging to the various pteridophyte groups have been essential for determining the ultrastructural peculiarities of each exospore type and enabling us to infer these peculiarities from the details that remain visible in sections of mature exospores.

The exospore of fossil spores offers similar potential advantages and disadvantages. Sections of the apertural area are especially informative. The main characteristics of this area may be observed even in badly preserved specimens. Direct comparisons with the aperture types of living groups are easy and supply reliable information on the affinities of source plants. The determination of the wall fine structure is more arduous and usually requires well-preserved sporoderm. As in extant spores, ultra-thin exospore sections often show few details, and differences in contrast that distinguish wall components are evident only in favourable cases. The observations necessitate cautious interpretation, but, with the help of data gained from the fine studies of Recent spores, they generally allow comprehension of the ultrastructural features of the walls being investigated.

B. Exospore features in fossil sphenopsids

The three sporomorphs studied here illustrate well the potential of fossil spore walls for identifying structural features, as well as deducing evolutionary relationships. The specimens selected for the micrographs (Figs. 1–24) are moderately well-preserved and show fairly sharp contrast variations in section.

The aperture characteristics are clearly visible in *Calamospora* and *Laevigatosporites* specimens in which the exospore fold delimiting the typical apertural slit is still apically closed by the very reduced but unbroken wall. These characteristics are less evident, but are recognizable in specimens having 'open' apertures with each side of the fold widely separated from the other, such as those of *Vestispora*. They also are quite recognizable in the apertural area section of *Elaterites triferens* illustrated by Kurmann and Taylor (1984: Fig. 2C), and it may be assumed that they are shared by all spores of ancient sphenopsids that have a protruding aperture.

FIGS. **9–13**. *Calamostachys tuberculata* (Fig. 9) and *Palaeostachys feistmantelii* (Figs. 10–13), La Houve terril, upper Westphalien D or first levels of Lorraine Stephanian. Strobilus and spores. **9**: detail of aperture, both sides separated and the exospore torn at the level of the especially well-developed narrow ridge (arrow heads) that initially bounded the apertural slit apex, the median zone sheets evident within septa, TEM, × 35,000. **10**: part of fructification, × 1. **11**: *Calamospora*, proximal face, LM, × 500. **12**: section of aperture, TEM, × 10,000. **13**: aperture widening inwards, three-layered peri-apertural area, wall with simpler structure outside this area, TEM, × 12,000.

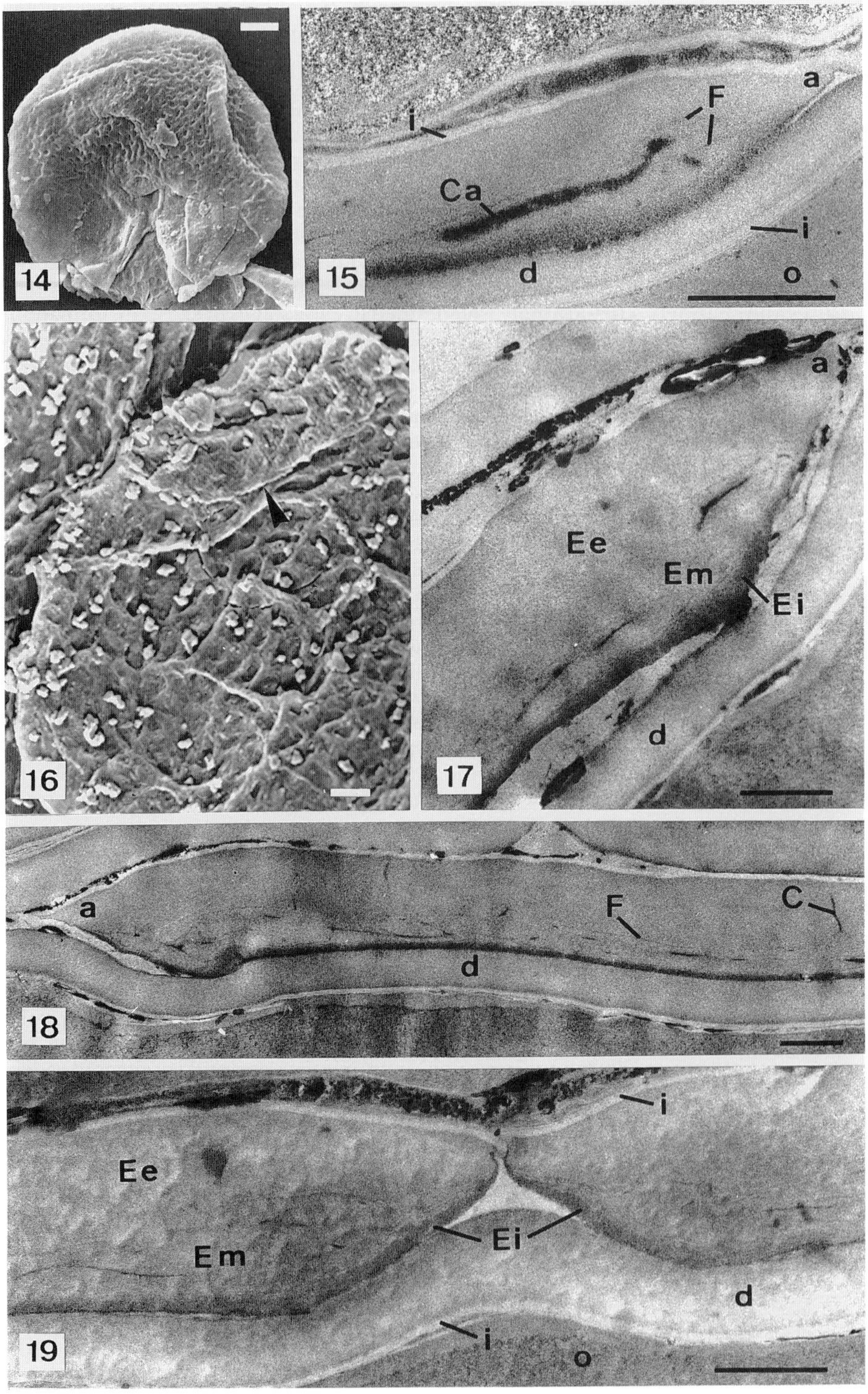
14
15
i
F
a
Ca
i
d
o
16
17
a
Ee
Em
Ei
d
18
a
F
C
d
19
i
Ee
Ei
Em
d
i
o

The exospore composition and ultrastructural details are readily discernible in sections due to the variations in contrast. In every case, the wall consists of two components. The usually darker one forms the thin, unbroken inner layer plus the separate, interconnected sheets extending above, and joining with, this layer (Fig. 25A–C). The inner layer is rather well-developed in *Calamospora*, less so in *Vestispora*, and even less in *Laevigatosporites*. In all three sporomorphs, it is thinner at the fold apex, particularly thick in the fold base region, and of a uniform thickness outside the apertural area. The separate sheets are laminate elements partly detached from the inner layer. They signify that this whole layer is probably composed of similar sheets that are closely fused together, eventually becoming indistinguishable. It is probable that the number of fused sheets is roughly correlated with that of separate sheets, and is relatively great in *Calamospora* (especially in the peri-apertural area), less in *Vestispora*, and reduced to very few (possibly one) in those specimens of *Laevigatosporites* that do not show any sheet separated from the inner layer. The distinct and fused sheets thus constitute a 'foliated substructure' varying in complexity among exospore types.

The second component of the exospore is amorphous sporopollenin probably analogous to that forming the outer part of the wall in extant spores. This material partly occupies the small spaces between the separate sheets of the substructure, and constitutes the unbroken, thick outer layer of the wall. No details are visible in the thick outer layer, except radial channels joining the exospore surface to subjacent cavities. These exospore cavities are obviously of two different sorts. In *Calamospora* and *Vestispora*, which have sheets distinct from the inner layer, all cavities are situated between the sheets; they are relatively large and elongated parallel to the sheets. In *Laevigatosporites*, which lacks separate sheets, the cavities are simple, minute perforations that are evenly located close to the inner layer. Both sorts of cavities presumably have different ontogenetic origins depending on the presence or absence of separate sheets in the substructure.

The foliated substructure thus appears to be the most significant component of the exospore. Ultrastructural studies of sporogenesis in living plants demonstrate that this substructure is completed early in the course of wall ontogeny, and does not undergo alterations in the arrangement and number of sheets after the outer layer begins to form. Consequently, in well-developed exospore specimens such as those discussed here, the substructure peculiarities are definitively fixed, and cannot depend on more or less incomplete wall maturation. Besides, from the data supplied by extant spores, it can be assumed that these peculiarities are reliable criteria indicating that *Calamospora*, *Vestispora* and *Laevigatosporites* are distinct sporomorphs, which probably belong to three related but distinct groups of plants. These peculiarities also indicate that important evolutionary modifications of exospore fine structure occurred among ancient sphenopsids. A substructure resembling that of the spores studied here is certainly shared by the exospore of other sphenopsid sporomorphs showing an apertural fold, although it was imperceptible in the *Elaterites triferens* specimens investigated by Kurmann and Taylor (1984), probably because the exospore was severely corroded and devoid of contrast variations. The three examples illustrated in this paper suggest that this foliated

Figs. 14–19. *Vestispora fenestra*, Tromborn, 1204.68 m, Westphalien D (Figs. 14–15) and *V. pseudoreticulata*, Sainte Fontaine 6, 1042.50 m, middle Westfalien C (Figs. 16–19). **14**: densely foveolate spore, SEM, × 1,000. **15**: side of open aperture (a, apertural fold apex), showing the exospore median zone with few sheets and intercalated cavities, one of them (Ca) particularly large (i, inner layer of the extra-exosporal wall applied to the exospore surface; o, outer layer of the same wall), TEM, × 40,000. **16**: part of large spore characterized by the complex reticulum and the operculum (arrow head) of the extra-exosporal wall, SEM, × 1,000. **17**: side of open aperture (a, apertural fold apex), consisting of exospore layers that are notably thicker than in the previous species, but have similar proportions and ultrastructural features, TEM, × 25,000. **18**: aperture side (a, apertural fold apex) and peri-apertural area, the exospore distal face below; the dimly visible sheets and elongate cavities of the median zone extend to a great distance from the aperture, TEM, × 16,000. **19**: both sides of distorted aperture (i, o: as in Fig. 15), TEM, × 30,000.

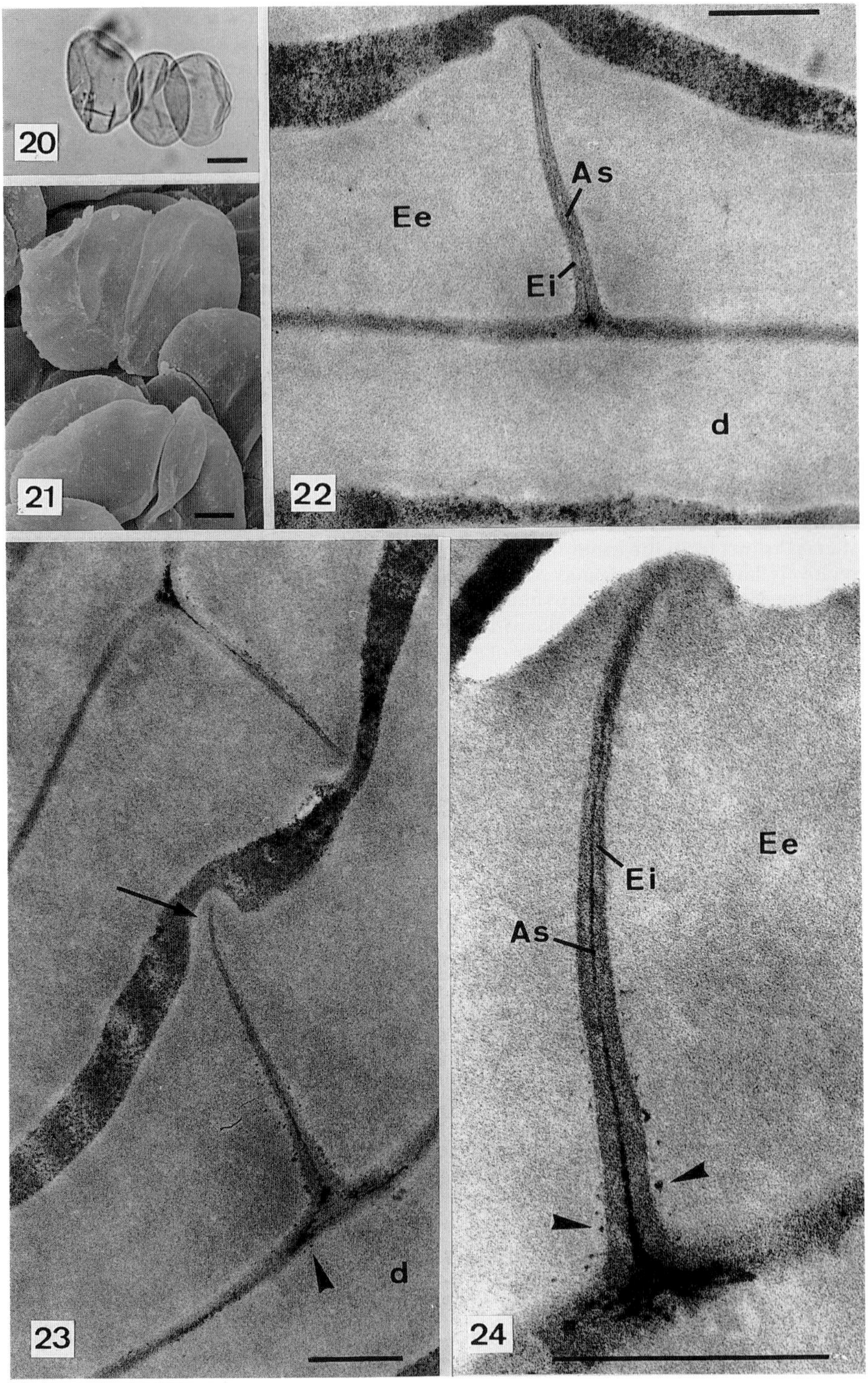
20
21
22
Ee
As
Ei
d
23
d
24
Ee
Ei
As

substructure has distinctive features in each sporomorph. It seems likely that these features not only distinguish the various sporomorphs, but also reflect relationships between sporomorphs and, consequently, between source plants. Studies of substructure characteristics in specimens of every sporomorph will be of much interest, and should supply new, significant information to palaeobotanists studying ancient sphenopsids.

C. Fossil sphenopsids and extant filicopsids

The most striking fact emerging from this study is the remarkable ultrastructural resemblance between the wall of fossil sphenopsids and the exospore of filicopsid spores (Lugardon, 1971, 1972, 1974). The extent of these similarities is underscored in Fig. 25, which compares schematic representations of the fossil spore walls (A–C) with three fern exospore types (E–G). In every case, the aperture is characterized by the protruding exospore fold, with a median, well-defined apertural slit closed outwards by very reduced wall layers. Also, the exospore includes a foliated substructure that forms the thin layer lining the spore cavity and may comprise distinct sheets that are especially abundant in the apertural area. Among living pteridophytes, these aperture and substructure features always occur together and are typical of the filicopsids.

Although Fig. 25 shows similar fine structure in fossil sphenopsids and living filicopsids, none of the three fossil exospore types described here precisely matches any fern exospore. *Calamospora* (A) and the exospore of Ophioglossaceae (E) have the most complex substructures. The separate, abundant sheets of the fossil spores do not reach the apex of the inner layer fold and fuse laterally with the outer layer a short distance from the aperture, whereas in Ophioglossaceae the sheets are fewer, 1 to 3 extend above the inner layer fold, and some sheets are distinct throughout the spore. *Vestispora* (B) and the wall of Gleicheniaceae (F) have simpler substructures. The sheets similarly extend from the middle and basal parts of the inner layer fold for some distance from the aperture, but their lateral ends join with the inner layer in *Vestispora*, whereas they are free and upturned in Gleicheniaceae. *Laevigatosporites* (C) and the exospore of ferns regarded as relatively advanced (G) show similar reduction of the substructure to a very thin inner layer without separate sheets, but they differ in the arrangement of the minute cavities in the innermost part of the outer layer. These cavities are fairly regularly aligned near the inner layer in *Laevigatosporites*, whereas they form a series of contiguous arcs in sections of fern exospores.

The obvious similarities of exospores in fossil sphenopsids and in the filicopsids, as well as the comparable, though not identical, variations of the wall ultrastructure within both groups, indicate a close relationship between these groups. They suggest that both lineages have a common origin. They also demonstrate that the spores of both groups have undergone evolutionary modifications that produced comparable variations in exospore substructure complexity. Studies of late Devonian and early Carboniferous spores (B. Lugardon, unpublished) suggest that these modifications largely occurred in the Devonian period, perhaps somewhat earlier and more rapidly in sphenopsid spores. The pattern of substructure evolution in the sphenopsids remains unknown. But the much more developed studies of fern spores clearly demonstrate that, in the filicopsids,

FIGS. **20–24**. *Laevigatosporites minimus*. La Houve, 409.13 m, first levels of Lorraine Stephanian. **20**: monolete spores, LM, × 500. **21**: part of a large group of spores, SEM, × 1,000. **22**: apertural area and distal face, the exospore inner layer transversely sectioned at the aperture, more-or-less obliquely elsewhere (the same in the following micrographs), TEM, × 30,000. **23**: aperture sections of two spores, the lower one with a higher apical ridge (arrow), probably indicating a closer vicinity to the proximal pole; punctiform cavities arranged in the innermost region of the outer exospore layer are abundant on both sides of the aperture, rarer (arrow head) on the wall distal face, TEM, × 25,000. **24**: detail of aperture showing the bilayered structure of the exospore, with the thin inner layer bounding the very narrow apertural slit, the absence of sheets separated from this layer, the outer layer including minute cavities (arrow heads) aligned close to the inner layer, TEM, × 75,000.

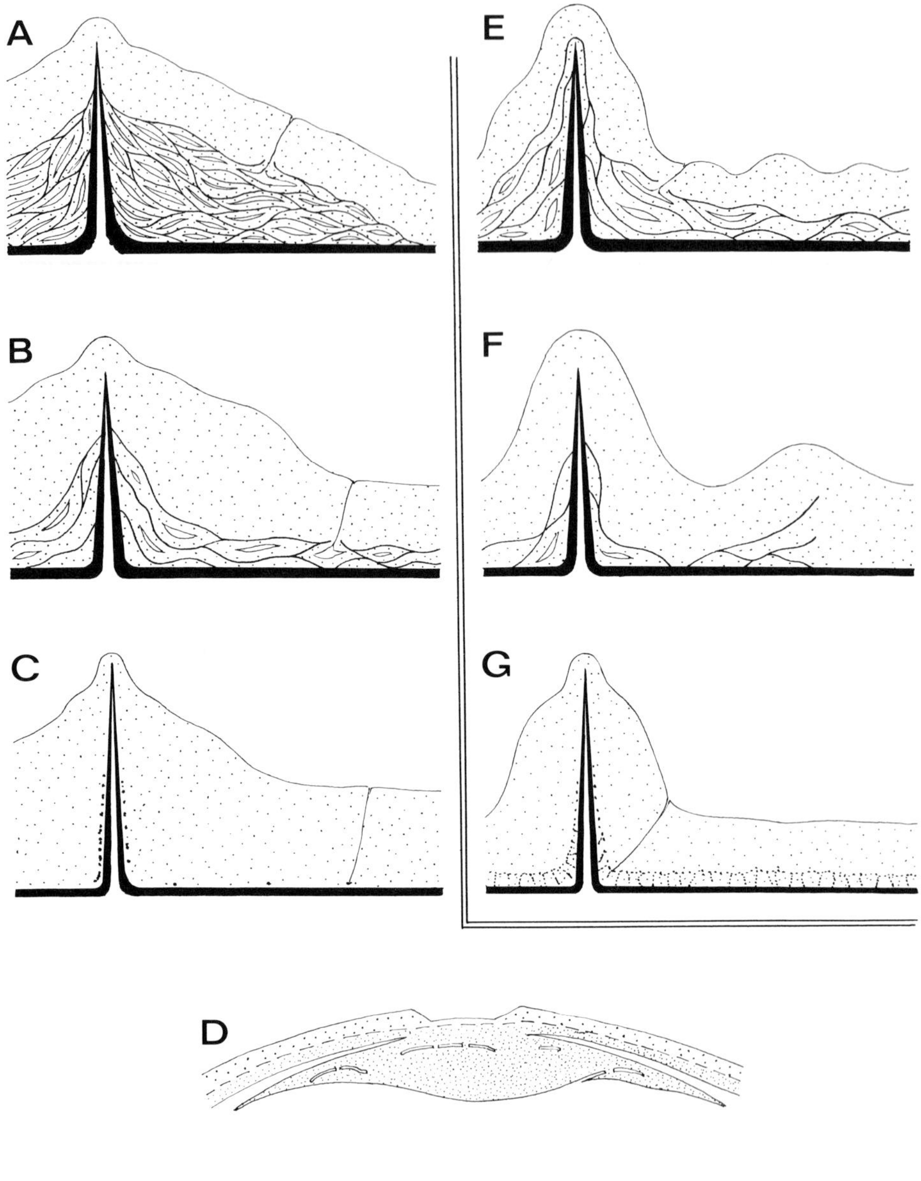

FIG. 25. Diagrams of exospore aperture and fine structure in sphenopsids (A–D) and filicopsids (E–G); **A**, *Calamospora*; **B**, *Vestispora* (the separate sheets joining to the inner layer some distance to the right of the diagram); **C**, *Laevigatosporites*; **D**, *Equisetum* (less enlarged); **E**, Ophioglossaceae; **F**, Gleicheniaceae; **G**, more advanced ferns.

exospore substructure becomes increasingly simplified from the groups regarded as primitive to the groups regarded as advanced.

D. Fossil and extant sphenopsids

Comparisons of the exospore ultrastructure in fossil and extant sphenopsids reveal significant differences in both wall composition and aperture configuration. The exospore of *Equisetum* (Lugardon, 1969) consists of two layers of nearly equal, uniform thickness, lacking prominent structural details such as channels or cavities. No formation similar to a foliated substructure can be seen. Ontogenetic studies show that the only structured components of the wall are lamellar elements aligned along the contact zone of the two layers and usually indistinct in mature spores (Uehara and Kurita, 1989; Lugardon, 1990). The aperture (Fig. 25D) consists of a small, much thinner area of the exospore, which is plain and inwardly reinforced with a thick, biconvex, lens-shaped mass of sporopollenin ('sub-apertural obturator') that is fastened to, and extends far beyond, the thin wall area. As shown in Fig. 25, no valid parallel can be drawn between the exospore of living *Equisetum* (D) and the wall of the fossil sphenopsids (A–C) investigated in this study. The differences appear so great that it is difficult to conceive that the two types of spore wall represent closely related plants. Nevertheless, it seems unacceptable to hypothesize that the extant Equisetaceae are not descended from the Carboniferous sphenopsids. Phylogenetic studies comparing fossil and living sphenopsids unanimously suggest close relationships (e.g. Stewart, 1983; Meyen, 1987; Bateman, 1991). Thus, the very peculiar characteristics of the exospore of extant Equisetaceae must reflect extensive evolutionary modifications of the wall that occurred since the Carboniferous, probably in several successive stages.

Conclusions

This study of exospore ultrastructure in fossil sphenopsids confirms the validity of several sporomorphs first distinguished using morphological features. Moreover, it reveals similarities to the exospore of filicopsids, while also demonstrating the structural singularity of the exospore in present-day Equisetaceae. The results reported here offer just an initial, incomplete survey of ultrastructural features of ancient sphenopsid spores, since they are concerned with the exospore of only three sporomorphs. But they show that these features are a potential source of additional, useful information. They allow us to hope that future fine-structure studies of spores preserved in Palaeozoic and Mesozoic fructifications will be beneficial in two respects. These investigations should solve the various issues raised by the spores of fossil and extant sphenopsids and, at the same time, contribute to progress in the knowledge of the phylogeny and palaeobiology of the sphenopsids.

Acknowledgements

We are grateful to Dr R.M. Bateman for his constructive comments and to P. Piquemal for assistance in photography and drawing.

References

Bateman, R.M. (1991). Palaeobiological and phylogenetic implications of anatomically-preserved archaeocalamites from the Dinantian of Oxroad Bay and Loch Humphrey Burn, southern Scotland. *Palaeontographica, Abt. B, Paläophytol.* **223**: 1–59.

Brousmiche, C. and Lugardon, B. (1990). Ressemblances inattendues des caractères ultrastructuraux entre quelques Arthrophytes fossiles et Filicophytes actuelles primitives. *Compt. Rend. Acad. Sci. Paris, Sér. 3, Sci. Vie* **310**: 873–879.

Good, C.W. (1975). Pennsylvanian-age calamitean cones, elater-bearing spores, and associated vegetative organs. *Palaeontographica, Abt. B., Paläophytol.* **153**: 28–99.

Good, C.W. (1977). Taxonomic and stratigraphic significance of dispersed spore genus *Calamospora*. In: R.C. Romans (editor). Geobotany, pp. 43–64, Plenum, New York.
Hartung, W. (1933). Die Sporenverhältnisse der Calamariaceen. *Arbeiten Inst. Paläobot.* **3**: 97–149.
Kurmann, M.H. and Taylor, T.N. (1984). Comparative ultrastructure of the sphenophyte spores *Elaterites* and *Equisetum*. *Grana* **23**: 109–116.
Lugardon, B. (1969). Sur la structure fine des parois sporales d'*Equisetum maximum*. *Pollen & Spores* **11**: 449–474.
Lugardon, B. (1971). Contribution à la connaissance de la morphogénèse et de la structure des parois sporales chez les Filicinées isosporées. Doctoral thesis, 257 pp. Paul Sabatier University, Toulouse.
Lugardon, B. (1972). La structure fine de l'exospore et de la périspore des Filicinées isosporées: 1, Généralités, Eusporangiées et Osmundales. *Pollen & Spores* **14**: 227–261.
Lugardon, B. (1974). La structure fine de l'exospore et de la périspore des Filicinées isosporées: 2. Filicales, Commentaires. *Pollen & Spores* **16**: 161–226.
Lugardon, B. (1990). Pteridophyte sporogenesis: a survey of spore wall ontogeny and fine structure in a polyphyletic plant group. In: S. Blackmore and R.B. Knox (editors). Microspores: Evolution and Ontogeny, pp. 95–120, Academic Press, London.
Meyen, S.V. (1987). Fundamentals of Palaeobotany. 432 pp. Chapman and Hall, London.
Remy, R. (1959). Die Sporen von *Cingularia typica* Weiss. *Monatsber. Deutsch. Akad. Wiss. Berlin* **1**: 257–261.
Remy, R. (1960). *Bowmanites nindeli* n. sp. *Monatsber. Deutsch. Akad. Wiss. Berlin* **2**: 122–125.
Remy, R. (1961). Beiträge zur Flora des Autunien III. *Monatsber. Deutsch. Akad. Wiss. Berlin* **3**: 441–336.
Serret, L. and Brousmiche, C. (1987). Quelques fructifications de Calamitaceae (Arthrophytes) récoltées dans le bassin houiller sarro-lorrain (organisation, sporae in situ). *Palaeontographica, Abt. B, Paläophytol.* **203**: 135–179.
Smith, A.H. and Butterworth, M.A. (1967). Miospores in the coal seams of the Carboniferous of Great Britain. *Special Pap. Palaeontol.* **1**: 1–324.
Steward, W.N. (1983). Paleobotany and the Evolution of Plants. 405 pp. Cambridge Univ. Press, Cambridge.
Storch, D. von (1980). *Sphenophyllum*-Arten aus drei intramontanen Karbonbecken: Pflanzen-geographische Besonderheiten im mitteleuropäischen Karbon. *Schriftenreihe. Geol. Wiss.* **16**: 171–273.
Tryon, A.F. and Lugardon, B. (1991). Spores of the Pteridophyta: Surface, Wall Structure, and Diversity Based on Electron Microscope Studies. 648 pp. Springer-Verlag, New York.
Uehara, K. and Kurita, S. (1989). An ultrastructural study of spore wall morphogenesis in *Equisetum arvense*. *Amer. J. Bot.* **76**: 939–951.
Zodrow, E.L. and Gao, Z. (1991). *Leeites oblongifolis* nov. gen. et sp. (Sphenophyllaean, Carboniferous), Sydney Coalfield, Nova Scotia, Canada. *Palaeontographica, Abt. B, Paläophytol.* **223**: 61–80.

Van Konijnenburg-van Cittert, J.H.A and Kurmann, M.H. (1994). Comparative ultrastructure of living and fossil matoniaceous spores (Pteridophyta). In: M.H. Kurmann and J.A. Doyle (editors). Ultrastructure of fossil spores and pollen, pp. 67–86, Royal Botanic Gardens, Kew.

5. COMPARATIVE ULTRASTRUCTURE OF LIVING AND FOSSIL MATONIACEOUS SPORES (PTERIDOPHYTA)

JOHANNA H.A. VAN KONIJNENBURG-VAN CITTERT[1] AND MARIE H. KURMANN[2]

[1]Laboratory of Palaeobotany and Palynology, University of Utrecht, Heidelberglaan 2, 3584 CS Utrecht, The Netherlands
[2]Palynology Unit, Herbarium, Royal Botanic Gardens, Kew, Richmond, Surrey, TW9 3AE, U.K.

Abstract

Extant and fossil matoniaceous *in situ* spores have been studied using light, scanning electron and transmission electron microscopy. Some developmental stages in the formation of the perispore have been observed in the extant genera *Matonia* and *Phanerosorus*. Mature spores of these two genera are almost identical in spore wall stratification. However, they differ in the exospore thickenings (interradial and apical) which are present in *Matonia* but are absent in *Phanerosorus*. Based on the presence or absence of exospore thickenings, three spore types (no thickenings, interradial thickenings, interradial and apical thickenings) were recognised in the fossil taxa. Transmission electron microscopy revealed that lamellae and cavities are present in the middle exospore layer adjacent to the aperture in all species. Spore ultrastructure supports the assessment that the Matoniaceae were more diverse during the Mesozoic.

Introduction

The Matoniaceae are characterised mainly by their sori which are arranged in two rows along the midrib on the lower side of the leaf. The sori consist of a ring of usually 6 to 12 sporangia surrounding a receptacle (or placenta) which is expanded into a large peltate indusium at its apex. Maturation of the sporangia is simultaneous and they dehisce as soon as they are mature.

At present, the family is restricted to the Malaysian Archipelago and consists of two genera, *Matonia* and *Phanerosorus*, each with one or two species. The ultrastructure of Recent matoniaceous spores has been studied by Tryon and Lugardon (1991). They showed that the spores of *Matonia* and *Phanerosorus* clearly indicate close relationships as seen in the organisation of the perispore and the ultrastructure of the exospore layers.

During the past, the Matoniaceae were much more widespread with an almost cosmopolitan distribution. The oldest fossils with matoniaceous affinities have been described from the Late Triassic of North America and Europe (Ash et al., 1982; Hirmer and Hörhammer, 1936). Hirmer and Hörhammer published a comprehensive study of the Matoniaceae in 1936. Since then, numerous new fossil species have been described based primarily on leaf shape (pinnate or bipinnate) and soral characters. Only a few studies examined ultrastructural features of the fossil spores (Litwin, 1984; Skog, 1988). Van Konijnenburg-van Cittert (1993) studied *in situ* spores of some 20 fossil species as well as representatives of the Recent Matoniaceae with light and scanning electron microscopy. She distinguished two spore types in the Recent taxa and three types in the fossils.

In this paper, we describe the results of a transmission electron microscopical study of the spore-types established by Van Konijnenburg-van Cittert (1993) and discuss the relevance of spore ultrastructure in the classification and evolution of the Matoniaceae.

Materials, Methods and Terminology

A. Materials

Fossil samples were prepared from compression specimens kept at the following institutions; British Museum (Natural History) London, Sedgwick Museum Cambridge, Natural History Museum Stockholm, Geological Museum Copenhagen, Palaeobotanical Institute Munich, and Institute of Palaeobotany and Palynology Utrecht. The preparations are housed in the Collections of the Institute of Palaeobotany and Palynology in Utrecht. Recent spore material was prepared from herbarium specimens housed at the following herbaria: British Museum (Natural History) (BM), Royal Botanic Gardens Kew (K), University of Leiden (L) and University of Utrecht (U).

B. Methods

Preparations of the fossil spores were made by selecting (if possible) one sporangium of each taxon, macerating it in Schulze's reagent (after cleaning in HF if necessary), and neutralising with ammonia. Spores of *Matonia* and *Phanerosorus* were removed from herbarium specimens, rehydrated and some samples acetolysed.

For light microscopy (LM), macerated fossil spores and acetolysed Recent spores were mounted in glycerine-jelly and sealed with paraffin wax. The slides were examined with a Leitz Ortholux microscope, and at least 50 spores per collection were measured.

For scanning electron microscopy (SEM), the samples were dehydrated in a graded acetone or ethanol series, placed on a standard specimen stub and air dried. They were coated with gold or platinium, viewed and photographed with a Camescan or a Hitachi S-2400.

For transmission electron microscopy (TEM), acetolysed and non-acetolysed spores were fixed in 2% glutaraldehyde containing 0.1% ruthenium red in 0.2 M cacodylate buffer for 1 hour and postfixed in 2% osmium tetroxide in 0.2 M cacodylate buffer for 3 hours. The spores were then sandwiched into 2% agar on milipore filters and dehydrated through a graded ethanol series. They were then transferred into acetone and embedded in Spurr's resin. Sections were cut with a Reichert ultramicrotome, stained in an LKB Ultrastainer with uranyl acetate and lead citrate and viewed in a Hitachi H-500 TEM.

C. Terminology

As far as possible, the descriptive terms used in this paper have been based on the pollen and spore terminology of Blackmore et al., 1992.

Descriptions and Discussion

A. Ultrastructure of Recent spores

Matonia R.Brown 1829

The genus *Matonia* is known from the Malaysian Archipelago (Malaya, Borneo, Sumatra, Ambon and New Guinea). There is still some discussion as to whether the genus is comprised of one or two species; the type-species *M. pectinata* Brown 1829 and *M. foxworthyi* Copeland 1908. The latter species was described by Copeland from herbarium specimens collected in Sarawak (Borneo) only. He stated that it differs from *M. pectinata* in its almost perpendicular, straight pinnules (falcate and arising at a more acute angle in

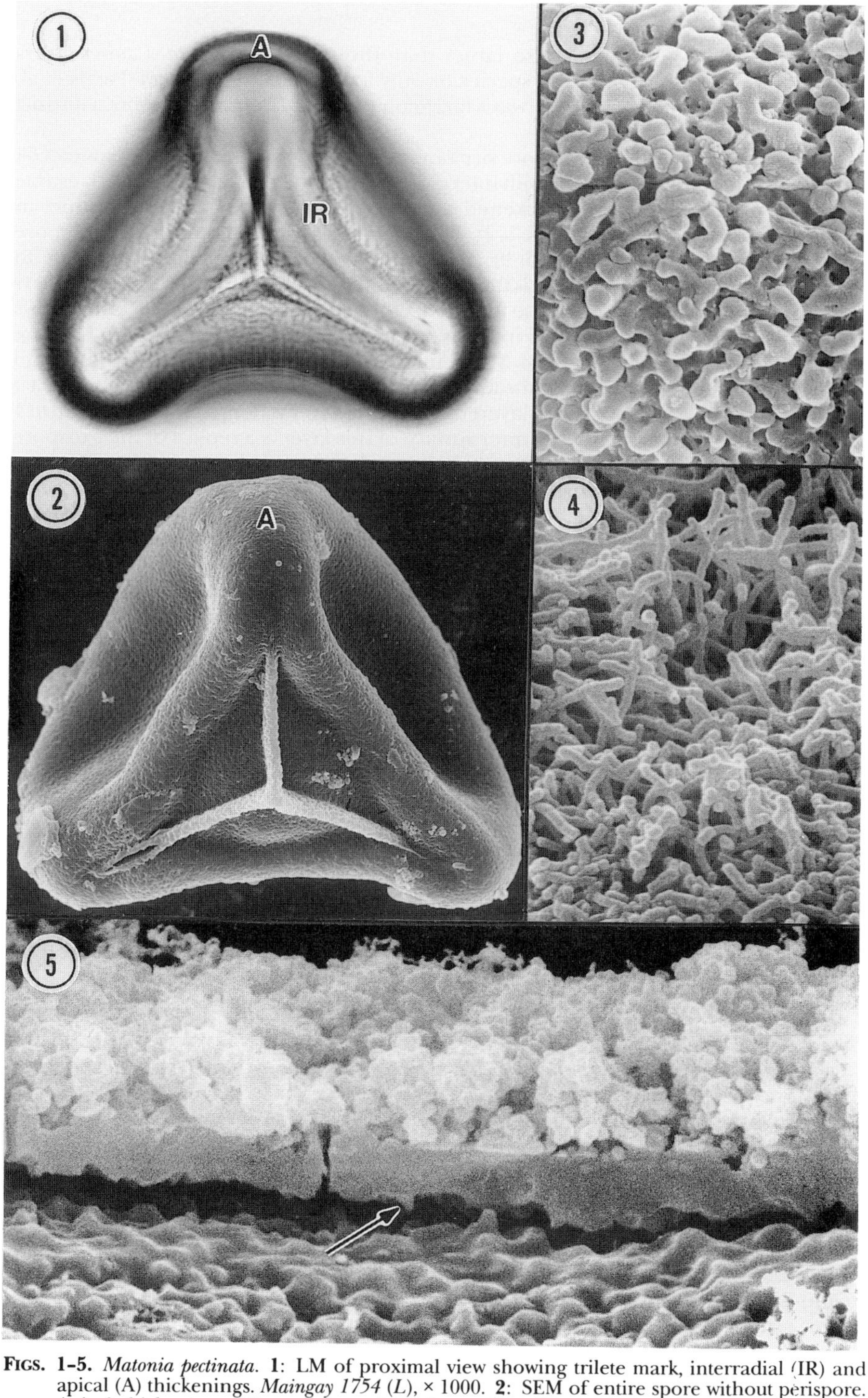

Figs. 1–5. *Matonia pectinata.* **1**: LM of proximal view showing trilete mark, interradial (IR) and apical (A) thickenings. *Maingay 1754* (*L*), × 1000. **2**: SEM of entire spore without perispore. Apical thickenings (A) pronounced. *Maingay 1754* (*L*), × 1000. **3**: Surface of immature perispore. *Maingay 1754* (*L*), × 10,000. **4**: Mature perispore showing details of rodlets. *Walker 130* (*U*), × 10,000. **5**: Fractured perispore lying on top of granular exospore. Arrow indicates a connective peg. *Lee & Badaruddin* (*K*), × 10,000.

M. pectinata) which are usually also larger than those of *M. pectinata*. Later authors doubted if it really was a different species or only a variety of *M. pectinata*. Parris et al. (1984) concluded that *M. foxworthyi* was a larger plant than *M. pectinata*, but not a distinct species.

Spore description (Figs. 1–12) — Spore size ranges from 50 to 75 μm (average diameter 64 μm). Shape is tetrahedral-globose with interradial thickenings subparallel to the laesurae (Figs. 1, 2, 6). The apices are thickened and these thickenings are more distinct in immature than mature spores (Figs. 1, 2). The aperture is trilete, laesurae ca. ¾ of spore radius, usually raised (Figs. 1, 2, 6, 9). The exospore has an irregularly granulate to verrucate surface (Figs. 5, 9, 11) and is three-layered. The inner and outer exospore layers are narrow and more electron dense in non-acetolysed spores. The middle layer shows characteristic cavities and lamellae in areas adjacent to the aperture (Figs. 10, 11). The exospore varies in thickness from 3 μm in the depressions between laesurae and interradial thickenings to 10 μm in the area of interradial thickenings. In the distal area it is about 6 μm. The perispore is two-layered. The inner layer is homogeneous and up to 1 μm thick (Figs. 5, 8, 9, 11, 12). The inner surface of this layer is a replica of the exospore (Figs. 5, 9, 11) and there are small connective pegs between the exo- and the perispore (Fig. 5, arrow). The outer perispore layer is up to 9 μm thick, and in mature spores it consists of rodlets (Figs. 4, 7, 11, 12). These rodlets can be up to 4 μm long and sometimes exhibit a translucent central area (Fig. 12, arrows). In immature spores, this layer is granular (Figs. 5, 9), then becomes verrucate (Fig. 3) and finally rodlets emerge (Fig. 4).

Discussion – The spores of the *M. pectinata* and *M. pectinata* var. *foxworthyi* differ only in small details, and the variation between taxa is no larger than within a taxon. Indeed, the only difference between the spores of *M. pectinata* and *M. pectinata* var. *foxworthyi* lies in the size of the perispore rodlets. In *M. pectinata* they are slightly shorter (up to 3 μm) and narrower (0.1 μm in diameter) than in *M. pectinata* var. *foxworthyi* (up to 4 μm long and 0.15–0.2 μm in diameter). In all other characters the spores are basically identical. Spore morphology supports the hypothesis that *M. foxworthyi* is not a separate species from *M. pectinata*.

Ultrastructurally, our findings closely compare with those of Tryon and Lugardon (1991). These authors also recognised a three-layered exospore with cavities in the middle exospore layer and a two-layered perispore.

It is interesting to note that Tryon and Lugardon (1991) specify a spore size in *M. pectinata* ranging from 50 to 58 μm, while Kremp and Kawasaki (1972) report 64–91 μm. Our measurements encompass both these with size ranges from 50 to 75 μm (64 μm average diameter). These reported differences might be due to differences in spore preparation, but may also reflect the number of specimens measured.

Phanerosorus Copeland 1908

Copeland (1908) made *Matonia sarmentosa* Baker into a new genus, *Phanerosorus*, based mainly on macromorphological characters. The fern with its pendant leaves and dichotomously branched leaflets is morphologically considerably different from *Matonia*, but the characters of sori, sporangia, indusium and spores are remarkably similar. Apart from *P. sarmentosus* known from Sarawak (Borneo), there is possibly one other species, *Phanerosorus maior* Diels (1932) from New Guinea and some small islands in the area. Most authors, however, are convinced that *P. maior* is not a different species but at best a variety of *P. sarmentosus* (Walker and Jermy, 1982, Parris et al., 1984).

Spore description (Figs. 13–22) — Spore size ranges from 50 to 67 μm (average diameter 60 μm). Shape is tetrahedral-globose (Figs. 16, 17). Aperture trilete, laesurae ca. ¾ of spore radius, usually raised. The exospore is three-layered (Figs. 13, 14) and has a verrucate surface (Figs. 17, 18). The inner and outer layers are narrow and electron dense. The middle layer shows cavities and lamellae adjacent to the aperture. The exospore is approx. 6 μm in thickness and there are no interradial or apical thickenings (Figs. 16, 17). The perispore is two-layered. The inner layer is homogeneous and up to 1 μm thick. The

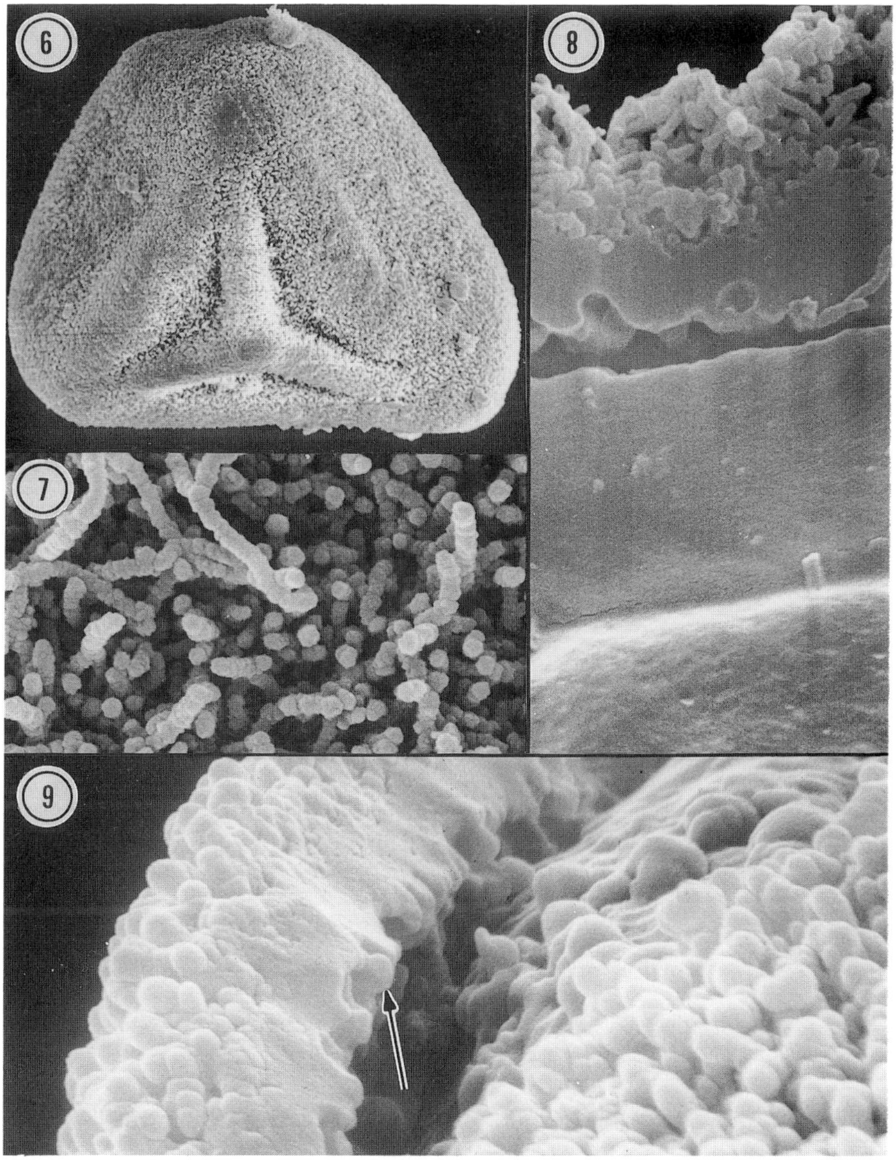

Figs. 6–9. *Matonia pectinata* var. *foxworthyi*. **6**: SEM of proximal surface, perispore present. *Kato et al. B.9574* (*L*), × 1000. **7**: Mature perispore showing details of rodlets. *Hallier 2460B* (*U*), × 10,000. **8**: Fractured sporoderm showing homogenous exospore and two layered perispore. *Hallier 2460B* (*U*), × 10,000. **9**: Fractured perispore of an immature spore. The outer perispore elements are beadlike. Arrow indicates connective peg. *Kato et al. B.9574* (*L*), × 10,000.

inner surface of this layer is a replica of the outer exospore surface and there are thin connective pegs between the exospore and the perispore, about 0.7–0.8 μm long (Fig. 22, arrow). The outer perispore layer is up to 4 μm in thickness and in mature spores consists

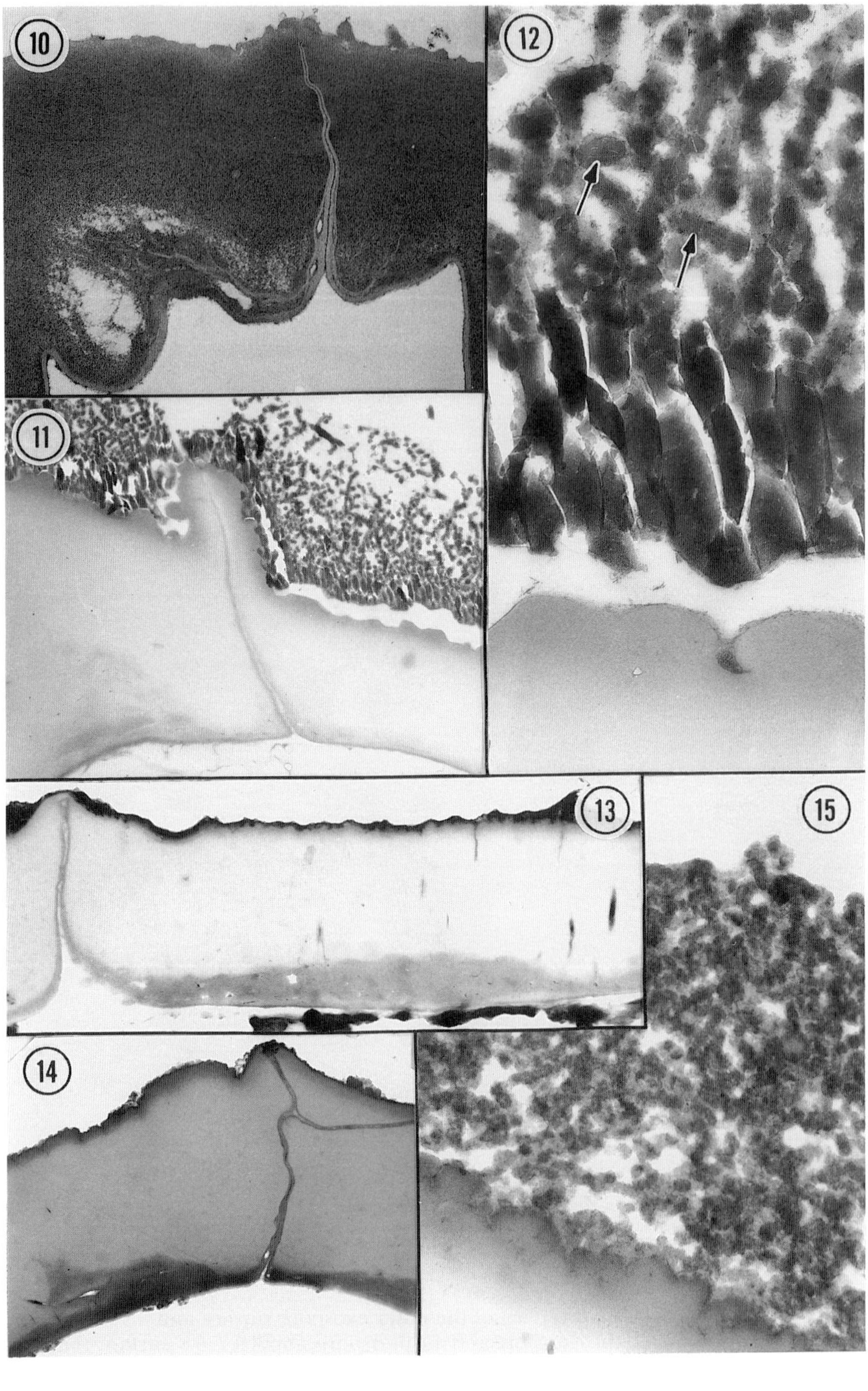
10
11
12
13
14
15

of rodlets (Fig. 21). In immature spores, this layer is granular (Fig. 19), and through the deposition of material rodlets are formed (Fig. 21). These rodlets are up to 3 μm long and 0.15–0.2 μm wide.

Discussion – We did not observe any distinctions between the spores of *Phanerosorus sarmentosus* and *P. maior* apart from a small size difference: *P. sarmentosus* spores have a mean diameter of 60 μm, while those of *P. maior* are usually 65–70 μm, but they overlap.

Our description of *Phanerosorus* spores differs from that by Tryon and Lugardon (1991) mainly in the size and structure of the perispore. They consider the perispore to be thin with only some rodlets in irregular masses, while we found it to be almost identical to the one of *Matonia* spores, both in thickness (3 μm or more) and in morphology (an inner, compact layer and an outer layer of rodlets). Moreover, on the inner side of the perispore the connective pegs, as discovered in *Matonia*, were observed.

Tryon and Lugardon (1991) recorded a much larger size range in *Phanerosorus* spores (35–80 μm), while Kremp and Kawasaki (1972) mention a larger size of 71–81 μm than we found. These varying figures are consistent with the ones observed in *Matonia pectinata*.

Comparison between Matonia and Phanerosorus spores – The mature spores of *Matonia* and *Phanerosorus* resemble each other in general shape, aperture, exospore and perispore stratification. Apart from small size differences (average equatorial diameter in *Matonia pectinata* is 64 μm, in *Phanerosorus sarmentosus* 60 μm), there are differences in the exospore structure. In *Matonia* there are interradial and apical thickenings in the exospore. The apical thickenings are more pronounced in immature spores. In *Phanerosorus* no exosporal thickenings have been observed, neither in immature nor mature spores. Despite the absence of interradial thickening in *Phanerosorus*, the middle exospore layer shows similar lamellae and cavities as observed in *Matonia* in the area of the interradial thickenings. The perispore in both genera is very elaborate and distinctive.

B. Ultrastructure of fossil spores

The oldest matoniaceous fossils known to date have been recorded by Ash et al. (1982) from the Late Triassic of the western USA. The family formed a major component of Late Jurassic and Early Cretaceous floras mainly in Laurasia (Hirmer and Hörhammer, 1936), but there are also reports from the Southern Hemisphere (Halle, 1913). A few records from the Late Cretaceous and the Early Tertiary of Europe (Givulescu et al., 1972), USA (Hall, 1965) and Australia (Tidwell and Skog, 1992) also exist. It appears that the family became extinct in the Northern Hemisphere during the Tertiary and eventually became restricted to the Malaysian Archipelago where it is found today.

Seven matoniaceous genera with fertile fronds have been described from the Mesozoic. Table 1 summarises the distinguishing features of these genera. Using light and scanning electron microscopy Van Konijnenburg-van Cittert (1993) described the *in situ* spores of some 20 species of Mesozoic Matoniaceae. Spores of these taxa share several features and can be described in general terms as follows: spore size ranges from

FIGS. **10–15.** *Matonia pectinata* (Figs. 10–12) and *Phanerosorus sarmentosus* (Figs. 13–15). **10**: TEM section through the suture of an acetolysed spore, showing the three-layered exospore. The inner exospore layer lines the aperture and the middle exospore layer shows lamellae and cavities adjacent to the aperture. *Haniff & Ness 8113* (*BM*), × 10,000. **11**: TEM section through the suture of a non-acetolysed spore showing a three-layered exospore and a two-layered perispore. Lamellae are present in the middle exospore layer adjacent to the aperture. *Haniff & Ness 8113* (*BM*), × 4000. **12**: Detail of perispore. Narrow inner, electron translucent, homogenous layer overlain by rodlets. Rodlets exhibit translucent central areas (arrows). *Haniff & Ness 8113* (*BM*), × 30,000. **13**: TEM section through aperture area in a non-acetolysed spore. The exospore is three-layered, and the middle exospore layer shows lamellae and cavities adjacent to the aperture. *Anderson S.27811* (*K*), × 5000. **14**: TEM through suture area of a spore showing similar cavities as seen in Fig 13. *Anderson S.27811* (*K*), × 5000. **15**: Tangential section through the perispore showing detail of outer rodlets. *Jermy 14466* (*L*), × 10,000.

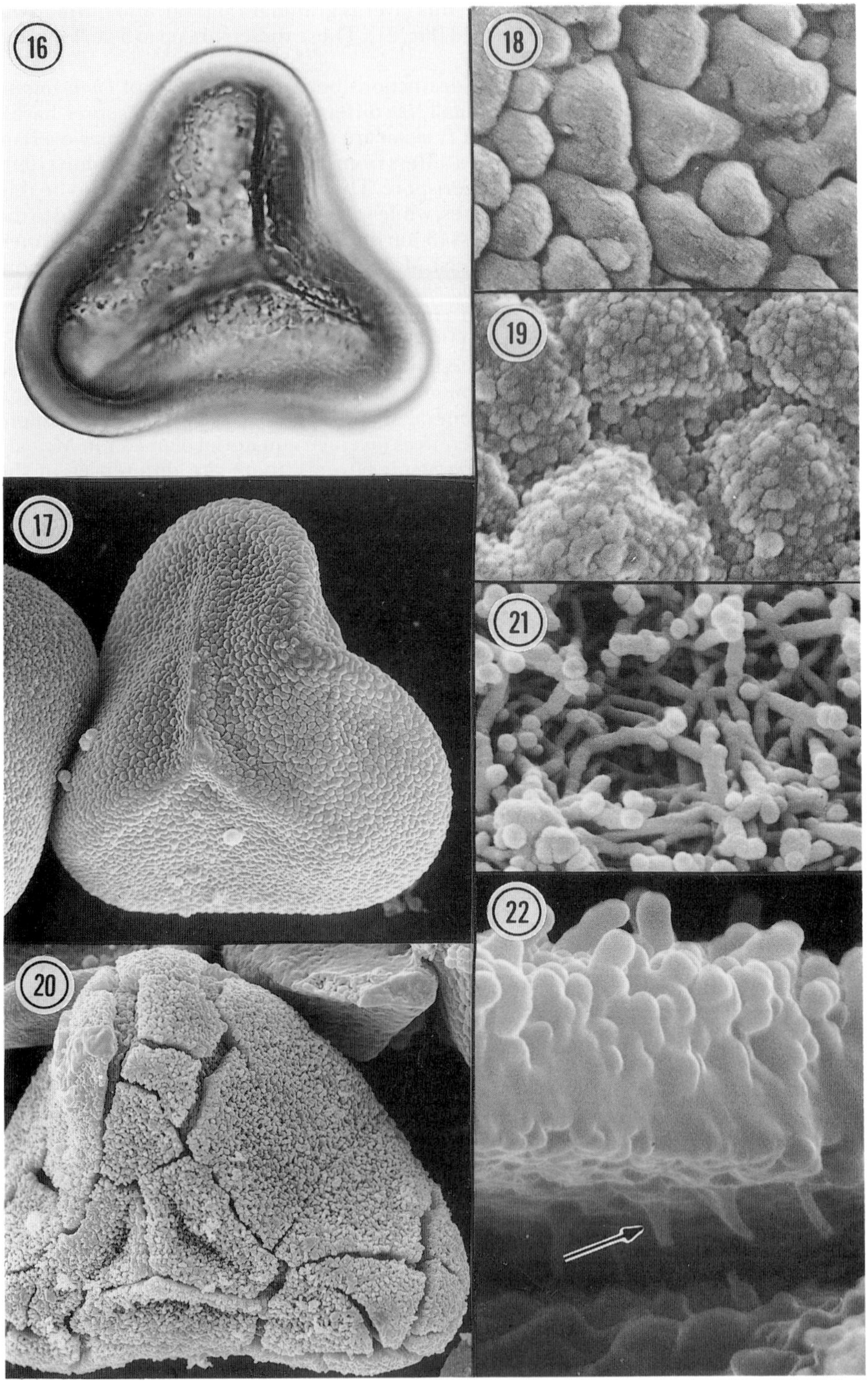
16
17
18
19
20
21
22

30–70 μm in equatorial view. Shape is rounded triangular in equatorial outline, often with interradial thickenings subparallel to the laesural radii. The aperture is trilete, laesurae are distinct and ca. ¾ of the spore radius in length. The exospore surface is psilate to faintly granulate. The perispore, if present, is thin and granulate.

Despite numerous common features, there is at least one distinct difference between the spores of the various taxa, namely the absence (non-valvate) or presence (valvate) of apical thickenings. Based on this character, Litwin (1985) concluded that spores of *Phlebopteris* form two distinct categories. Van Konijnenburg-van Cittert (1993) also recognised these two main spore types in the Matoniaceae. Moreover, she was able to further divide the non-valvate type based on the presence or absence of interradial thickenings.

Type 1 – The spores in this group have no apical thickenings (non-valvate). They usually range from 30 to 50 μm in diameter. The laesurae are raised and extend up to ¾ of the spore radius. The exospore thickness ranges from 0.5–3 μm and a perispore was usually not observed. In some taxa, the spores possess interradial thickenings which are subparallel to the laesurae (type 1a), in others these interradial thickenings are absent (type 1b).

Type 1a consists of the following taxa: *Matonia brownii, Nathorstia latifolia* (Figs. 28–31), *N. pectinata, Phlebopteris affinis* (Figs. 32–34), *P. angustiloba, P. minutifolius, P. polypodioides, P. smithii* and *Piazopteris branneri* (Figs. 23–27) (Van Konijnenburg-van Cittert, 1993). Transmission electron microscopical investigations of three representatives of this type (*Piazopteris branneri* (Figs. 26, 27), *Nathorstia latifolia* (Figs. 30, 31) and *Phlebopteris affinis* (Fig. 34)) showed that they all contain lamellae and/or cavities in the middle exospore layer adjacent to the laesurae, in the area of the interradial thickenings. The exospore is up to twice the thickness in the area of the interradial thickenings compared to the rest of the spore (Table 2).

Type 1b differs from type 1a in the absence of interradial thickenings and consists of the following taxa: *Delosorus heterophyllus, Nathorstia alata, N. galleyi.* Unfortunately, no material of this type was available for transmission electron microscopy. However, Skog (1988) figured a fractured spore of *Delosorus heterophyllus* showing an inner, compact exospore and an outer, granular deposit forming the perispore.

Type 2 – The spores of this type have apical thickenings (valvate spores), are frequently larger than the ones of type 1, ranging from 50–70 μm in diameter, and the exospore is thicker (2–6 μm). Interradial thickenings are always present and a perispore is frequently observed. The taxa belonging to this group are: *Matonia braunii* (Figs. 35–40), *Matonidum cingulatum, M. goeppertii* (Figs. 41–44), *Phlebopteris hirsuta, P. indica, P. muensteri* (Figs. 45–49) and *Selenocarpus muensterianus* (Figs. 50–53). Transmission electron microscopical investigations of four representatives of this type (*Matonia braunii* (Figs. 38–40), *Matonidium goepperti* (Figs. 40, 44), *Phlebopteris muensteri* (Figs. 47–49) and *Selenocarpus muensterianus* (Figs. 52, 53)) again revealed the lamellae and/or cavities in the middle exospore layer in the area of the interradial thickenings adjacent to the laesurae. In this area, the exospore is twice the thickness of the distal wall precisely as in type 1. However, the interradial thickenings appear much more prominent in type 2 as the exospore in the distal region is up to four times the thickness of that in spores of type 1. A thin, darker staining perispore layer consisting of granules has been observed in most of species classified in this group.

Figs. 16–22. *Phanerosorus sarmentosus.* **16**: LM of proximal side showing trilete mark. *Jermy 14466* (*L.*) × 1000. **17**: SEM of spore without perispore showing proximal exospore surface and trilete mark. *Jermy 14466* (*L*), × 1000. **18**: High magnification of Fig. 17, gemmate to verrucate exospore surface. × 10,000. **19**: Deposition of perispore initiated by appearance of granules. *Jermy 14466* (*L*), × 10,000. **20**: SEM of proximal surface of mature spore. Perispore is cracked. *Jermy 14466* (*L*), × 1000. **21**: High magnification of Fig. 20, showing perispore rodlets. × 10,000. **22**: Wall fracture displaying perispore and connective pegs on its inner surface (arrow). *Jermy 14466* (*L*), × 10,000.

TABLE 1. Distinguishing features in fossil matoniaceous genera.

Genus and References	Leaf morphology	Sorus shape	Indusium	Sporangia shape	Stratigraphic range
Delosorus[1]	bipinnate	circular	present	ovoid	Early Cretaceous
Matonia[2]	pinnate	circular	present	ovoid	Early Jurassic–Late Cretaceous
Matonidium[3]	pinnate	circular	present	ovoid	Middle Jurassic–Early Cretaceous
Nathorstia[4]	pinnate	circular	?present	wedge-shaped	Cretaceous
Phlebopteris[3]	pinnate	circular	absent	ovoid	Late Triassic–Early Cretaceous
Piazopteris[5]	bipinnate	circular	absent	ovoid	Early Jurassic–Early Cretaceous
Selenocarpus[3]	pinnate	crescent-shaped	absent	ovoid	Early Jurassic

[1]Skog, 1988; [2]Harris, 1980; [3]Hirmer and Hörhammer, 1936; [4]Krassilov, 1964; [5]Lorch, 1967.

In *Selenocarpus muensterianus* (Figs. 50–52), the apical thickenings are less pronounced than in the other taxa of type 2, the exospore being only around 2 μm in this area. Furthermore, the spores are considerably smaller (average diameter of 38 μm) than the other type 2 taxa. These size characteristics are more similar to the type 1a spores and therefore this taxon could be considered an intermediate between the types 1a and 2.

It is worth noting that the spores of various species in the same genus do not exhibit the same spore type. For example, spores of *Phlebopteris affinis* fall into type 1a, while those of *P. muensteri* belong to type 2. The same applies for various species in the genera *Matonia* and *Nathorstia.*

Conclusions

The characteristic features of the spores of the various taxa are summarized in Table 2. In the Recent Matoniaceae we have recognised two different types of spores based on thickenings in the exospore. In *Matonia pectinata,* both interradial and apical exospore thickenings are present, while in *Phanerosorus* these thickenings are absent. Within the fossil spores, we were able to recognise three different spore types based on that same character: no exospore thickenings (type 1b), interradial exospore thickenings only (1a) and both interradial and apical exospore thickenings (type 2). The fossil spores are more diverse, both in terms of size and structure, than the living spores. The valvate and non-valvate spore types exhibited in the extant Matoniaceae were already in existence in the Mesozoic.

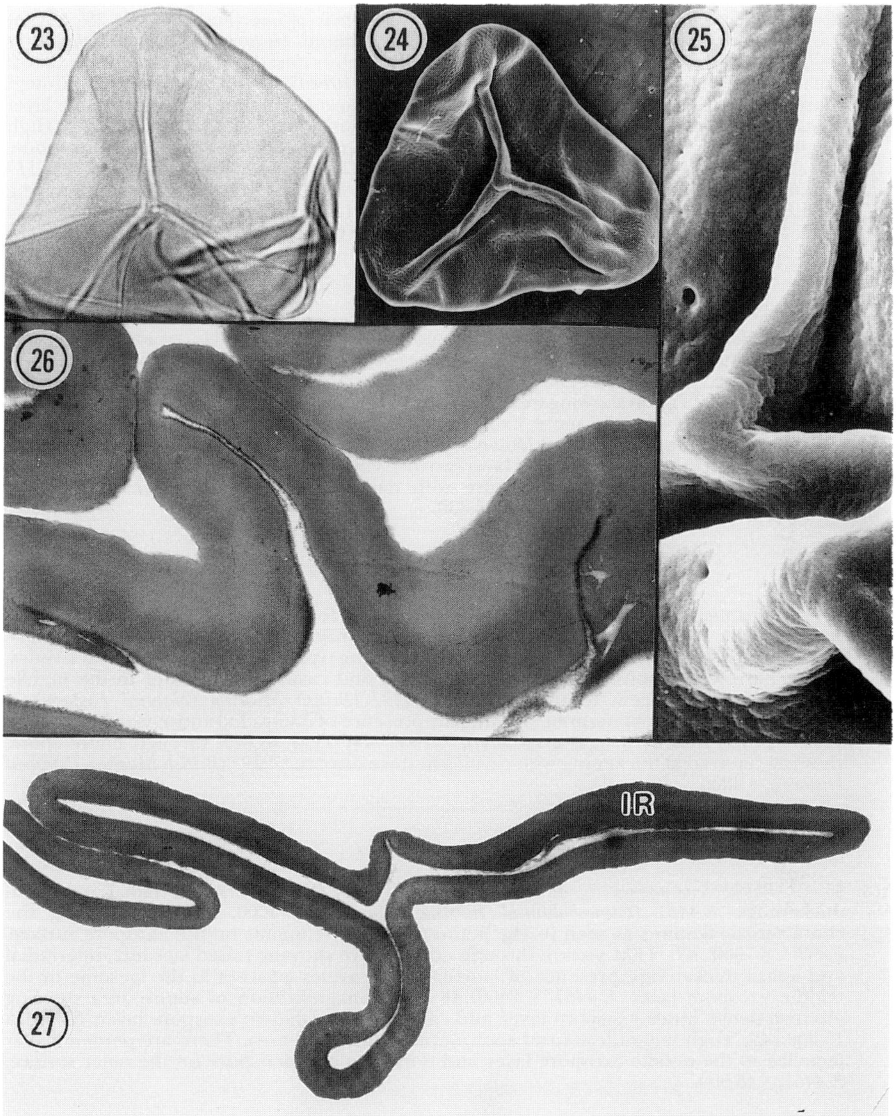

FIGS. 23–27. *Piazopteris branneri*. **23**: LM of proximal face, showing trilete mark and relatively thin sporoderm, *V.42327* (*British Museum* (*Natural History*)), × 1000. **24**: SEM of proximal surface showing raised laesura, no interradial thickenings are present. *V.42323* (*British Museum* (*Natural History*)) courtesy of C.R. Hill, ×1000. **25**: High magnification of Fig. 24. Courtesy of C.R. Hill, × 5300. **26**: TEM through the aperture region. Aperture is lined by the dark staining inner exospore layer, and there are cavities in the middle exospore layer adjacent to the aperture. *V.42324* (*British Museum* (*Natural History*)), × 15,000. **27**: TEM through an entire spore showing interradial thickening (IR). *V.42324*, × 4200.

The fossil spores of type 1a are smaller than those of type 2, and the absence of apical thickenings is combined with thinner exospores and the lack of a perispore (Table 2). Despite this combination of features, we do not consider type 1a an immature form of

FIGS. 28–34. *Nathorstia latifolia* (Figs. 28–31) and *Phlebopteris affinis* (Figs. 32–34). **28**: LM of proximal area showing trilete mark. *S.101203* (*Natural History Museum, Stockholm*), × 1000. **29**: SEM of a group of spores. Trilete mark is slightly raised and spore surface is almost psilate with the exception of some pitted areas. *S.101203*, × 1000. **30**: Section through sporoderm in the aperture region. Inner exospore layer lining the aperture, middle layer containing lamellae and outer layer almost electron translucent. *S.101203*, × 10,000. **31**: High magnification of the proximal exospore showing inner, more electron-dense layer and middle layer transversed by lamellae. *S.101203*, × 30,000. **32**: LM of spore showing many folds. *S.113* (*Institute of Palaeobotany and Palynology, Utrecht*), × 1000. **33**: SEM of proximal surface showing rounded-triangular outline, slightly raised aperture and slight interradial thickenings. *S.113*, × 1000. **34**: TEM through suture. Cavities in the middle exospore layer. *S.113*, × 15,000.

FIGS. 35–40. *Matonia braunii*. **35**: LM of spore showing laesura, apical and interradial thickening. *S.3009* (*Institute of Palaeobotany and Palynology, Utrecht*), × 1000. **36**: SEM of proximal face, with raised laesure, pronounced apical and interradial thickenings. *S.3009*, × 1000. **37**: High magnification of Fig. 36 showing exospore surface structure with perispore remnants. *S.3009*, × 10,000. **38**: TEM of entire spore. Exospore is thickened in the interradial and apical areas. *V.59751* (*British Museum* (*Natural History*)), × 3000. **39**: Aperture with laesura clearly raised and lamellae in the middle exospore layer (arrow). *V.31922* (*British Museum* (*Natural History*)), × 15,000. **40**: Section through aperture with darker staining outer layer representing remnants of the perispore. *V.59751*, × 15,000.

FIGS. 41–44. *Matonidium goeppertii*. **41**: SEM of two partly folded spores. Exospore surface is covered with remnants of perispore. *V.45252* (*British Museum* (*Natural History*)), × 1000. **42**: TEM section through the aperture displaying typical matoniaceous characteristics. Inner, narrow exospore layer lines the aperture and cavities are present in the middle exospore layer adjacent to the aperture. *39297* (*British Museum* (*Natural History*)), × 10,000. **43**: LM of proximal area, showing the presence of interradial and apical thickenings. *52605* (*British Museum* (*Natural History*)), × 1000. **44**: TEM section through entire spore, showing interradial thickening with cavities in the exospore. *39295* (*British Museum* (*Natural History*)), × 4200.

FIGS. 45–49. *Phlebopteris muensterii*. **45**: LM of spore showing trilete mark, interradial and apical thickenings. *E.4461* (*Palaeobotanical Institute, Munich*), × 1000. **46**: SEM showing the characteristic features as seen in Fig. 5 and perispore remnants on the exospore surface. *E.4461*, × 1000. **47**: TEM section through entire spore showing raised laesurae, interradial and apical thickenings, presence of lamellae and cavities adjacent to the laesurae in the middle exospore layer. *E.4461*, × 2400. **48**: High magnification of suture area showing electron-dense inner exospore layer and cavities in the middle exospore layer. *E.4461* × 15,000. **49**: High magnification of sporoderm in the distal area. There are perpendicular lamellae in the middle exospore layer and remnants of a perispore on the outer surface. *E.4461*, × 15,000.

FIGS. 50–53. *Selenocarpus muensterianus*. **50**: LM of spore showing interradial and slight apical thickenings. *1964 XX* (*Palaeobotanical Institute, Munich*), × 1000. **51**: SEM of spore showing rounded-triangular outline and nearly psilate exospore surface. *1964 XX*, × 1000. **52**: TEM section through entire spore showing raised laesura, interradial thickenings and cavities in the exospore adjacent to the aperture. *1964 XX*, × 3000. **53**: High magnification of sporoderm in the suture area. Note lamellae and cavities in the exospore and presence of perispore remnants on the exospore surface. *1964 XX*, × 30,000.

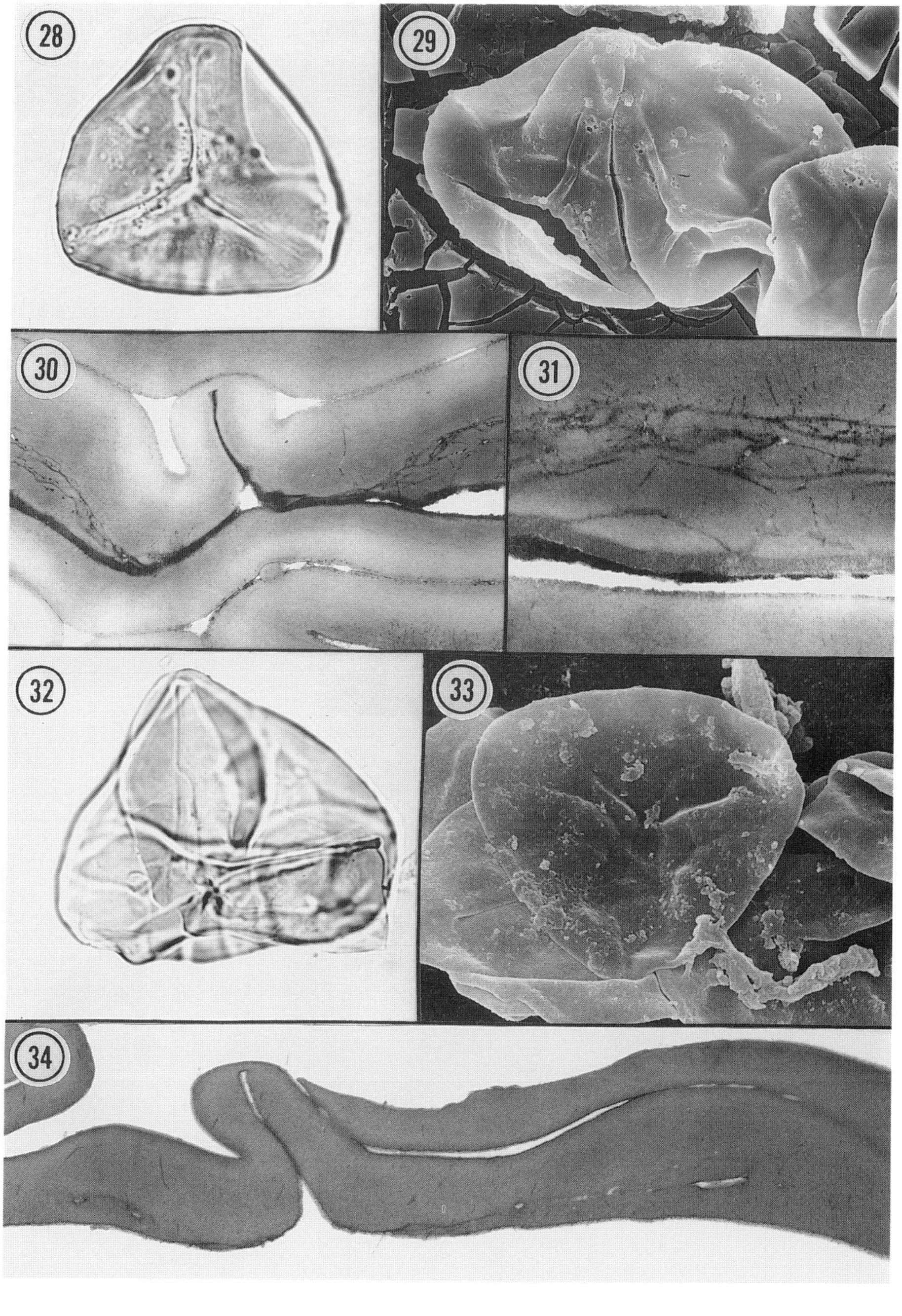
28
29
30
31
32
33
34

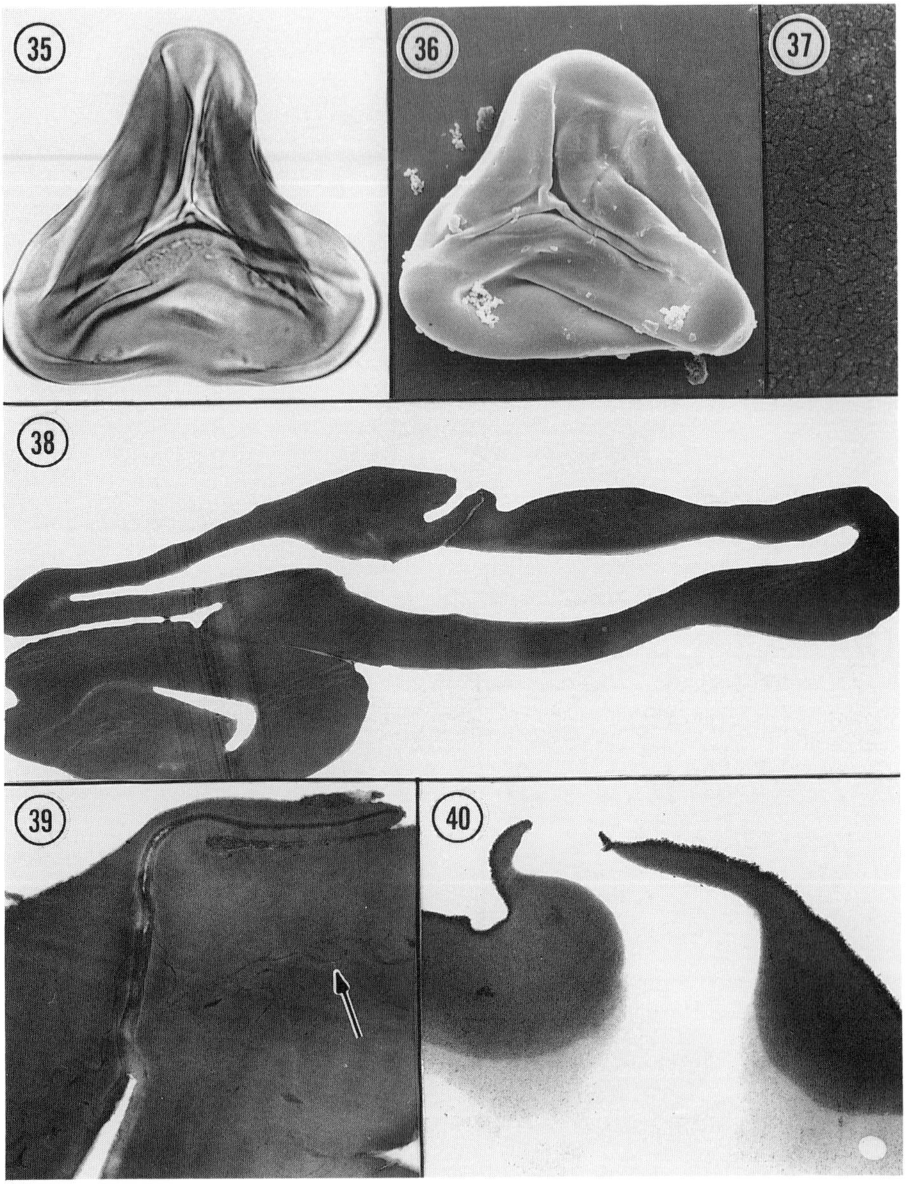
35
36
37
38
39
40

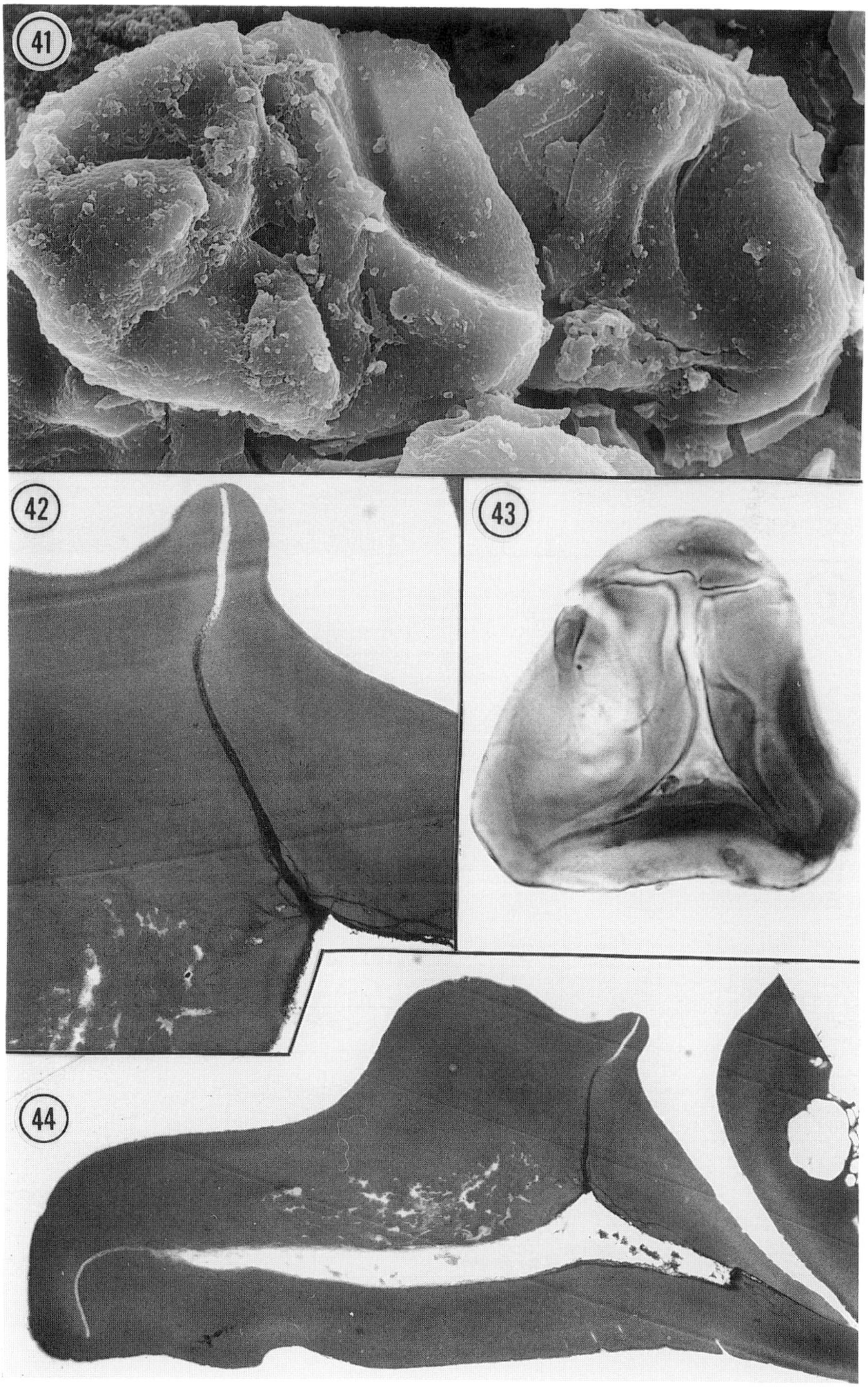
41
42
43
44

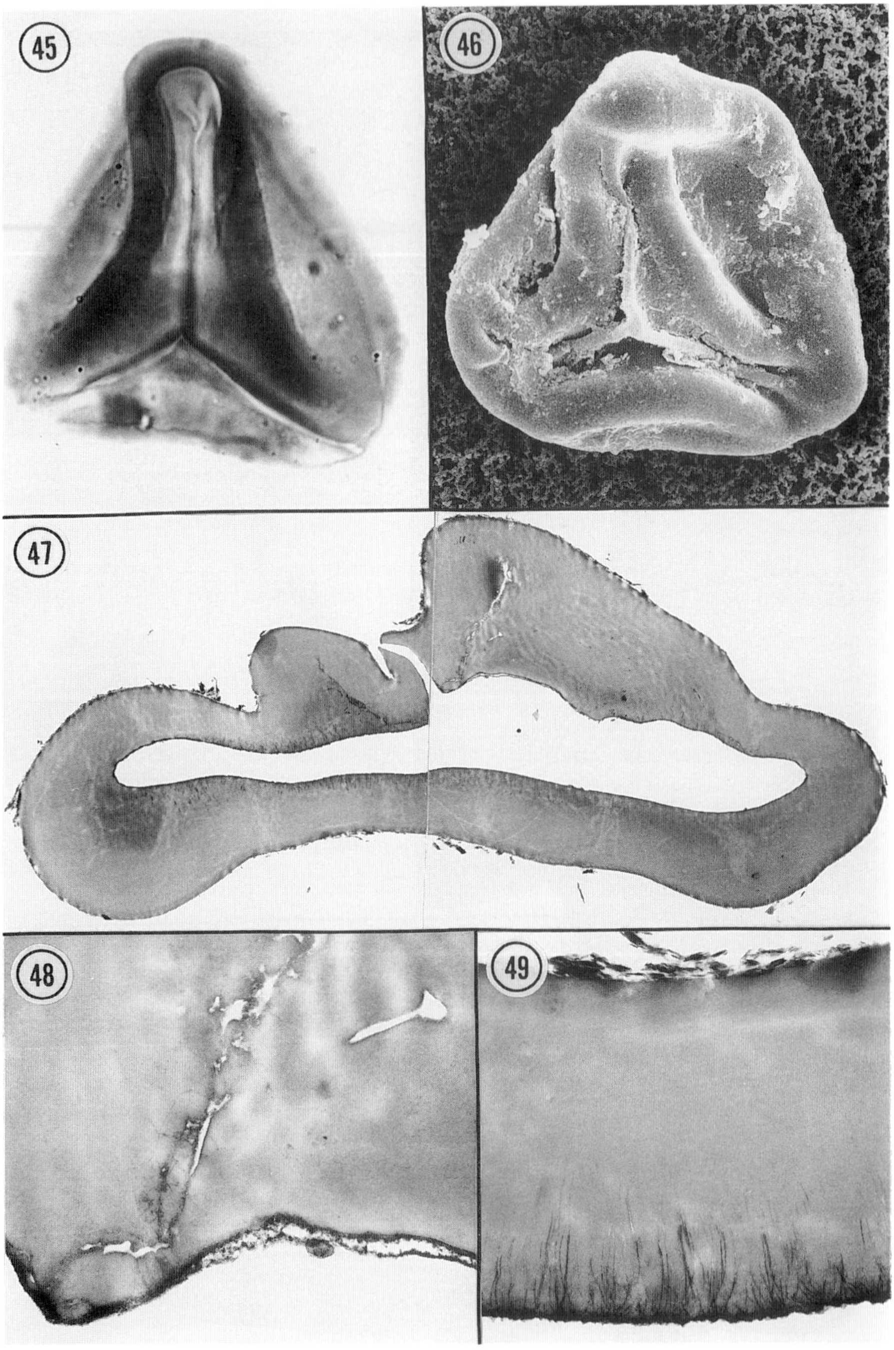
45
46
47
48
49

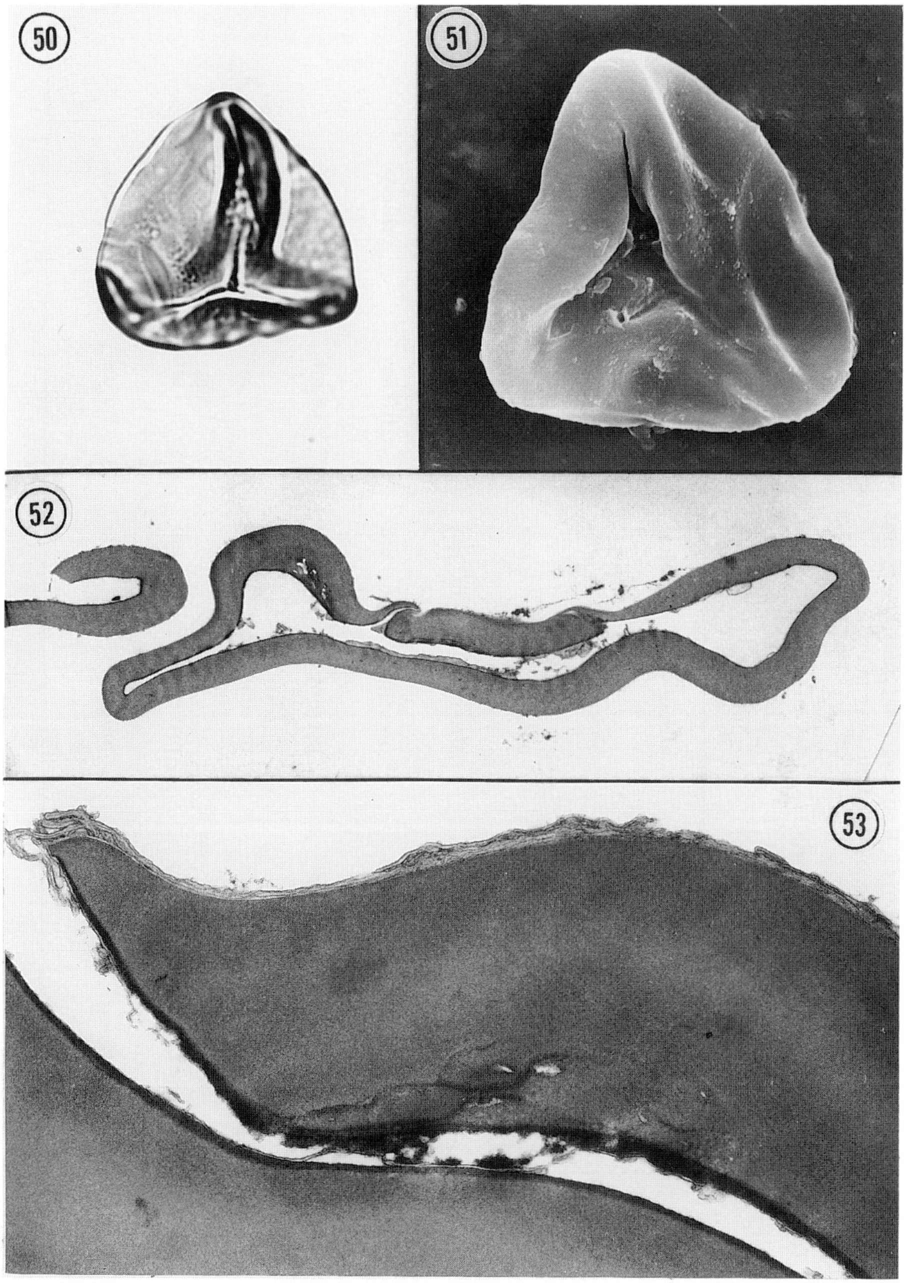
50
51
52
53

TABLE 2. Summary of the characteristic features in matoniaceous spores (measurements are averages, equatorial diameters are based on LM and exine thickness on TEM measurements).

Taxon	Equatorial diameter	Apical thickness	Exospore thickness		Perispore	Spore type
			distal	interradial		
Nathorstia latifolia (Figs. 28–31)	38 μm	absent	1.0 μm	2.3 μm	absent	1a
Phlebopteris affinis (Figs. 32–34)	45 μm	absent	0.5 μm	1.0 μm	absent	1a
Piazopteris branneri (Figs. 23–27)	38 μm	absent	1.0 μm	2.0 μm	absent	1a
Phanerosorus sarmentosus (Figs. 13–22)	60 μm	absent	6.0 μm	6.0 μm	present	1b
Matonia pectinata (Figs. 1–12)	64 μm	present	6.0 μm	10.0 μm	present	2
Matonia braunii (Figs. 35–40)	50 μm	present	1.9 μm	3.8 μm	present	2
Matonidium goeppertii (Figs. 41–44)	55 μm	present	3.8 μm	7.8 μm	present	2
Phlebopteris muensteri (Figs. 45–49)	62 μm	present	3.3 μm	8.3 μm	present	2
Selenocarpus muensterianus (Figs. 50–53)	38 μm	present	1.0 μm	2.2 μm	present	2

type 2. The apical thickenings in the spores of type 2 are much more pronounced in immature specimens than in mature ones.

The development of the outer perispore is initiated by the deposition of a granular layer on the outside of the inner, compact perispore. This layer becomes verrucate and through the addition of further material rodlets emerge. The presence of such elaborate, two-layered perispores in the spores of the extant species is unusual for these presumably less derived ferns. The first fossil matoniaceous spores with perispores occur in the Lower Jurassic. This is relatively shortly after the first appearance of matoniaceous fossils which have been recorded from the Late Triassic.

All spores, extant and fossil, examined with transmission electron microscopy show lamellae and/or cavities in the area adjacent to the laesurae. This character is not correlated with the presence or absence of interradial thickenings. When apical thickenings are present, the exospore is twice the thickness of the distal one in these regions.

Acknowledgements

The authors wish to thank Dr C.R. Hill for his help and advice. JHAVK-VC is most grateful to the Hanneke van Holk Fonds (Dutch Association of University Women) for a grant which enabled her to visit the collections in London (The Natural History Museum), Cambridge (Sedgwick Museum), Zürich (ETH) and Munich (Palaeobotanical Institute). Furthermore, she is thankful to the Leiden Herbarium, the Natural History Museum at Stockholm and the Geological Museum at Copenhagen for the opportunity to borrow material. MHK gratefully acknowledges financial support from the Board of Trustees, RBG, Kew. Thanks are also due to P. Hoen for help with the SEM work, and Drs K.L. Alvin and A.F. Tryon for reviewing the manuscript.

References

Ash, S.R., Litwin, R.J. and Traverse, A. (1982). The Upper Triassic fern *Phlebopteris smithii* (Daugherty) Arnold and its spores. *Palynology* **6**: 203–219.

Blackmore, S., Le Thomas, A., Nilsson, S. and Punt, W. (1992). Pollen and Spore Terminology. First Concept. 47 pp, Onderwijs Media Instituut, Univ. Utrecht, The Netherlands.

Copeland, E.B. (1908). New genera and species of Bornean ferns. *Philipp. J. Sci., C.* **3**: 343–351 (1908 publ. 1909).

Diels, L. (1932). Matoniacea nova papuasica. *Notizbl. Bot. Gart. Berlin-Dahlem*, **11**: 311–312.

Givulescu, R., Neagu, T. and Dragastan, O. (1986). *Matonidium goeppertii* Schenk in der Oberen Kreide der Inneren Flysch Zone der Ostkarpaten Rumäniens. *Acta Palaeobot.* **26**: 29–32.

Hall, J.W. (1965). Remains of a Cretaceous member of the Matoniaceae. *Amer. J. Bot.* **52**: 638–639 (abstr.).

Halle, T.G. (1913). Some Mesozoic plant-bearing deposits in Patagonia and Tierra del Fuego and their floras. *Kungl. Svenska Vetensk. Acad. Handl.* **51** (**3**): 1–58.

Harris, T.M. (1980). The Yorkshire Jurassic fern *Phlebopteris braunii* (Goeppert) and its reference to *Matonia* R. Br. *Bull. Brit. Mus.* (*Nat. Hist.*), *Geol.* **33**: 295–311.

Hirmer, M. and Hörhammer, L. (1936). Morphologie, Systematik und geografische Verbreitung der fossilen und rezenten Matoniaceen. *Palaeontographica, Abt.B, Paläophytol.* **81**: 1–70.

Krassilov, V.A. (1964). On the genus *Nathorstia* (Ferns). *Paleontol. Zhurn.* **2**: 115–120. (in Russian)

Kremp, G.O.W. and Kawasaki, T. (1972). The spores of the extant Pteridophytes. 398 pp. Hirokawa Publishing Company, Tokyo.

Litwin, R.J. (1985). Fertile organs and *in situ* spores of ferns from the Late Triassic Chinle Formation of Arizona and New Mexico, with discussion of the associated dispersed spores. *Rev. Palaeobot. Palynol.* **44**: 101–146.

Lorch, J. (1967). A Jurassic flora of Makhtesh Ramon, Israel. *Israel J. Bot.* **16**: 131–155.

Parris, B.S., Jermy, A.C., Camus, J.M. and Paul, A.M., (1984). The Pteridophyta of Gunung Mulu National Park. In: A.C. Jermy (editor). Studies on the flora of Gunung Mulu National Park, Sarawak, pp. 145–233, Forest Department, Kuching, Sarawak.

Skog, J.E. (1988). Reassignment of *Aspidium heterophyllum* to a new genus in the family Matoniaceae. *Amer. J. Bot.* **75**: 1120–1129.

Tidwell, W.D. and Skog, J.E. (1992). Two new fossil matoniaceous stem genera from Tasmania, Australia. *Rev. Palaeobot. Palynol.* **70**: 263–277.

Tryon, A.F. and Lugardon, B. (1991). Spores of the Pteridophyta. 648 pp. Springer-Verlag, New York.

Van Konijnenburg-van Cittert, J.H.A. (1993). A review of the Matoniaceae based on *in situ* spores. *Rev. Palaeobot. Palynol.* **78**: 235–267.

Walker, T.G. and Jermy, A.C. (1982). The ecology and cytology of *Phanerosorus* (Matoniaceae). *Fern Gaz.* **12**: 209–213.

Foster, C.B. and Balme, B.E. (1994). Ultrastructure of *Teichertospora torquata* (Higgs) from the Late Devonian: oldest saccate palynomorph. In: M.H. Kurmann and J.A. Doyle (editors). Ultrastructure of fossil spores and pollen, pp. 87–97, Royal Botanic Gardens, Kew.

6. ULTRASTRUCTURE OF *TEICHERTOSPORA TORQUATA* (HIGGS) FROM THE LATE DEVONIAN: OLDEST SACCATE PALYNOMORPH

CLINTON B. FOSTER[1] AND BASIL E. BALME[2]

[1]Australian Geological Survey Organisation, GPO Box 378, Canberra, Australia 2601.
[2]The University of Western Australia, Department of Geology, Nedlands, Australia, 6009.

Abstract

Teichertospora Balme 1988 was created for trilete miospores with triangular amb, clearly defined circular intexine, and endoreticulate exoexine. It was based on the type species *Teichertospora cuvierica* Balme 1988, from the Frasnian of Western Australia. *Teichertospora cuvierica* is now regarded as a junior subjective synonym of *Auroraspora torquata*, known from Late Devonian (late Frasnian-Famennian) assemblages from North America and western Europe. As a consequence, the type of *Teichertospora* is *T. torquata* (Higgs) McGregor and Playford 1990.

Serial ultratome sections and TEM studies of well preserved individual specimens have demonstrated that *T. torquata* is protosaccate. The protosaccus, however, seems relatively fragile and may be missing in poorly preserved specimens. In other specimens a pseudocavum develops as the result of damage during polar compression. The exine is two layered with an ?homogeneous innermost layer (intexine) which is, except at the laesurae, in contact with the outer layer (exoexine). Over the polar areas the exoexine is alveolate; for the remainder of the grain the exoexine comprises the filigree-like structure of the protosaccus. This complex exinal structure predates by at least 30 million years the development of protosaccate and eusaccate pollen grains among Carboniferous gymnosperms. The natural affinities of *T. torquata*, as indicated by wall ultrastructure, remain uncertain as it demonstrates features shared with spores of certain lycopsids and pollen of the progymnosperms (solid inner layer, proximal germinal area) and gymnosperms (protosaccate exoexine).

Introduction

Teichertospora Balme 1988 was created for trilete, monosaccate palynomorphs from Late Devonian (Frasnian) assemblages from the Carnarvon Basin, Western Australia. The genus is monotypic and characterised by an orthotriangular amb; circular intexine, often with a continuous marginal fold; and an endoreticulate exoexine (Figs. 1–7). Balme (1988) noted that no undoubtedly saccate forms had previously been recorded from pre-Carboniferous strata, and drew attention to similarities between *Teichertospora* and younger Palaeozoic monosaccate genera such as *Felixipollenites* Millay and Taylor 1974 and *Plicatipollenites* Lele 1964.

Teichertospora cuvierica Balme 1988 was designated as the type of *Teichertospora*, but McGregor and Playford (1990) have demonstrated that on morphological criteria this species is a junior subjective synonym of *Auroraspora torquata* Higgs 1975, previously known from low palaeolatitude Late Devonian (late Frasnian through Famennian) strata

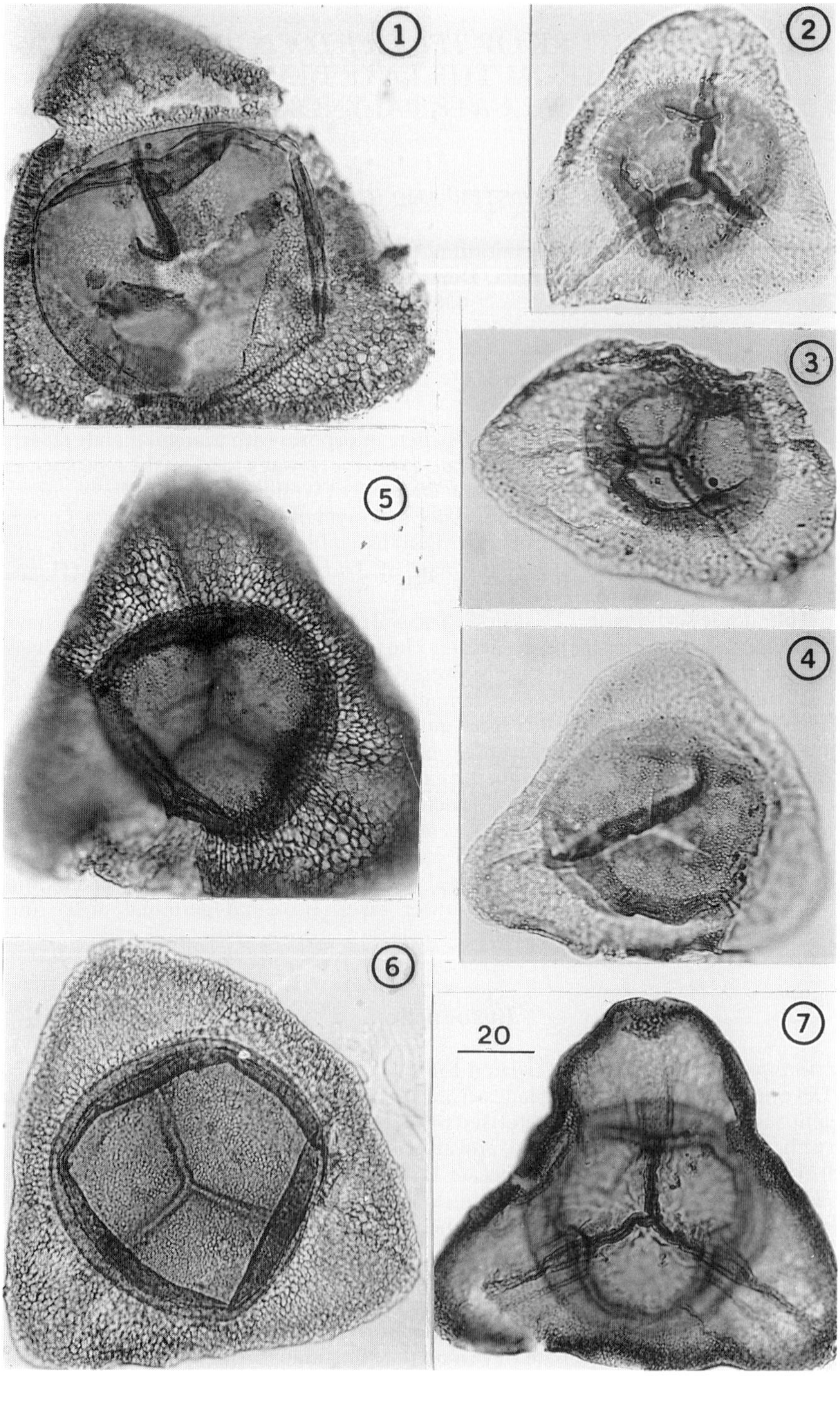
1
2
3
4
5
6
7
20

in western Europe and North America (see McGregor and Playford, 1990, text-fig. 2). However, because other members of *Auroraspora* Hoffmeister, Staplin and Malloy 1955 lack an endoreticulate exoexine, generic reassignment of *A. torquata* to *Teichertospora* was adopted. McGregor and Playford (1990) discussed the morphological variability of the exoexine which ranges from clearly endoreticulate (saccate) to specimens in which the endoreticulum is scarcely visible or even apparently nonexistent. Similar variability is also evident in our material and is attributed to preservational differences (see Figs. 1–7).

The saccate construction of *Teichertospora* led Balme (1988) to speculate on the affinities of its parent plant. He suggested that it may have belonged to either a member of the ancestral stock of the Cordaitanthales or, alternatively, to a cryptogam, perhaps a lycopsid. If the latter, the endoreticulate structure of the exoexine is an example of convergence, with the later pollen of the cordaitanthaleans and gymnosperms possessing an analogous structure. The aim of this study is to elucidate the ultrastructure of *T. torquata*, and to assess the implications of the structure in terms of its possible botanical affinities.

Material and Methods

Specimens of *T. torquata* were picked from palynological residue recovered from samples, supplied by Western Australian Petroleum Pty Ltd, of core 12 (420–430 m), Cape Cuvier No. 1 Well, Carnarvon Basin, Western Australia. This was the original type locality for *T. cuvierica*. The interval, an equivalent of the Gneudna Formation, is dated as Frasnian on the evidence of marine invertebrate faunas, and McGregor and Playford (1990) believed that palynological evidence indicated a late, or latest, Frasnian age.

Plant microfossils were strew mounted in resin and those selected for sectioning were cut from the strew mounts and re-embedded in Araldite ready for ultratome sectioning, using either a diamond or a glass knife. Sections of approximately 90 nm thickness were skilfully prepared by Ms Ruo Lun Shen of the Electron Microscopy Unit, Research School of Biological Sciences, Australian National University, and studied at the same facility using either a Zeiss 109 or an Hitachi 600 TEM instrument. The resultant TEM photographic images were computer enhanced to remove differently exposed backgrounds, and to make the contrast of multiple exposures more uniform. This was done at the Australian Geological Survey Organisation (AGSO) through the careful work of John Creasy and Christian Thun. The images (uncompressed) used in this paper occupy 2–6 megabytes of computer storage space. They can be stored in TIFF format and accessed electronically using E-mail (cfoster@bmr.gov.au).

Observations

Figures 8–14 show a series of serial sections through an uncompressed specimen of *T. torquata*. Two exinal layers are identified. The innermost layer, or intexine, seems mostly to lack any recognisable structure (see Fig. 11) except beneath the laesurae; it is sometimes folded (see below), and tightly appressed to the second layer (exoexine). In the polar regions the exoexine is alveolate; distally it is in immediate contact with the intexine, but at the proximal pole in some specimens the exinal layers are separated to form a small vestibulate cavum (Figs. 8, 12–14). Away from the poles, but still in contact

FIGS. 1–7. Optical photomicrographs of *Teichertospora torquata* (Higgs) showing preservational variation; specimens from Cape Cuvier No. 1 Well, Carnarvon Basin, Western Australia. Late Devonian. Scale bar in micrometres. **1**: Ruptured specimen showing cavum developed as a result of intexinal folding. **2–4**: Smaller specimens with partially degraded, fine endoreticula. **5–7**: Specimens with well developed endoreticula, with lumina of varying sizes. 5. Designated holotype for *T. curvierica* Balme 1988, now assigned to *T. torquata* (UWA No. 105037).

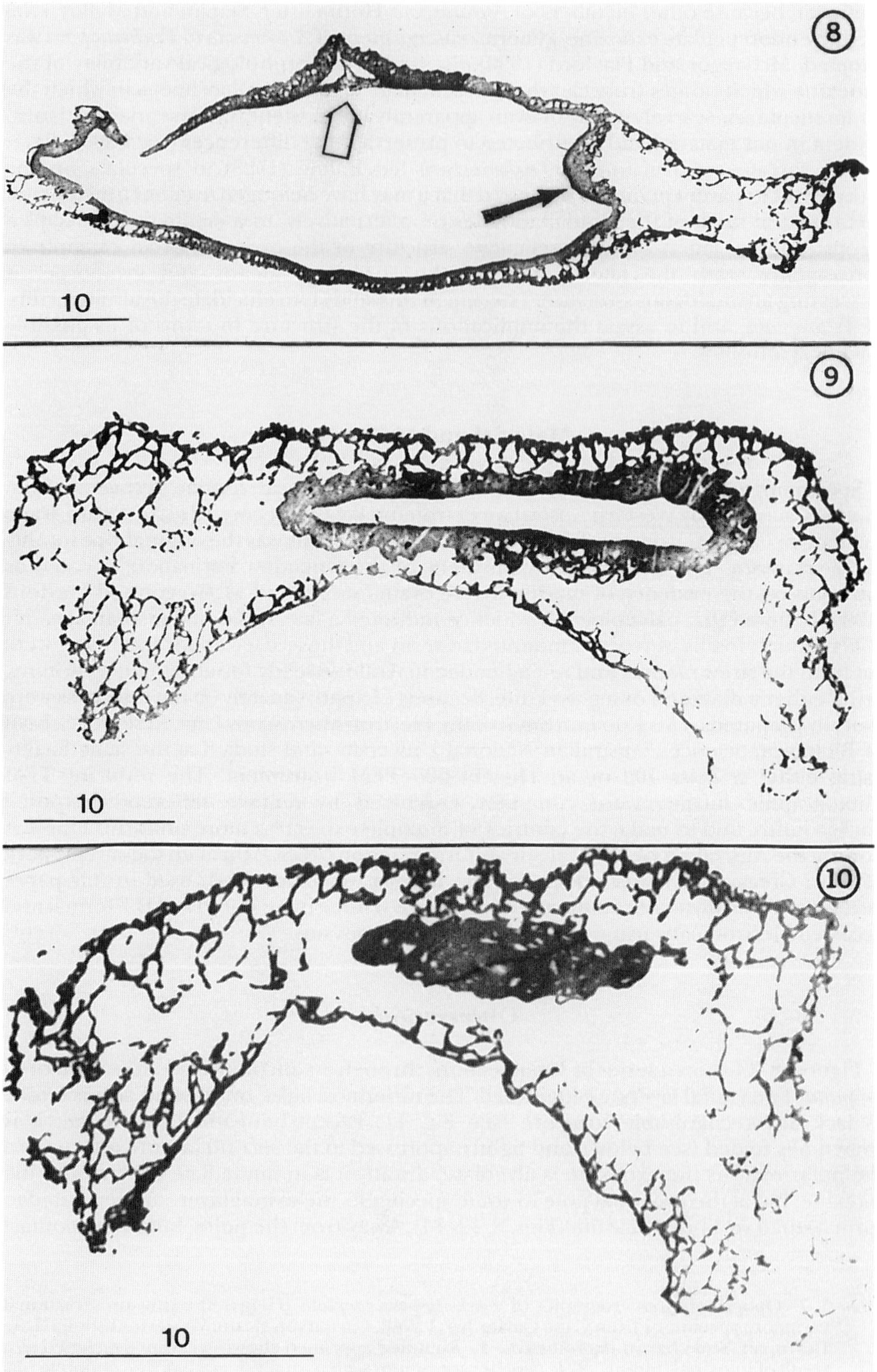
8
10
9
10
10
10

with the intexine, the exoexine is further differentiated (or expanded) to form the filigree-like endoreticulate structure of the saccus. At the laesurae both layers show special features: here the intexine seems less robust and appears to break down more easily, and as noted above, a separation of exinal layers may occur. Both these features are possibly an aid to germination. The descriptive terms of intexine, exoexine, and protosaccus conform to those used when describing palynomorphs using standard optical microscopy techniques. Selected TEM sections of highly compressed specimens are shown in Figs. 15–17. Some of these specimens lack endoreticulate structure of the saccus, and we consider this to be the result of differences in preservation (see Figs. 2–4). Why there is an admixture of compressed and uncompressed specimens within the type assemblage is uncertain. Perhaps the degree of compression is related to variation in sedimentary grain size and consequent differences in compaction of the host rock. The significance of the observed ultrastructural features is discussed below.

Interpretations

Our initial sections appeared to support the earlier conclusions of Balme (1988), based on optical microscopy, that the elements of the endoreticulum were detached from the intexine and that the miospore was eusaccate (cf. Figs. 1, 7). This contrasts with the protosaccate exinal structure of many Palaeozoic and Early Mesozoic pollen types, in which the filigree-like internal structures of the exoexine remain in contact with the intexine (see e.g. Scheuring, 1974; Foster, 1979). Protosaccate grains lack a cavum (or cava) within the expanded exoexine, and in this respect differ significantly from the eusaccate pollen of living Pinaceae and of certain fossil groups such as the Cordaitanthales and Callistophytales (see Taylor and Taylor, 1987). The initial results were nevertheless unexpected, as studies after 1988 had led us to suspect that *T. torquata* would be protosaccate. Such cavate structure could be either genuine, or an artefact of sectioning, or the result of degradation during fossilisation.

More specimens of *T. torquata* were selected and sectioned. Figures 8–10 show a series of ultrathin sections through a single, uncompressed specimen. Early sections show a clear protosaccus, comparable to the saccus architecture of Permian protosaccate grains, but with progressive sectioning an apparent cavum, resembling that illustrated by Balme (1988), was seen within the saccus. We now consider that the cavum is formed by the tearing of the endoreticulum from the intexine; this occurs when the intexine undergoes folding as a result of compression during fossilisation (Fig. 18).

Development of intexinal folds is common among many Carboniferous and Permian saccate pollen genera; *Potonieisporites* Bharadwaj 1954 and *Plicatipollenites* are well-known examples. The folds result from mechanical compensation, in polar compression, of grains in which the distal hemisphere is more strongly inflated than the proximal surface, so that the length of the intercept of the polar axis and the equatorial plane is shorter on the proximal side than the distal. The amb of specimens in polar compression is therefore determined by the equatorial diameter (ED), and the number and distribution of the intexinal folds depends on ED and the relative lengths of the proximal and distal polar axis intercepts. In *T. torquata* the exoexine is torn away from the intexine presumably as it folds (Figs. 8, 13). Mechanical tearing, as proposed here, is not as obvious in the few published sections of Late Palaeozoic protosaccate grains (see Foster, 1979), and it may reflect the relative fragility of the endoreticulum in *T. torquata*.

FIGS. 8–10. Transmission electron micrographs (TEM) of serial sections through an uncompressed specimen of *T. torquata* (TEM block 1). Scale bars in micrometres. **8**: Mid section, showing attachment of alveolate exoexine over polar areas. Solid arrow indicates infolded intexine and the adjacent pseudocavum developed by tearing of the protosaccate exoexinal layer. Note the differentiation of intexine beneath the laesura (open arrow). **9**: Early section intersecting the intexine and protosaccate exoexine of the saccus. **10**: Initial section from block 1. Note cavum developed through tearing of protosaccus.

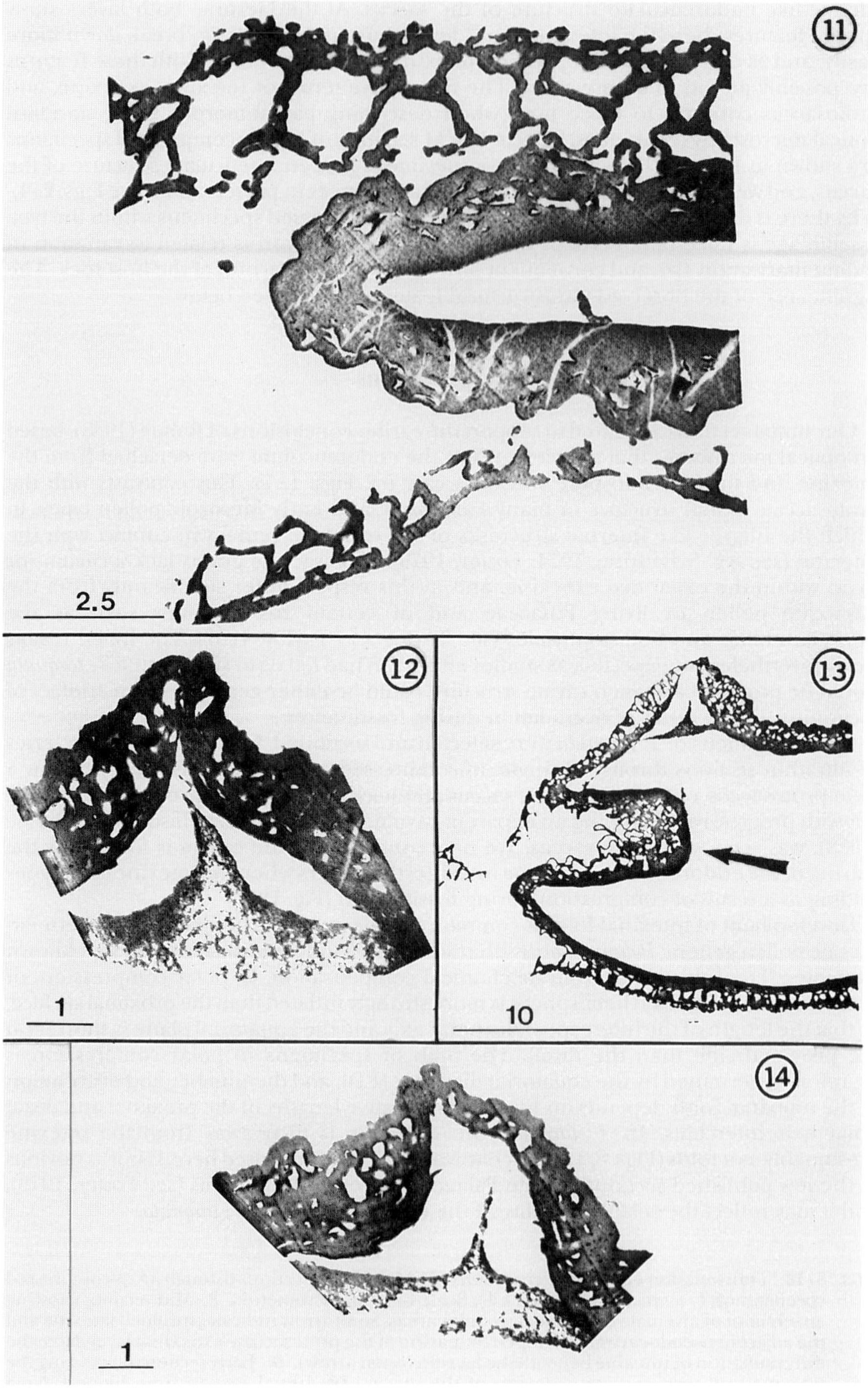
11
2.5
12
1
13
10
14
1

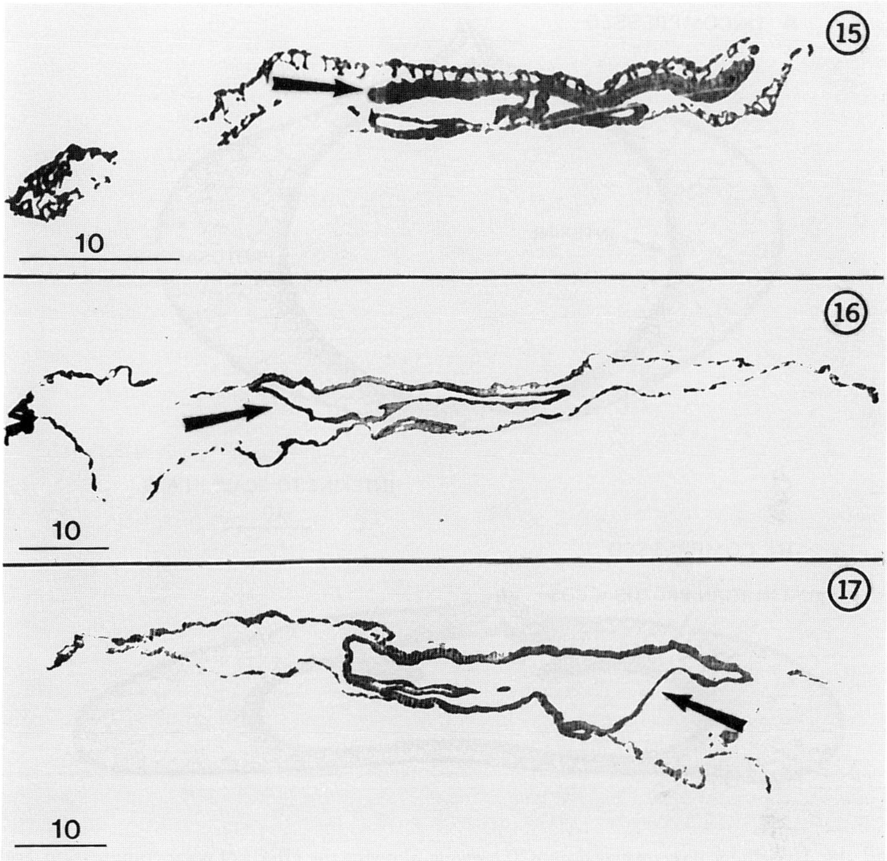

Figs. 15–17. TEM sections through poorly preserved examples of *T. torquata* showing the range of variation of preservation of the endoreticulum of the protosaccus, from visible (**15**: block 4), to non-existent (**16**: block 8; **17**: block 5). All specimens are highly compressed. Scale bars in micrometres. Arrow indicates intexinal layer.

We have shown that *T. torquata* is protosaccate; does it share any other features with geologically younger gymnosperm grains? The innermost layer (here called intexine) of gymnosperm pollen is characteristically lamellate (see discussion in Foster and Price, 1981, p. 179; Foster, 1979, Pl. 27, fig. 1e). There is no unequivocal evidence of multiple layering at high magnification in *T. torquata*, but on some specimens (not figured), very poorly defined 'lineations' are evident in the intexinal layer. However, lamellae within the inner layer are not always recognisable: their apparent absence can be effected either by

Figs. 11–14. Detailed TEM sections of *T. torquata* from TEM block 1 (also see Figs. 8–10). Scale bars in micrometres. **11**: Detail of intexine and protosaccus. **12**: Alveolate exoexine over laesura, note the differentiation of the intexine beneath the laesura, presumably involved in germination (see also Fig. 14). **13**: Infolded intexine (arrowed) with adjacent torn exoexine; alveolate exoexine over polar areas clearly displayed. **14**: Showing details of exinal layers at the laesura.

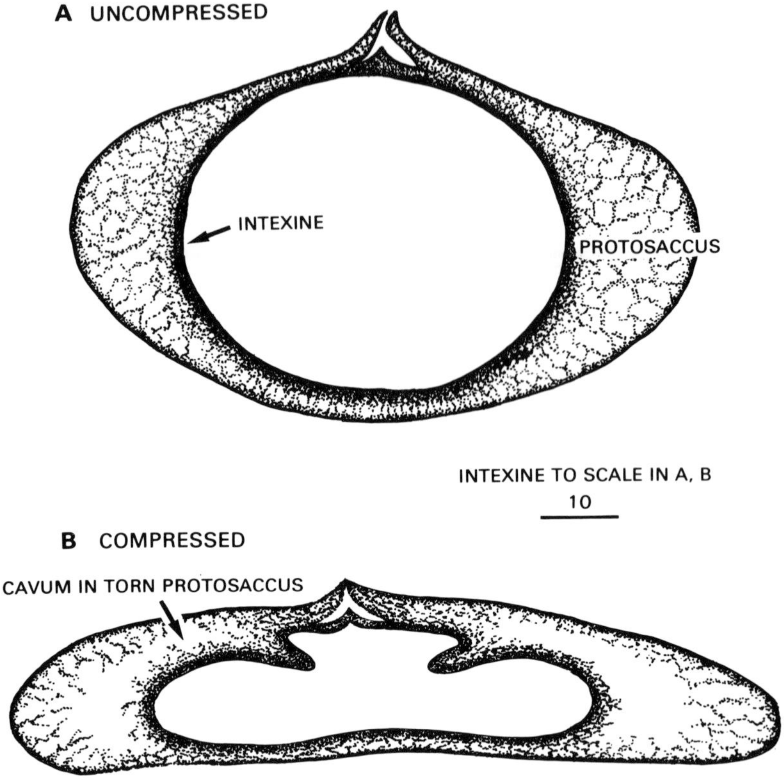

Fig. 18. Hypothetical reconstruction of *T. torquata*, showing the effects of polar compression, and the consequent development of the pseudocavum caused by tearing of the protosaccus. Intexinal layer of each at the same scale. Scale bar in micrometres.

preparation techniques (TEM sections too thick) and/or the level of ontogenetic development of individual specimens (see Taylor and Daghlian, 1980).

Natural Affinities: Ultrastructural Evidence

Taylor and Taylor (1987) have provided a useful review of the known natural affinities of dispersed miospores for which ultrastructural data are available. Their paper was based principally on numerous studies of T. N. Taylor and his collaborators, who had noted previously (Taylor et al., 1987) that the majority of Permian/Triassic grains from southern hemisphere deposits were protosaccate and suggested that this was "a basic pattern for early saccate pollen types". The term protosaccate may itself imply an ancestral or analogous architectural form relative to the eusaccate grains of some extant pinophytes, such as *Pinus strobus* (see Taylor et al., 1987, Pl. 3, fig. 4). It is nonetheless true that certain Late Carboniferous pollen types, such as the cordaitanthalean genera *Felixipollenites*, *Sullisaccites* Millay and Taylor 1974, and *Florinites* Schopf, Wilson and Bentall 1944 are eusaccate, as was the pollen of the callistophyte seed fern *Callandrium*.

Because of the presence of a cavum in type specimens of *Teichertospora*, Balme (1988) initially suggested the possibility that it derived from ancestral cordaitanthalean stock. However, from evidence presented here, the cavum of *Teichertospora* results only from mechanical compression, and so further comparisons with eusaccate cordaitanthalean pollen are less secure.

Millay and Taylor (1976) discussed the ultrastructural differences between Devonian *Rhabdosporites* Richardson 1960, the miospores of the progymnosperm *Tetraxylopteris*, and the Carboniferous gymnosperm pollen *Felixipollenites*. *Rhabdosporites* is cavate, without an endoreticulum, and the exoexine is completely detached from the intexine except in the area of the proximal pole. During early stages of development, grains of *Felixipollenites* have a cavum developed only within the equatorial expansion of the exoexine, which is structurally continuous with the intexine both proximally and distally. This type of exine structure gives the miospore a degree of rigidity that characterises all later saccate pollen. However, as grain expansion occurs in *Felixipollenites* the saccus separates from the distal surface of the intexine (= corpus of Taylor and Daghlian, 1980). For specimens of *Teichertospora* examined in this study, the exinal layers are attached over both the polar surfaces (except immediately beneath the laesurae), and in this respect differ from the progymnosperm miospores of *Rhabdosporites* and *Geminospora* Balme 1960 (see Figs. 8, 15–17). They are more comparable to pollen of the Carboniferous cordaitanthaleans and later gymnosperms. The strongly delineated trilete aperture in both the intexine and exoexine of *T. torquata* (Figs. 1–7) suggests that germination occurred through the proximal pole, a feature which is often considered to indicate "prepollen" among saccate morphotypes (see Millay and Taylor, 1974, 1979). This criterion alone, however, is too inconclusive to determine systematic affinity. For example, among Permian gymnosperm protosaccate grains, such as *Barakarites rotatus* Bharadwaj and Salujha 1964, germination seems, from structural evidence, equally possible from either pole (see Foster, 1983). The current evidence of proximal germination for *T. torquata* is consistent with either a cordaitanthalean or cryptogam origin. Figures 12–14 show that the intexine at the proximal pole is less robust than elsewhere and appears to have broken down; presumably this is linked to a germinal mechanism. We note also that the separation of these exinal layers in the germinal area is similar to that shown by gulate megaspores of lycopsid origin (e.g. Taylor and Taylor, 1988).

The development of a lamellate intexine is common in grains of gymnospermous origin. However, only very few of the present sections showed faint 'lineations' within the intexine: for the most part, the intexine, except at the proximal pole, appears unstructured. But as noted above, the absence of lamellae requires cautious interpretation because both preservational and ontogenetic factors, as well as technical factors (e.g. TEM section thickness) can affect their detection. Superficially *Teichertospora* appears to show a stacked-lamellae layer similar to that of the eusaccate cordaite grains of *Felixipollenites* (see Millay and Taylor, 1974). However, we interpret this as unexpanded protosaccus and therefore a part of the exoexine. Minor supra-exoexinal elements are evident in our TEM sections, and they may have been recognised as grana in Higg's (1975) original description.

Conclusions

Teichertospora torquata shows structural features that are shared with the northern hemisphere cordaitanthaleans (dipolar attachment of exinal layers, endoreticulate exoexine), the seed plants of the Gondwanan Late Palaeozoic (protosaccate exoexine), and certain fossil and living lycopsid spores. This may of course simply reflect the limitations of current studies. Ultrastructure alone is therefore not enough to determine the systematic affinities of these grains (see also Taylor et al., 1987). What is undeniable, and extremely interesting, is that *T. torquata* exhibits the oldest protosaccate construction yet determined. It is worth noting here that other Late Devonian miospores, such as those assigned to *Archaeoperisaccus* Naumova 1953, have been described as saccate, but currently available evidence from transmitted-light microscopy indicates that such specimens are probably alveolate rather than endoreticulate, and that any sculptural elements are supra-exinal or on the outer surface of the exoexine (for summary see Jansonius and Hills, 1976, cards 157–158; 1982, cards 3941–3942). *Teichertospora torquata* precedes other endoreticulate pollen grains such as those assigned to *Florinites* spp. from the Late Carboniferous (Namurian; Clayton et al., 1977) by 30 million years, and by 65 million from its morphologically closest genus, *Plicatipollenites*. It is distinguished from the latter by its strongly delineated trilete aperture, uniquely subtriangular amb, and less robust saccus construction. It is structurally more complex than any known progymnosperm spore and could be, as suggested earlier (Balme, 1988), a prime candidate as an ancestral gymnosperm pollen. Continuing ultrastructural studies of both *in situ* miospores and *sporae dispersae* will undoubtedly provide firmer evidence of plant affinity and evolutionary pathways.

Computer enhancement techniques have aided in ultrastructural interpretation and presentation, but most importantly in image storage, retrieval, and handling. Images can be stored in various computer formats (e.g. TIFF) and accessed electronically using E-mail.

Acknowledgements

We thank our referees for their critical review and suggestions to improve the manuscript.

References

Balme, B.E. (1988). *Teichertospora* n. gen., an endoreticulate miospore from the Late Devonian of Western Australia. *Senckenberg. Leth.* **69**: 159–169.

Clayton, G., Coquel, R., Doubinger, J., Gueinn, K.J., Loboziak, S., Owens, B. and Streel, M. (1977). Carboniferous miospores of western Europe: illustration and zonation. *Meded. Rijks. Geol. Dienst.* **29**: 1–71.

Foster, C.B. (1979). Permian plant microfossils of the Blair Athol Coal Measures, Baralaba Coal Measures, and basal Rewan Formation of Queensland. *Publ. Geol. Surv. Queensland* **372**, *Palaeontol. Pap.* **45**: 1–244.

Foster, C.B. (1983). *Jugasporites* Leschik 1956, a Late Palaeozoic operculate pollen genus. *Mem. Assoc. Australas. Palaeontologists* **1**: 327–338.

Foster, C.B. and Price, P.L. (1981). Exine infrastructure of *Praecolpatites sinuosus* (Balme et Hennelly) Baradwaj et Srivastava 1969, and *Marsupipollenites triradiatus* Balme et Hennelly 1956. *Palaeobotanist* **28–29**: 177–187.

Higgs, K. (1975). Upper Devonian and Lower Carboniferous miospore assemblages from Hook Head, County Wexford, Ireland. *Micropaleontology* **21**: 393–419.

Jansonius, J. and Hills, L.V. (1976). Genera file of fossil spores. Cards 157–158. University of Calgary Geology Department, Calgary.

Jansonius, J. and Hills, L.V. (1982). Genera file of fossil spores — supplement. Cards 3941–3942. University of Calgary Geology Department, Calgary.

McGregor, D.C. and Playford, G. (1990). Morphology and distribution of the miospore *Teichertospora torquata* comb. nov. in the Upper Devonian of Euramerica and Australia. *Palynology* **14**: 7–18.

Millay, M.A. and Taylor, T.N. (1974). Morphological studies of Paleozoic saccate pollen. *Palaeontographica Abt.B., Paläophytol.* **147**: 75–99.

Millay, M.A. and Taylor, T.N. (1976). Evolutionary trends in fossil gymnosperm pollen. *Rev. Palaeobot. Palynol.* **21**: 65–91.
Millay, M.A. and Taylor, T.N. (1979). Paleozoic seed fern pollen organs. *Bot. Rev.* **45**: 301–375.
Scheuring, B.W. (1974). "Protosaccate" Strukturen, ein weitverbreitetes Pollenmerkmal zur frühen und mittleren Gymnospermenzeit. *Geol. Paläontol. Mitt. Innsbruck* **4**: 1–30.
Taylor, T.N. and Daghlian, C.P. (1980). The morphology and ultrastructure of *Gothania* (Cordaitales) pollen. *Rev. Palaeobot. Palynol.* **29**: 1–14.
Taylor, T.N. and Taylor, E.L. (1987). The ultrastructure of fossil gymnosperm pollen. *Bull. Soc. Bot. France, Actual. Bot.* **134**: 121–140.
Taylor, T.N., Delevoryas, T. and Hope, R.C. (1987). Pollen cones from the Late Triassic of North America and implications on conifer evolution. *Rev. Palaeobot. Palynol.* **53**: 141–149.
Taylor, W.A. and Taylor T.N. (1988). Ultrastructural analysis of selected Cretaceous megaspores from Argentina. *J. Micropalaeontol.* **7**: 73–87.

Osborn, J.M. and Taylor, T.N. (1994). Comparative ultrastructure of fossil gymnosperm pollen and its phylogenetic implications. In: M.H. Kurmann and J.A. Doyle (editors). Ultrastructure of fossil spores and pollen, pp. 99–121, Royal Botanic Gardens, Kew.

7. COMPARATIVE ULTRASTRUCTURE OF FOSSIL GYMNOSPERM POLLEN AND ITS PHYLOGENETIC IMPLICATIONS

JEFFREY M. OSBORN [1] AND THOMAS N. TAYLOR [2]

[1] **Division of Science, Northeast Missouri State University, Kirksville, MO 63501, U.S.A.**
[2] **Department of Plant Biology and Byrd Polar Research Center, The Ohio State University, Columbus, OH 43210, U.S.A.**

Abstract

Numerous studies have been carried out on fossil gymnosperm pollen detailing its overall appearance in transmitted light and, to some extent, its surface morphology as revealed by scanning electron microscopy. Fewer investigations, however, have addressed the fine structural nature of fossil pollen using transmission electron microscopy. Pollen wall ultrastructure from grains preserved within intact reproductive organs (*in situ*) has been documented from gymnospermous taxa of the following orders: Lyginopteridales, Medullosales, Callistophytales, Cordaitales, Glossopteridales, Voltziales, Coniferales, Caytoniales, Corystospermales, Cycadales, Bennettitales, and Pentoxylales. Several dispersed (*sporae dispersae*) pollen types are also known from a few of these groups as well as the Gnetales. Moreover, a number of other taxonomically enigmatic palynomorphs (both *in situ* and dispersed taxa) have been described at the ultrastructural level, and are currently classified as *incertae sedis*. Pollen morphology and ultrastructure of these groups have played an important role in recent phylogenetic studies, which have been based on a number of reproductive and vegetative characters. Although many palynological characters are considered conservative evolutionary features, they can also reflect several other biological and physical phenomena. For example, features associated with pollen ontogeny as well as pollination syndromes may be important influences, while the most prominent abiotic factors to consider are mode of preservation, degree of diagenesis, and a suite of preparation protocols. These features are discussed as they relate to phylogenetic interpretations and relevance of several salient palynological characters, including overall exine infrastructure, nexine organization, and saccus type and internal composition.

Introduction

The use of both extant and fossil pollen as a systematic tool in evaluating phylogenetic relationships among seed plants is well documented and extends back to the early days of palynology (e.g., Wodehouse, 1928). Since that time, palynological characters have become increasingly important for phylogenetic analyses, with investigations of pollen becoming broader in both number and type. The majority of studies on fossil pollen has involved the analysis of grains by light microscopy (LM). However, micromorphological and ultrastuctural investigations employing scanning and transmission electron (SEM and TEM) reveal structural features of phylogenetic significance otherwise not available by LM.

Many excellent studies have been carried out on both extant and fossil gymnosperm pollen detailing its overall appearance in transmitted light (e.g., Van Konijnenburg-van Cittert, 1971) and, to some extent, its surface morphology as revealed by SEM (e.g.,

Krassilov, 1977). Fewer investigations, however, have addressed the fine structure of gymnosperm pollen using TEM. For extant groups, those TEM studies that have been undertaken have principally focused on describing exine stratification in mature pollen, although several have examined the various stages of sporoderm ontogeny. Moreover, only a small number of investigations, as well as taxa, have been conducted on pollen ultrastructure of extant gymnosperms (for review see Kurmann, 1992). With regard to fossil taxa, although twenty orders of extinct gymnosperms are recognized (Taylor and Taylor, 1993), and fossil members of the four extant orders are also known, data on pollen fine structure for some groups are either entirely lacking or come from the evaluation of a single taxon.

The existing studies of fossil gymnosperm pollen can be categorized into two distinct groups, based on the types of grains that are investigated. These include palynomorphs that are isolated from sediments in a dispersed condition (*sporae dispersae*), and those grains that are recovered from within intact, megafossil reproductive organs (*in situ*). Numerous contributions have examined *sporae dispersae* grains, which have been recovered by standard acid maceration techniques, and addressed the nature of such palynomorphs as seen in transmitted light. Stratigraphic palynology has been useful in illustrating palynomorph distribution, diversity, and occurrence throughout the geologic column (i.e., biostratigraphy). However, one of the most significant constraints in studying dispersed grains is the fact that in most cases nothing is known about their parent plants. It is possible in some instances to compare *sporae dispersae* taxa with *in situ* grains (e.g., Thomas, 1987; Willard, 1989), and to evaluate overall grain morphology, surface sculpture, and the nature of the aperture or haptotypic mark in order to infer systematic relationships (e.g., Ward et al., 1989). Despite the fact that more than 50 dispersed taxa have been examined at the fine structural level (Table 1), for most, their taxonomic affinities remain equivocal.

Investigations that focus on *in situ* gymnosperm grains are typically afforded the opportunity to assess the systematic position of the plants that produced the pollen (Table 2). In a few cases, despite having complete morphological and/or anatomical data on the entire reproductive organ as well as information regarding pollen ultrastructure, the affinities continue to remain problematic (e.g., *Lasiostrobus*; Taylor, 1970; Table 2). Furthermore, pollen preserved *in situ* is of paramount importance in evaluating fossil gymnosperms from a biological perspective (Taylor et al., in press). For example, studies of *in situ* gymnosperm pollen have been instrumental in addressing several developmental and reproductive parameters such as microsporogenesis (Taylor, 1990), pollen germination and pollen tube growth (Rothwell, 1972), and the possible functional roles of the exine and sacci in pollination (Taylor and Zavada, 1986).

Tables 1 and 2 indicate the fossil gymnosperm taxa which are now known at the ultrastructural level. Although several new taxa have been investigated and others have been re-examined (Tables 1, 2) since the last review of fossil gymnosperm pollen ultrastructure was published (Taylor and Taylor, 1987), we will not review this topic here. Rather, the intent of this paper is to focus on the relative phylogenetic significance of several ultrastructural characters present in fossil gymnosperm pollen.

Ultrastructural Characters

In recent years, phylogenetic systematics (cladistics) has become a prominent component in palynology. An area that has received a significant amount of attention is the proposed phylogenetic relationships among extinct and extant seed plants as well as the question of angiosperm origins as elucidated by cladistic investigations (e.g., Crane, 1985; Doyle and Donoghue, 1986). These studies have been based on a number of both vegetative and reproductive features including several palynological characters. In addition to the four pollen characters used by Crane (1985) and the six employed by

TABLE 1. Selected gymnosperm taxa for which *sporae dispersae* pollen has been studied with transmission electron microscopy [1]

PTERIDOSPERMOPHYTA	
MEDULLOSALES (*Schopfipollenites*-type)	
Schopfipollenites sp. [2]	Pn
Schopfipollenites sp. [3]	Pn
GLOSSOPTERIDALES	
Protohaploxypinus spp. [4]	Pm
CONIFEROPHYTA	
CORDAITALES	
Florinites sp. [2]	Pn
VOLTZIALES	
Cheirolepidiaceae	
(circumpolloid grains)	
Circulina sp. [5]	Tr
Classoidites glandis [5,6]	Tu
Classopollis classoides [7]	Ju
C. harrisii [5]	Li/Rh
C. minor [8]	Ju
CONIFERALES	
Podocarpaceae	
Rugubivesiculites rugosus [9]	Cr
GNETOPHYTA	
Equisetosporites spp. [10]	Cr
Ephedripites sp.1 [11]	Cr
INCERTAE SEDIS	
Teichertospora torquata [12]	Dv
Nanoxanthiopollenites mcmurrayii [13,14]	Pn
Cannanoropollis janakii [15]	Pm
Marsupipollenites triradiatus [16]	Pm
Platysaccus leschikii [15]	Pm
Plicatipollenites malabarensis [15]	Pm
Praecolpatites sinuosus [16]	Pm
Protohaploxypinus limpidus [15]	Pm
Striatopodocarpites phaleratus [15]	Pm
INCERTAE SEDIS (CONT.)	
Bharadwajipollenites wielandii [17]	Tr
Equisetosporites chinleana [17,18,19]	Tr
Granamonocolpites luisae [17]	Tr
cf. *G. asymmetricus* [9]	Cr
Triadispora bölchii [20]	Tr
Unnamed bisaccate sp. [18]	Tr
Lueckisporites virkkiae [21]	Li
Lunatisporites noviaulensis mollis [21]	Li
Ovalipollis notabilis [21]	Li
O. ovalis [21]	Li
Araucariacites sp. [22]	Ju
A. hungaricus [8]	Ju
A. v. *granulatisporites* [6]	Cr
A. australis [23]	Cr
Cycadopitys sp. [8]	Ju
Inaperturopollenites limbatus [22]	Ju
I. ex gr. *hiatus* [8]	Ju
Spheripollenites scabratus [6]	Ju
Balmeiopsis limbatus [23]	Cr
Clavabisvesiculites sp. [9]	Cr
Cyclusphaera psilata [24,25]	Cr
Eucommiidites sp. 1 [11]	Cr
E. sp. 2 [11]	Cr
E. sp. [18]	Ju
E. sp. [26]	Cr
Granabivesiculites inchoatus [18]	Cr
G. cf. *inchoatus* [9]	Cr
G. sp. [18]	Cr
Granamultivesiculites sp. [9]	Cr
Monosulcites sp. 1 [11]	Cr
M. sp. [9]	Cr
Oculopollis maximus [27]	Cr
Punctamultivesiculites inchoatus [18]	
Verrumonocolpites conspicuus [18]	Cr
Unnamed (vestigial saccate) sp. [18]	Cr

[1]Age; Dv=Devonian, Cr=Cretaceous, Ju=Jurassic, Li=Liassic, Pn=Pennsylvanian, Pm=Permian, Rh=Rhaetian, Tr=Triassic, and Tu=Turonian.

[2] Pettitt, 1966; [3] Abadie et al., 1978; [4] Osborn, 1991; [5] Médus, 1977a; [6] Kedves and Párdutz, 1973; [7] Rowley and Srivastava, 1986; [8] Kedves, 1985; [9] Zavada and Dilcher, 1988; [10] Osborn et al., 1993; [11] Trevisan, 1980; [12] Foster and Balme, 1994; [13] Taylor, 1980; [14] Taylor, 1982; [15] Foster, 1979; [16] Foster and Price, 1981; [17] Zavada, 1990; [18] Zavada, 1984; [19] Pocock and Vasanthy, 1988; [20] Scheuring, 1976; [21] Scheuring, 1974; [22] Kedves and Párdutz, 1974; [23] Zavada, 1992; [24] Taylor et al., 1987; [25] Zavada, 1987; [26] Doyle et al., 1975; [27] Médus, 1977b.

TABLE 2. Selected gymnosperm taxa for which *in situ* pollen has been studied with transmission electron microscopy.

PTERIDOSPERMOPHYTA
- LYGINOPTERIDALES
  - *Crossotheca* sp. [1]
  - *Phacelotheca pilosa* [2]
  - *Potoniea illinoiensis* [3,4]
  - *P. carpentieri* [4]
  - *Schopfiangium varijugatus* [5]
- MEDULLOSALES (*Schopfipollenites*-type)
  - *Aulacotheca iowensis* [6,7,8]
  - *Bernaultia formosa* [9]
  - *B. sclerotica* [9]
  - *Boulayatheca fertilis* [10]
  - *Codonotheca caduca* [6,7,8]
  - *Dolerotheca* sp. [8,11]
  - *Halletheca reticulata* [8,11,12]
  - *Rhetinotheca patens* [4]
  - *R. tetrasolenata* [8]
  - *Schopftheca boulayoides* [8]
  - *Sullitheca dactylifera* [4]
- MEDULLOSALES (*Parasporites*-type)
  - *Parasporotheca leismanii* [1,4]
- CALLISOPHYTALES
  - *Idanothekion callistophytoides* [13,14]
- GLOSSOPTERIDALES
  - *Arberiella* sp. (*Protohaploxypinus*-type & *Striatopodocarpites*-type) [15]
- CORYSTOSPERMALES
  - *Pteruchus africanus* [16]
  - *P. dubius* [16,17]
  - *P. papillatus* [16]
  - unnamed sp. [18]
- CAYTONIALES
  - *Caytonanthus arberi* [19,20,21]
  - *C. kockii* [20]

CONIFEROPHYTA
- CORDAITALES (*Florinites*-type)
  - *Cordaianthus* sp. [13,14]
- CORDAITALES (*Sullisaccites*-type)
  - *Cordaianthus* sp. [13,14]
- CORDAITALES (*Felixipollenites*-type)
  - *Gothania lesliana* [13,14,22]
- VOLTZIALES
  - Voltziaceae
    - *Darneya peltata* [23]
    - *Sertostrobus laxus* [23,24]
    - *Willsiostrobus denticulatus* [23]
    - *W. cordiformis* [23,25]
    - *W. rhomboides* [23,25]
  - Cheirolepidiaceae (*Classopollis*-type)
    - *Hirmeriella* (=*Cheirolepidium*) *muensteri* [26]
    - *Classostrobus comptonensis* [27]
    - *Pseudofrenelopsis* sp. [28]
- CONIFERALES
  - Pinaceae
    - *Pinus* sp. [29]
  - Podocarpaceae
    - *Millerostrobus pekinensis* [30]
    - *Moreno fertilis* [31]
    - *Squamostrobus tigrensis* [32]
    - *Trisacocladus tigrensis* (*Trisaccites*-type) [33]
  - Taxodiaceae
    - *Drumhellera kurmanniae* [34]
    - *Elatides williamsonii* [35]
    - *Metasequoia milleri* [36]

CYCADOPHYTA
- CYCADALES
  - *Androstrobus balmei* [37]
- BENNETTITALES (=CYCADEOIDALES)
  - *Cycadeoidea dacotensis* [19,38]
  - *Leguminanthus siliquosus* [39]
- PENTOXYLALES
  - *Sahnia laxiphora* [40]

INCERTAE SEDIS
- *Erdtmanispermum balticum* (*Eucommiidites*-type) [41]
- *Erdtmanitheca texensis* (*Eucommiidites*-type) [41]
- *Lasiostrobus polysacci* [42,43]
- *Melissiotheca longiana* [44]

---

[1] Millay et al., 1978; [2] Meyer-Berthaud and Galtier, 1986; [3] Stidd, 1978; [4] Taylor, 1982; [5] Stidd et al., 1985; [6] Taylor, 1976a; [7] Taylor, 1976b; [8] Taylor, 1978; [9] Taylor and Rothwell, 1982; [10] Kurmann and Taylor, 1984; [11] Millay and Taylor, 1976; [12] Taylor, 1971; [13] Millay and Taylor, 1974; [14] Millay and Taylor, 1976; [15] Zavada 1991a; [16] Zavada and Crepet, 1985; [17] Taylor et al., 1984; [18] Osborn and Taylor, 1993; [19] Osborn, 1991; [20] Pedersen and Friis, 1986; [21] Zavada and Crepet, 1986; [22] Taylor and Daghlian, 1980; [23] Taylor and Grauvogel-Stamm, in prep.; [24] Taylor and Taylor, 1987; [25] Taylor, 1988; [26] Pettitt and Chaloner, 1964; [27] Taylor and Alvin, 1984; [28] Taylor and Alvin, in prep.; [29] Osborn and Stockey, in prep.; [30] Taylor et al., 1987; [31] Del Fueyo et al., 1990; [32] Archangelsky and Del Fueyo, 1989; [33] Baldoni and Taylor, 1982; [34] Serbet and Stockey, 1991; [35] Kurmann, 1991; [36] Rothwell and Basinger, 1979; [37] Hill, 1990; [38] Taylor, 1973; [39] Ward et al. 1989; [40] Osborn et al., 1991; [41] Pederson et al., 1989; [42] Taylor, 1970; [43] Taylor and Millay, 1977; [44] Meyer-Berthaud, 1989.

Doyle and Donoghue (1986), some of which overlap, a number of other characters have also been incorporated into other phylogenetic analyses of both gymnosperms and angiosperms (Table 3). The list of characters in Table 3 is not totally inclusive, and the degree to which it can be exclusively used is dependent upon the taxonomic rank at which a phylogenetic analysis is conducted (i.e., within a genus, a family, a division, etc.).

Nevertheless, it is clear that an array of morphological and fine structural features are available for evaluation. With both general access to TEM and a growing number of taxa being studied, the focus of many investigations has been the desire to identify a large number of characters and, as a result, infer homologies. Although many palynological characters are considered conservative features in terms of evolutionary change and manifestations of phylogeny, they can also reflect several other biological and physical phenomena. For example, events during pollen ontogeny as well as many facets associated with the reproductive biology of a taxon (e.g., pollination) may be important influences on observed fine structure. The most prominent abiotic factors to consider are mode of preservation, degree of preservational alterations (e.g., diagenesis), preparation protocols for TEM, and nature of the ultrathin sectioning plane. In the present paper, we address these features as they relate to phylogenetic interpretations and relevance of three salient palynological characters, including overall exine infrastructure, nexine organization, and saccus type and internal composition.

A. Gross exine infrastructure

A vast terminology has been developed for pollen wall ultrastructure based principally on the structural aspects of the sporoderm, as opposed to ontogenetic or functional features. Of the many systems of pollen wall nomenclature proposed, those of Faegri and Iversen (1989) and Erdtman (1969) are most widely recognized (see Zavada, 1984). Comparisons and discussions of these two schemes have received a significant amount of attention (e.g., Zavada, 1984) and will not be elaborated here. Briefly, however, the sporoderm as interpreted by Faegri and Iversen (1989) consists of three major layers: an outer ektexine and middle endexine (exine collectively), both composed of resistant sporopollenin, and an inner intine primarily consisting of cellulose. The sporopolleninous ektexine and endexine exhibit differential stainability in both LM and TEM preparations (see below). Erdtman (1969) also recognizes a three-layered sporoderm, composed of an outer sporopolleninous sexine and nexine (exine collectively) and an inner pecto-cellulosic intine. In Erdtman's classification, the wall layers are recognized based on their topographic positions.

Exine stratification is generally categorized into several general types, including homogeneous (atectate), tectate-alveolate ("honeycomb-like" type and "spongy" type), tectate-granular, intectate-granular, tectate-columellate, semitectate-columellate, and intectate-columellate (e.g., Doyle et al., 1975; Kurmann, 1992). However, researchers have differing views concerning interpretations of such terms as "alveolate" and "spongy" exines and, in practice, use these differently. In other cases a new set of terms may be created. The range of terminology is cumbersome and indeed problematic when attempting to address homologous wall layers. For example, similar names applied to structurally different layers by multiple authors may incorrectly imply homology. Therefore, the phylogenetic implications of how exine stratification is interpreted and described are critical, and necessitate consideration of all variables that may influence the observed structural features of the pollen wall.

The exine clearly undergoes several major changes in ultrastructure during sporoderm ontogeny (Figs. 1–3). Relatively few studies have documented developmental sequences for fossil gymnosperm pollen (see Taylor, 1990). Those that have been published have resulted from examinations of multiple specimens of *in situ* grains. However, it is possible that the exine might be erroneously described if one were to only examine grains at a particular stage of development that were contained within a single pollen sac. For example, *Classopollis*-type grains isolated from the cones of the cheirolepidiaceous

TABLE 3. Selected palynological characters used in phylogenetic analyses

GROSS MORPHOLOGICAL FEATURE
Heterospory
Dispersal Unit
Tetrads acalymmate/calymmate
Shape
Size
Symmetry
Aperturate/inaperturate
Aperture type
Aperture position
Sulcus-pollen tube present/absent
Saccate/nonsaccate
MICROMORPHOLOGICAL FEATURE
Striate/nonstriate
Supratectal ornament present/absent
Surface sculpture
Sculpture attached/detached
Sculpture density
Sculpture detail
Aperture membrane sculpturing
ULTRASTRUCTURAL FEATURE
Gross exine ultrastructure (=stratification)
Exine partitioning
Sexine ultrastructure
Exine tectate
Tectum ultrastructure
Nexine ultrastructure
Nexine thickness over aperture
Endexine present/absent
Endexine present/absent in extra-apertural areas
Endexine thickness
Foot layer present/absent
Saccus internal composition

conifer *Classostrobus* (Taylor and Alvin, 1984) may have been characterized as either intectate-granular (Fig. 1), or as tectate-alveolate or tectate-columellate lacking a supratectal ornament (Fig. 2) if only immature grains had been sectioned. However, identification of additional ontogenetic stages demonstrated that the mature pollen of this taxon is in fact complex, consisting of a four-zoned, tectate-collumellate exine with supratectal spinules (Fig. 3).

It is also important to sample pollen sacs from different positions along the whole reproductive organ (e.g., distal vs proximal ends of a cone), as the organ may not exhibit synchronous pollen development. When possible, it is also advantageous to section entire pollen sacs, rather than mechanically or chemically macerating these structures in order to obtain pollen. The former approach permits identification of intra-locular structures such as resistant tapetal membranes and orbicules, which are important in documenting a complete ontogenetic series (Taylor et al., in press).

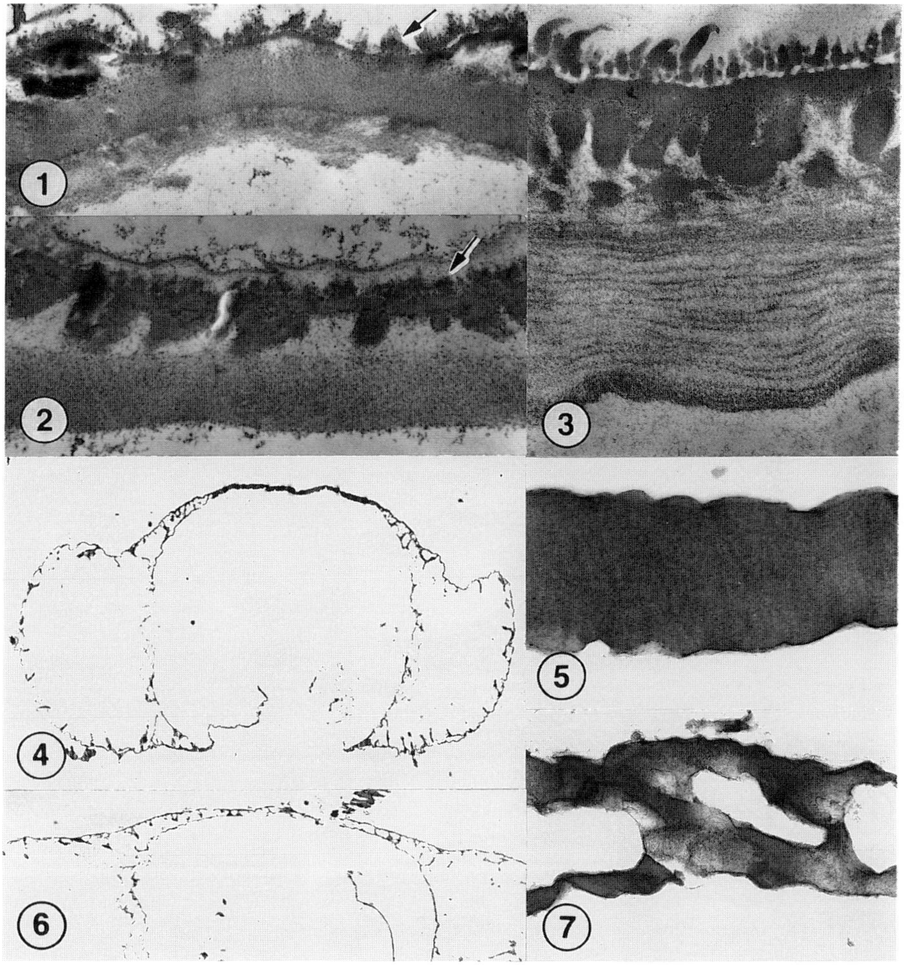

FIGS. 1–7. Exine infrastructure. **1**: Section through the cryptopore region of an immature *Classopollis* pollen grain. Note the homogeneous, non-lamellate nexine and dark staining, discontinuous sexine elements (arrow), × 10,000. **2**: Section through an unspecialized region of an immature *Classopollis* grain. This grain is older than the one illustrated in Fig. 1 and shows a better developed sexine and non-lamellate nexine. Note that the sexine consists of immature columellae and a superficially granular tectum (arrow) that results from poorly developed supratectal spinules, × 15,000. **3**: Section through an unspecialized region of a mature *Classopollis* grain. Note the well-defined lamellae within the nexine and the distinct columellae, homogeneous tectum, and supratectal spinules of the sexine, × 20,000. **4**: Median transverse section through a permineralized corystosperm pollen grain showing a homogeneous proximal wall. Note how the exine layers separate and gradually grade into an alveolate structure where the sacci are attached, and that the endoreticulations within the sacci are only attached to the outer sacci walls, × 1,060. **5**: Detail of the proximal wall from the permineralized corystosperm grain illustrated in Fig. 4 showing a homogeneous ultrastructure, × 31,500. **6**: Lateral transverse section through a permineralized corystosperm pollen grain showing an alveolate proximal wall, × 1,060. **7**: Detail of a laterally sectioned proximal wall from a permineralized corystosperm grain showing an alveolate fine structure, × 31,500.

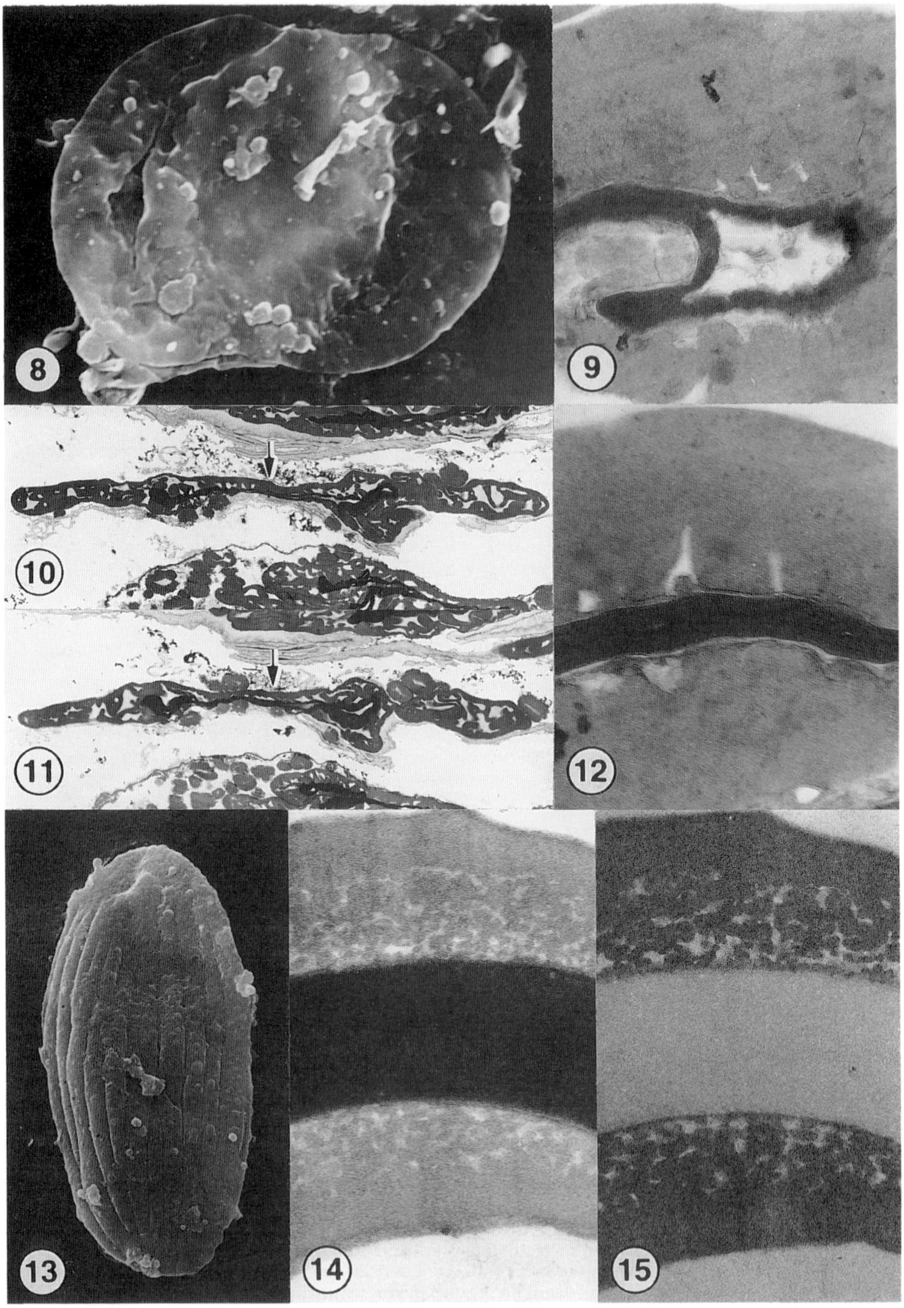
8
9
10
12
11
13
14
15

Two related characters to note regarding developmental patterns are tectum thickness and exine partitioning (i.e., sexine thickness relative to nexine thickness). Zavada and Gabarayeva (1991) have recently shown in the tectate-granular exine of extant *Welwitschia* that the majority of the thick, homogeneous tectum observed in mature grains results from the developmental compaction and fusion of sporopolleninous granules. Consequently, the granular infratectal layer and the homogeneous tectum exhibit markedly different thicknesses during ontogeny. It is possible that this pattern could also occur in fossil groups; if immature grains are exclusively sampled, then exine descriptions and phylogenetic interpretations would be skewed.

The plane of section through a pollen grain and the nature of grain preservation are also key factors concerning interpretations of infrastructure. This is especially important with regard to studies of saccate grains. For example, bisaccate pollen of the Mesozoic seed fern group Corystospermales shows a distinctly different fine structure when sectioned medially as opposed to laterally. The proximal wall (cappa) in median sections is homogeneous (Figs. 4–5) and becomes tectate-alveolate in organization laterally (Figs. 6–7) where the sacci are attached (see below). The ability to detect these differences is the result of the permineralized nature of the fossils and the fact that multiple grains were individually serially sectioned (Osborn and Taylor, 1993). Until these structurally preserved specimens were examined, corystosperm pollen was known only from two studies based on compressed *Pteruchus* pollen organs. Principally due to preservation of the *Pteruchus* specimens, Taylor et al. (1984) were unable to detail the fine structure of the cappa, while Zavada and Crepet (1985) described the proximal wall to be ultrastructurally homogeneous with a few small lacunae. A similar type of organization has been detected in the bisaccate grains of another Mesozoic seed fern, *Caytonanthus* (Fig. 8; Caytoniales). Although pollen organs of *Caytonanthus* are also known only from compression specimens, serial sectioning (Osborn, 1991) indicates that the previously described alveolate exine (Pedersen and Friis, 1986; Zavada and Crepet, 1986) is prominent only in lateral wall regions (Fig. 11), and grades into a homogeneous infrastructure near the median area (Fig. 10), as in corystosperm pollen.

Preservation type may also affect interpretations of nonsaccate pollen. For instance, pollen ultrastructure of *Sahnia* (Pentoxylales) is known only from compression specimens, with the exine having a tectate-granular ultrastructure (Osborn et al, 1991). Grains exhibit a wide range of tectum and infratectum thicknesses resulting from differential degrees of granule packing (Figs. 9, 12). Although strata thickness in a tectate-granular exine may be a developmental phenomenon, as noted above for *Welwitschia,* it is clearly the result of preservational compression in *Sahnia* pollen (Figs. 9, 12). In fact, in some grain regions (Fig. 9), as well as in entire grains, the sexine shows a homogeneous ultrastructure.

In addition to mode of preservation, several other preservational influences (diagenesis) and the effects of temperature, chemical maceration, and electron microscopy preparation techniques are known to alter exine fine structure in both fossil and extant pollen (e.g., Sengupta, 1977; Niklas, 1980; Kedves, 1985; Kedves and Kincsek,

FIGS. **8–15**. Exine infrastructure. **8**: Proximal surface of a mature *Caytonanthus* pollen grain, × 3,000. **9**: Section of a compressed *Sahnia* pollen grain showing a dark-staining nexine and a light-staining sexine. Note that the sexine appears almost entirely homogeneous at left and tectate-granular at right, × 23,250. **10**: Near median transverse section through a compressed *Caytonanthus* pollen grain showing homogeneous sexine infrastructure (arrow), × 3,300. **11**: Lateral transverse section through the same compressed *Caytonanthus* pollen grain illustrated in Fig. 10 showing an alveolate sexine infrastructure (arrow), × 3,300. **12**: Section through a compressed *Sahnia* pollen grain showing a tectate-granular infrastructure. Note the thick, homogeneous tectum and weakly defined granular infratectum due to compression and compaction of granules, × 40,000. **13**: Polyplicate surface of a dispersed ephedroid palynomorph, × 2,700. **14**: Transverse section of a dispersed ephedroid palynomorph showing a darkly stained nexine and a lightly stained, tectate-granular sexine, × 31,500. **15**: Same exine region shown in Fig. 14, but from a different ultrathin section; note the lightly stained nexine and the darkly stained sexine, × 31,500.

1989; Rowley et al., 1990). For example, it is possible that the permineralized corystosperm grains noted above have been altered due to preservational phenomena, because the rocks in which they were preserved are known to have been thermally altered (see Osborn and Taylor, 1993). Nevertheless, the exine of these grains is still thought to be homogeneous in medial positions of the proximal wall (Figs. 4–5); this is based on comparisons with other Mesozoic bisaccates (*Caytonanthus*) that have a similar cappa fine structure but have not been subjected to the same thermal effects.

Regarding TEM preparation techniques, fossil exines have variable affinities for heavy metal stains. This can result in a number of anomalous conditions. In the majority of fossil gymnosperm grains examined to date, the sexine is light-staining and the nexine is dark-staining (e.g., Figs. 3, 9–12, 22, 24; also see discussion below). However, in some cases the exine may exhibit an opposite staining reaction (Figs. 14–15). The reasons for this are unclear, but the resultant sections afford the opportunity to detect exine strata at different levels of contrast and clarity. Such staining patterns have also been detected in fossil angiosperm pollen (Daghlian, pers. comm.).

B. Nexine organization

Most discussions of nexine organization have centered around proposed relationships between early and putatively primitive, extant angiosperms and the various gymnosperm groups to which they are suggested to be most closely related. Although the cladistic studies of Crane (1985) and Doyle and Donoghue (1986) differ with regard to several features, they both link angiosperms into a clade of highly derived seed plants, informally designated as "anthophyes", that also includes Bennettitales, Gnetales, Pentoxylales, and perhaps *Eucommiidites* pollen-producing plants (Pedersen et al., 1989). Many extant angiosperms regarded as primitive have monosulcate, tectate-granular pollen, as do all of the gymnospermous anthophyte groups. This has proven to be a problem concerning the investigation of many Early Cretaceous *sporae dispersae* grains and their identification as either angiospermous or gymnospermous (e.g., Zavada, 1984; Ward et al., 1989). Consequently, the distinction between pollen of these two groups has been primarily based upon both the appearance and structure of the nexine (=basal layer *sensu* Zavada, 1991b), specifically the nexine 2 layer or the endexine (Blackmore and Barnes, 1987).

The two layers of the nexine are divided into an the outer nexine 1 (=footlayer) and an inner nexine 2 (=endexine) based on their differential staining properties in light microscopy (e.g., with fuchsine B and auramine O; Faegri and Iversen, 1989). However, these two layers do not routinely show the same distinguishing affinities for stains used in TEM. As noted above, the nexine of most fossil gymnosperm pollen stains uniformly darker than the overlying sexine (Figs. 3, 9–12, 14, 22, 24), but opposite staining reactions may also take place (Fig. 15) and both the nexine and the sexine may appear homogeneous (Figs. 16–17). It is also possible that preparation artifacts, resulting from excessive staining for TEM, can lead to misidentification of nexine layers. In some cases, residual stains accummulate at the interface between the exine and the lumen of the pollen grain and superficially resemble a dark-staining inner layer (Fig. 16). This can be detected as an artifact if multiple sections from multiple grains are prepared using a range of stain regimes (Figs. 16, 19).

The endexine of extant gymnosperms is structurally lamellate at maturity, while the endexine of extant angiosperm pollen usually lacks lamellae except in apertural regions (e.g., Van Campo, 1971; Doyle et al., 1975). This character alone, however, may not be a good criterion for distinguishing between dispersed gymnospermous and angiospermous pollen for several reasons. For example, Blackmore and Barnes (1987) and Blackmore and Crane (1988) have suggested that differences in endexine structure of dispersed pollen may be indicative of ontogenetic stages rather than true structural dissimilarity. The timing of deposition of endexine materials differs between gymnosperms and angiosperms. The majority of endexine is deposited during the tetrad stage in gymnosperm pollen, whereas deposition of this wall layer is generally not initiated until

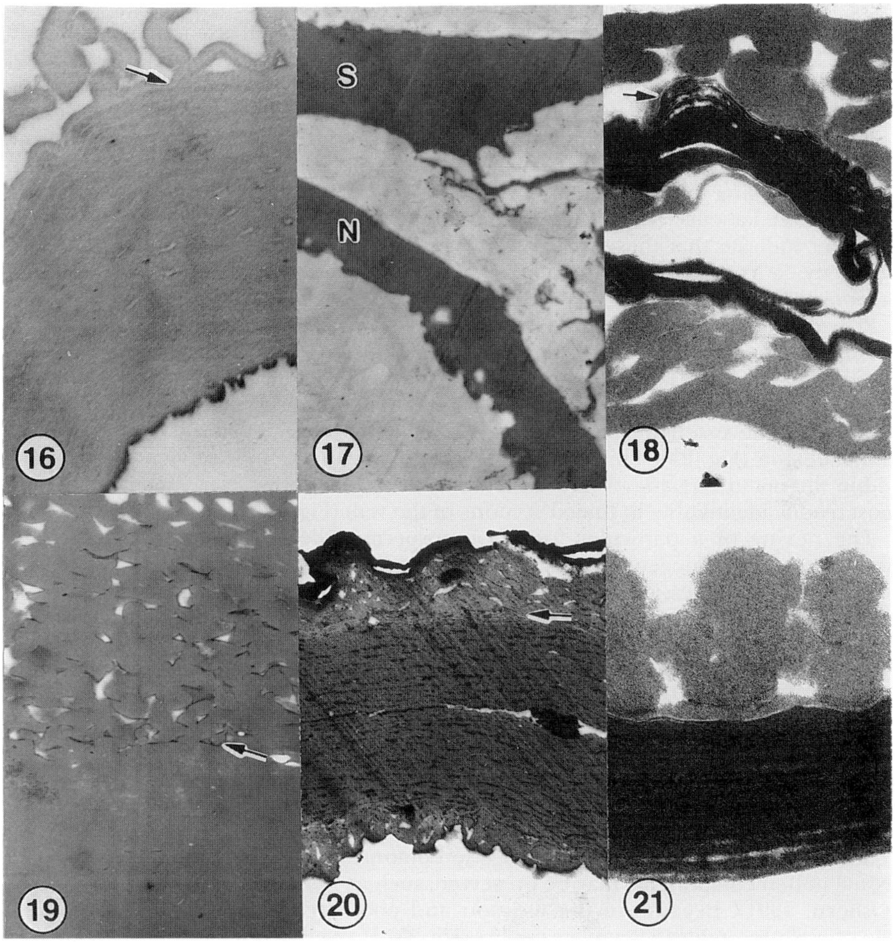

Figs. 16–21. Nexine organisation. **16**: Section through the exine of a mature *Schopfipollenites* pollen grain showing a distinctly lamellate nexine (below arrow) and the lower portion of an alveolate sexine (above arrow), × 44,000. **17**: Section of the proximal wall near the aperture of a mature *Gothania* grain showing a non-lamellate nexine (N). Note also that the nexine and sexine (S) have the same affinity for TEM stains and appear homogeneous, × 22,500. **18**: Section through a mature *Caytonanthus* pollen grain showing a dark-staining nexine and light-staining sexine. Note that lamellae are only detectable in folded regions of the wall (arrow), × 20,100. **19**: Portion of an immature *Schopfipollenites* pollen grain showing a homogeneous, non-lamellate nexine (below arrow) and an alveolate sexine (above arrow), × 6,000. **20**: Section through an unspecialized exine region of an immature *Gothania* grain showing well-defined lamellae within the nexine (below arrow), × 18,000. **21**: Exine of an immature *Caytonanthus* grain showing a distinctly lamellate, dark-staining nexine, × 40,000.

the free-spore phase in angiosperms (e.g., Blackmore and Crane, 1988). It is clear from developmental studies of *in situ* fossil gymnosperm pollen that nexine lamellae also form at different stages. For instance, well-defined lamellations in the nexine of *Classostrobus* (Taylor and Alvin, 1984) and *Schopfipollenites* (Medullosales; e.g., Taylor, 1982) are only detectable in sections of mature pollen (Figs. 3, 16). However, when immature pollen of these two taxa are examined, nexine lamellae are notably absent (Figs. 1–2, 19). Interestingly, a contrasting situation has been observed in several taxa with *in situ* saccate pollen. Here, sections of immature pollen of *Gothania* (Cordaitales) and *Caytonanthus* exhibit well-defined nexine lamellae (Figs. 20–21) when compared with mature grains (Figs. 17–18). This is similar to what has been observed during ontogeny of the saccate pollen from several extant conifers. In the free-spore phase, nexine lamellae become stretched and tightly appressed to each other, especially during saccus expansion, thereby inhibiting their detection as individual units (Kurmann, 1990). Despite the fact that tripartite lamellae occur in the immature pollen of *Gothania* and *Caytonanthus*, this does not indicate that these grains were lamellate throughout the earliest phases of ontogeny; a complete developmental sequence for these taxa has not yet been determined (Taylor and Daghlian, 1980; Osborn, 1991).

Preservation may also play a part in the ability to detect lamellae within the nexine. For example, in the case of several dispersed ephedroid grains (Fig. 13) the majority of grains sectioned lack nexine lamellae altogether (Osborn et al., 1993), although some may show an occasional lamella near a wall furrow (Fig. 22). However, sections of a single grain that has undergone extensive preservational folding reveal lamellations throughout the entire nexine (Fig. 23). This condition was also important in documenting lamellae within the mature nexine of *Caytonanthus,* as described above. Here, the lamellae are most readily identifiable in folded sections of the wall (Fig. 18).

The nexine of a particular fossil gymnosperm pollen grain may, therefore, be ultrastructurally homogeneous (i.e., lack tripartite lamellae) due to either developmental stretching and appression, or preservational compression and compaction. However, without detailed developmental data documenting a full range of intermediate stages, the degree to which ontogeny versus preservation affect the mature nexine is unclear. Consequently, the presence of lamellae provides useful information, although their absence within the nexine is clearly uninformative. For instance, one might be inclined to characterize the pollen of the bennettitalean genus *Cycadeoidea* as non-lamellate, because nexine lamellae have never been observed despite the preparation of multiple grains (Fig. 24). This assessment might inaccurately link the group closer to angiosperms, because the nexine of angiosperm pollen is characterized as non-lamellate except in the apertural regions of some grains.

Studies of *in situ Cycadeoidea* pollen also demonstrate that in some instances intra-exinal pollen components may be preserved, such as remnants of the cellulosic intine (Osborn, 1991). Because of the location and poor preservation of this material, it superficially resembles a lamellate nexine (Fig. 25). If a well-defined nexine had not been identified in these grains as a continuous, dark-staining layer consistent in thickness (Figs. 24–25), then the non-exinous layer (Fig. 25) could conceivably have been described as nexine. Intine preservation in other *in situ* and dispersed pollen grains from both gymnosperms and angiosperms is also possible, and it is advisable to employ appropriate preparation protocols (e.g., acid macerations, acetolysis) to chemically extract the oxidized wall layer (Zavada, pers. comm.) in order to preclude misidentification.

C. Saccus type and internal composition

Extensions of the sporoderm, formed by separation of wall layers, are known from an array of pteridophyte and gymnosperm groups, as well as several angiosperm taxa. For the most part, these extensions are categorized by the degree of wall separation (i.e., the amount of space enclosed by the separated sexine and nexine layers). Spores and pollen grains with only slight separation of the exine are described as having a cavea (=cavum) or a camera, while those with a more extensively separated wall are said to have a pseudosaccus. The pseudosaccate condition is known in many lycopsid microspores (e.g., *Endosporites*; Brack and Taylor, 1972), while most angiosperms that have space in their exines are said to be caveate (e.g., Compositae; Blackmore et al., 1984). Both pseudosacci

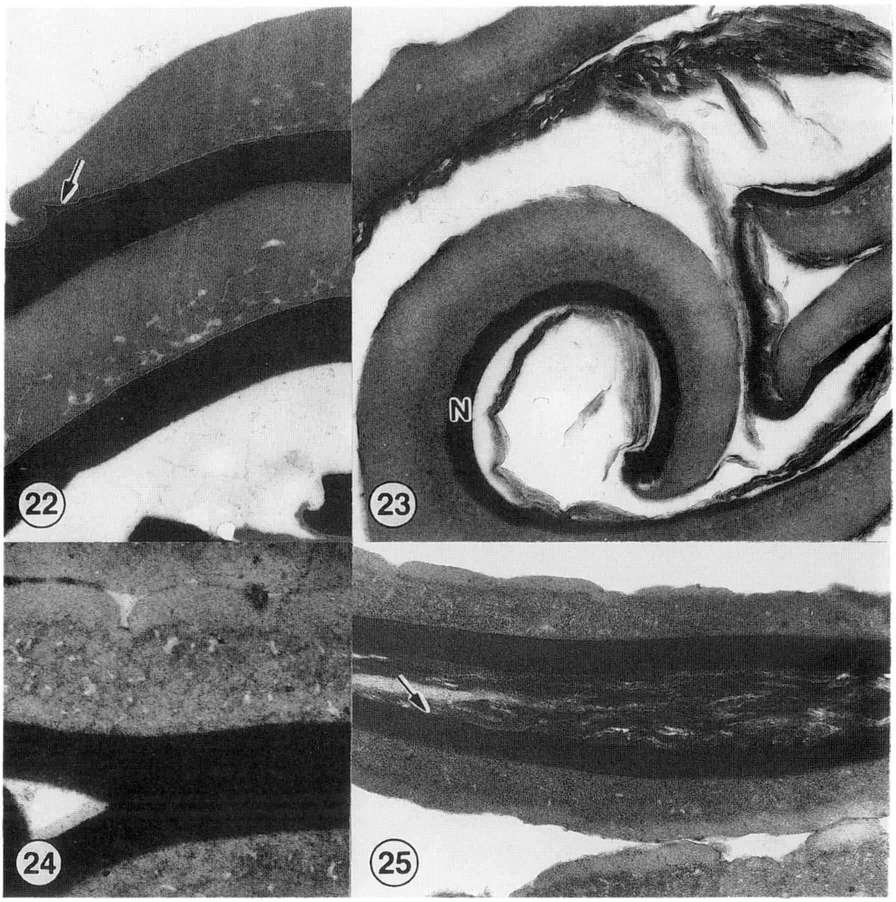

Figs. 22–25. Nexine organisation. **22**: Section of a dispersed ephedroid palynomorph showing a dark-staining nexine and a light-staining sexine. Note that the nexine is almost entirely homogeneous except for a single lamella (arrow) that can be detected near the sporderm furrow, × 31,300. **23**: Transverse section through a highly folded, dispersed ephedroid grain. Note that the individual lamellae within the nexine (N) can only be detected in folded regions of the grain, × 25,100. **24**: Section of a permineralized *Cycadeoidea* pollen grain showing a darkly stained, homogeneous nexine and a lightly stained, tectate-granular sexine. Note the absence of nexine lamellae, × 40,000. **25**: Transverse section through a *Cycadeoidea* grain showing 'pseudolamellae' within the lumen of the pollen grain. Note that these structures are not part of the nexine, as the lower boundary of the non-lamellate nexine (arrow) is well-defined and continuous, × 25,100.

and caveae are completely hollow. A saccus, on the other hand, is typically formed by a relatively extensive separation of exine layers, and is characterized as variably filled with a network of sporopolleninous plates, or endoreticulations (e.g., Traverse, 1988).

Saccate pollen occurs principally in gymnosperms. Four orders of seed ferns (Pteridospermophyta) are known to have saccate pollen, including Callistophytales, Glossopteridales, Corystospermales, and Caytoniales; the pollen of *Parasporotheca* (Medullosales) is also described as having vestigial sacci. Saccate pollen is also prominent in the Coniferophyta, and is known from Cordaitales, Voltziales (Voltziaceae), and Coniferales (Pinaceae and Podocarpaceae). Moreover, the pollen of one angiosperm

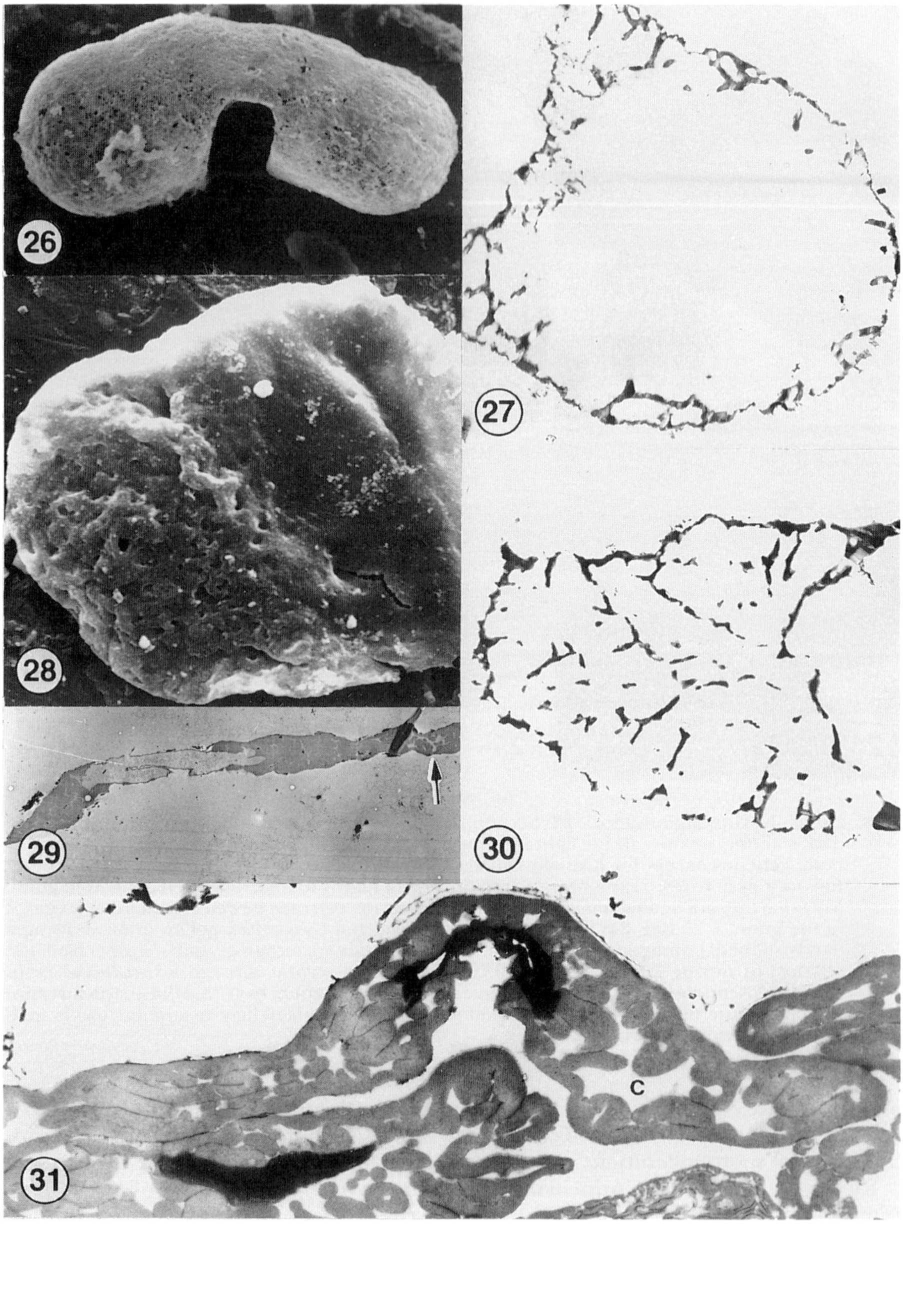
26
27
28
29
30
31
c

family (Lactoridaceae) has been characterized as saccate (Zavada and Taylor, 1986a; Zavada and Benson, 1987). However, the sacci in this family lack any type of conspicuous internal structure (i.e., an endoreticulum), and Crane (1990) has further suggested that the "saccus-like structures" of *Lactoris* as well as the caveae of other angiosperms are not homologous with the true sacci of various gymnospermous clades.

The saccate grains of gymnosperms can be further classified with regard to both the number of sacci and nature of their attachment to the corpus. Monosaccate pollen may have a single saccus that is attached to the proximal wall, superficially resembling the pseudosaccate condition of some pteridophyte spores, as in *Gothania* (Fig. 32, *Felixipollenites*-type cordaitalean grains; Taylor and Daghlian, 1980). However, distal separation of the monosaccus in *in situ* pollen of *Gothania* has been suggested to be a developmental phenomenon, because the majority of grains show both proximal and distal attachment of the saccus in earlier ontogentic stages (Millay and Taylor, 1974; Taylor and Daghlian, 1980). By comparison, the monosaccus of *Cordaianthus* (*Florinites*- and *Sullisaccites*-type grains), a geologically younger cordaitean genus, is consistently found attached to both the proximal and distal poles of the grain, thereby forming a single, continuous saccus that laterally encircles the corpus (Millay and Taylor, 1974).

The bisaccate condition is characterized by two independent sacci that are distinctly restricted in their attachment to the corpus, and is known from most seed ferns as well as other saccate members of the Coniferophyta. Here, the sacci are typically found transversely attached (Figs. 4, 26), although they may also occur in a more distally inclined fashion (Fig. 33). A third type of saccus configuration is somewhat intermediate between the monosaccate and bisaccate conditions; grains of this type have been referred to as having a transitional saccus (Traverse, 1988) or a girdling saccus (Osborn, 1991; Zavada, 1991a). For example, the *sporae dispersae* genus *Protohaploxypinus*, presumed to have been produced by a number of glossopteridalean seed ferns, has what appears to be two relatively large sacci transversely attached to the taeniate corpus. However, upon closer examination it is apparent that these two transverse bladders are continuous over the distal surface of the grain via slight lateral attachment (Fig. 35). Examinations of ultrathin sections with TEM confirm this organization (e.g., Osborn, 1991; Zavada, 1991a). A number of other enigmatic saccate conditions are also known from several fossil gymnosperm taxa. *In situ* pollen of the problematic species *Lasiostrobus polysacci* has 3–8 subequatorial projections that have been termed sacci; however, these structures lack well-defined endoreticulations (Taylor and Millay, 1977). Another example includes the dispersed taxa *Lueckisporites virkkiae* and *Lunatisporites noviaulensis mollis*, which, in addition to having two prominent, transversely attached sacci, have elongate bladders attached to the proximal surface paralleling the long axis of the corpus (Scheuring, 1974).

The ultrastructural determination of the type of saccate condition exhibited by a particular pollen grain is based on several factors. First, as noted above the majority of *Gothania* grains resemble the pollen of *Cordaianthus* in that the monosaccus is both proximally and distally attached to the corpus. Identification of a limited number of ontogenetically more mature grains (Fig. 32) indicates that the saccus is in fact distally

Figs. 26–31. Saccus structure. **26**: Equatorial view of a permineralized, bisaccate corystosperm, × 800. **27**: Median transverse section through the saccus of a permineralized corystosperm pollen grain showing eusaccate condition. Note that the endoreticulations are discontinuous between the outer saccus wall (right) and the saccus floor, × 2,600. **28**: Proximal surface of a compressed *Pteruchus* pollen grain, × 7,000. **29**: Oblique section through a compressed *Pteruchus* grain showing sacci that superficially appear to be completely infilled with endoreticulations. Note that some space (arrow) within the right saccus can be detected, × 2,000. **30**: Lateral transverse section through the saccus of a permineralized corystosperm grain showing continuous endoreticulations extending between the outer saccus wall (left) and the saccus floor, thereby superficially appearing 'protosaccate', × 2,100. **31**: Oblique section through a compressed *Caytonanthus* pollen grain. Note that the left saccus appears extensively infilled with endoreticulations, but that a distinct cavity (C) can be detected in the right saccus. The darkly stained layer represents the nexine in the corpus region, × 9,284.

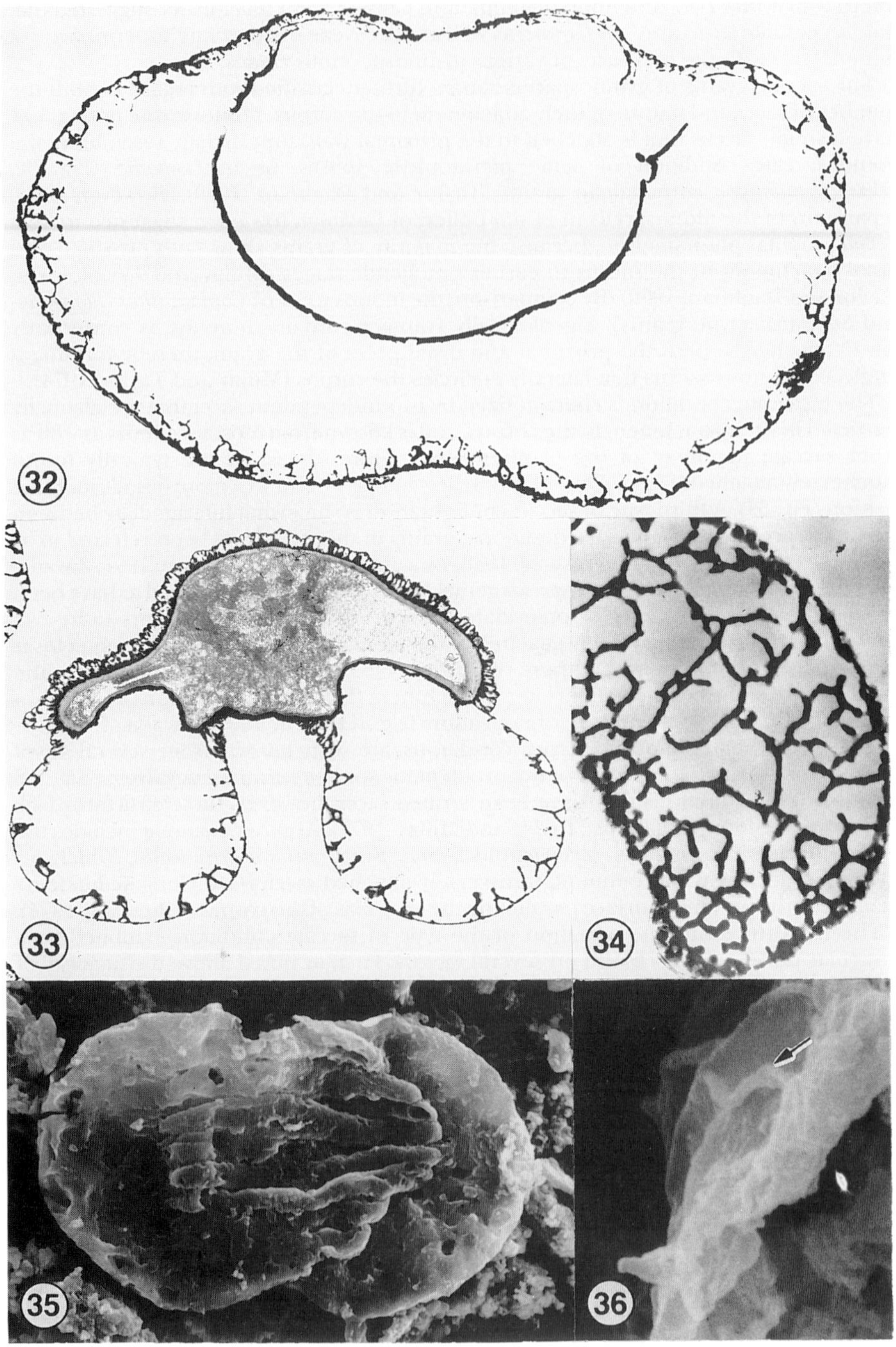
32
33
34
35
36

separated from the corpus. Secondly, sacci are believed to function in several ways, principally during pollination and pollination-related events. It is clear that sacci efficiently increase overall grain sizes without significantly increasing grain weight, and thereby provide for greater dispersal by wind. Others have suggested that sacci have a number of additional functions, which include: to harmomegathically maintain volumetric continuity; to physically ensure that the distal aperture is oriented against the nucellus to maximize siphonogamy; and to reduce competition with other pollen grains by maximally occupying the micropylar space, thereby physically preventing conspecific grains from entering that space (e.g., Chaloner, 1976; Zavada and Taylor, 1986b; Zavada, 1991a).

Perhaps a more salient aspect of saccus ultrastructure to address in a phylogenetic context is the nature of the internal sporopolleninous units, or endoreticulations. The degree to which gymnospermous sacci are filled with endoreticulations has become an important topic of discussion and phylogenetic speculation. Grains in which the endoreticulations are continuous between the corpus and the saccus walls (i.e., completely infill the saccus cavity) have been referred to as "protosaccate" (Scheuring, 1974). By comparison, when the saccus cavity is almost entirely hollow and the endoreticulations are restricted in attachment to the outer saccus wall (i.e., discontinuous), the grain is said to be "eusaccate" (Scheuring, 1974). The phylogenetic implication of the term 'protosaccate' as pleisiomorphic was the impetus for Meyen (1987) to refer to grains of this type as "quasisaccate", because eusaccate grains geologically precede protosaccate forms in the fossil record. However, if Foster and Balme's (1994) report of protosaccate grains from the Late Devonian is accurate, then Meyen's stratigraphic reservations would be less relevant. Nevertheless, descriptions of protosaccate and eusaccate grains from different major groups of gymnosperms have prompted speculation as to whether or not these two types of sacci are homologous (Crane, 1990). Doyle (1987–1988) has suggested that the protosaccate condition is indeed derived from the eusaccate type, specifically with reference to Caytoniales and Callistophytales.

In our opinion, the phylogenetic utility of this character is poor, and the degree to which the internal composition of the saccus can be homologized in seed plants is equivocal at best. First, the majority of *in situ* grains for which the protosaccate condition has been described has come from compression specimens. It is clear that this type of preservation has significant effects on saccus structure. For example, until recently ultrastructural data on corystosperm pollen have come exclusively from three compressed *Pteruchus* species (Taylor et al., 1984; Zavada and Crepet, 1985). These compressed grains have been important in providing information on a variety of features, especially surface morphology (Fig. 28). When ultrathin sectioned, the sacci appear to be extensively filled with endoreticulations (Fig. 29). However, discovery of permineralized *in situ* corystosperm pollen, as noted above, has provided the opportunity to make comparisons of grains preserved differently (Osborn and Taylor, 1993). The three-dimensionally preserved grains unequivocally demonstrate that corystosperm pollen is eusaccate (Figs. 4, 26). Therefore, although the sacci of compressed *Pteruchus* grains superficially appear to be protosaccate, the extensive infillings in fact represent the discontinuous endoreticulations that have been preservationally compressed.

The ultrathin section plane has also been shown to be important with regard to

FIGS. 32–36. Saccus structure. **32**: Equatorial section of a permineralized, mature *Gothania* pollen grain showing the monosaccus attached only to the proximal wall, × 333. **33**: Median section through an extant, bisaccate *Pinus* pollen grain showing the eusaccate condition, × 1,000. **34**: Lateral section through the saccus of an extant *Pinus* grain showing an extensive endoreticulum extending throughout the saccus, × 5,000. **35**: Proximal view of a dispersed *Protohaploxypinus* grain showing the taeniate cappa and girdling saccus, × 1,000. **36**: Saccus floor of a permineralized *Pinus* pollen grain showing endoreticulation attachment scars (arrow); the saccus cavity is to left, × 2,000.

interpretations of saccus structure. For example, serial sections through the same permineralized corystosperm grains indicate that although the grains are definitively eusaccate, as observed in medial sections (Figs. 4, 27), if lateral sections are examined the grains morphologically appear protosaccate (Fig. 30). This structure results from the sacci tapering off laterally where they approach attachment to the corpus. This sectioning phenomenon is also convincingly illustrated in comparable medial and lateral sections through the sacci of extant *Pinus* pollen (Figs. 33–34) as well as compressed grains of *Caytonanthus* (Fig. 31).

It is also possible that the ontogenetic age of pollen grains may contribute to their characterization as protosaccate. For example, in the dispersed protosaccate grains of *Triadispora*, the sacci are relatively small (10–12 μm wide) and grains are often found in tetrads, a condition suggesting dispersal at this stage (Scheuring, 1976). The possibility exists that these grains were superficially protosaccate only, because either the sacci had not yet fully expanded or they were developmentally destined to remain small in size. In extant *Pinus* pollen (e.g., Dickinson and Bell, 1970; Willemse, 1971), endoreticulations separate from the corpus wall while grains are in the late tetrad stage; however, *Pinus* sacci clearly expand during subsequent developmental stages to attain relatively large sizes. This is also illustrated in the mature pollen of Middle Eocene *Pinus*, where the corpus wall lining the saccus floor shows separation scars indicating the former attachment sites of endoreticulations (Fig. 36). Another indication that saccus size may be important in whether or not the endoreticulations tear away from the nexine is the fact that the only extant genus with protosaccate pollen, *Dacrydium* (Podocarpaceae), is also characterized by small sacci (Pocknall, 1981; Médus et al., 1989).

Finally, endoreticulation size may also play a role in the morphological determination of a particular grain as 'eusaccate' or 'protosaccate'. For example, corystosperm grains may be 'eusaccate' because of the delicate nature of their endoreticulations (Figs. 27, 30) which would conceivably be easily ruptured, or separated, from the underlying nexine during saccus expansion. Although serial sections through *Caytonanthus* grains indicate that they are eusaccate, the sacci are small and still densely filled with thick endoreticulations (Figs. 10–11, 31). Despite the fact that the sacci of *Caytonanthus* are smaller and more extensively filled than those of corystosperm grains, the relative robustness of *Caytonanthus* endoreticulations may preclude their separation from the corpus wall to the degree observed in other grains.

Conclusions

The present paper underscores a conservative approach in assessing the relative emphasis that should be placed on the morphology and ultrastructure of fossil pollen. Based on the above discussion, we encourage less 'homologizing' of particular exine features in fossil pollen, especially when those data are taken from published micrographs in the literature. For example, an important consideration concerns interpretations of the nexine, specifically the endexine (nexine 2) and the ectexinous footlayer (nexine 1). Nexine layers of fossil gymnosperm grains are highly variable in their affinities for TEM stains, and clearly in the presence or absence of lamellations. In fact, it is distinctly possible that the endexine *per se* is absent from many fossil grains. In an analogous system, in many extant pollen types the underlying intine frequently acts as support for the endexine and, consequently, both are lost following acetolysis (Blackmore and Crane, 1988). It is not unlikely that this loss could also occur during lithification in the case of fossil grains, and thereby calls into question the utility of such characters as "endexine present or absent" (e.g., Doyle and Hotton, 1991).

Although the primary emphasis here has been on developmental and functional effects on pollen form, as well as geological and methodological influences, the phylogenetic element may also be important. One of the most challenging aspects in the study of fossil pollen ultrastructure is gauging the relative importance of potentially phylogenetically significant data while taking into account, and either appropriately weighing-in or filtering-out, the other influences that may magnify or mask characters that are valuable in phylogenetic analyses. Therefore, when only a few, random sections from a single grain or a small number of grains have been studied, it is objectionable to make phylogenetic inferences about various ultrastructural characters observed in fossil pollen.

Acknowledgements

This investigation represents a portion of a dissertation submitted by the first author in partial fulfillment of the requirements for the Ph.D. degree from The Ohio State University (OSU). This study was supported in part by a Doctoral Dissertation Improvement Grant from the National Science Foundation (NSF; BSR-9016397), a student scholarship from the American Association of Stratigraphic Palynologists, and a Presidential Fellowship and a Graduate Student Alumni Research Award both from the Graduate School at OSU to JMO, as well as NSF grant DPP-9118314 to TNT. We thank Ruth A. Stockey (Botany Department, University of Alberta, Canada) for providing fossil pine material.

References

Abadie, M., Lachkar, G., Masure, E. and Taugourdearu-Lantz, J. (1978). Observations nouvelles sur le G. *Schopfipollenites* R. Potonié et Kremp 1954. *Ann. Mines Belg.* **2**: 125–132.

Archangelsky, S. and Del Fueyo, G. (1989). *Squamastrobus* gen. n., a fertile podocarp from the Early Cretaceous of Patagonia, Argentina. *Rev. Palaeobot. Palynol.* **59**: 109–126.

Baldoni, A.M. and Taylor, T.N. (1982). The ultrastructure of *Trisaccites* pollen from the Cretaceous of southern Argentina. *Rev. Palaeobot. Palynol.* **38**: 23–33.

Blackmore, S. and Barnes, S.H. (1987). Embryophyte spore walls: origin, development, and homologies. *Cladistics* **3**: 185–195.

Blackmore, S. and Crane, P.R. (1988). The systematic implications of pollen and spore ontogeny. In: C.J. Humphries (editor). Ontogeny and Systematics, pp. 83–115, Columbia University Press, New York.

Blackmore, S., Van Helvoort, H.A.M. and Punt, W. (1984). On the terminology, origins and functions of caveate pollen in the Compositae. *Rev. Palaeobot. Palynol.* **43**: 293–301.

Brack, S.D. and Taylor, T.N. (1972). The ultrastructure and organization of *Endosporites*. *Micropaleontology* **18**: 101–109.

Chaloner, W.G. (1976). The evolution of adaptive features in fossil exines. In: I.K. Ferguson and J. Muller (editors). The Evolutionary Significance of the Exine, pp. 1–14, Academic Press, London.

Crane, P.R. (1985). Phylogenetic analysis of seed plants and the origin of angiosperms. *Ann. Missouri Bot. Gard.* **72**: 716–793.

Crane, P.R. (1990). The phylogenetic context of microsporogenesis. In: S. Blackmore and R.B. Knox (editors). Microspores: Evolution and Ontogeny, pp. 11–41, Academic Press, London.

Del Fueyo, G., Archangelsky, S. and Taylor, T.N. (1990). Una nueva Podocarpacea fertil (Coniferal) del Cretácico inferior de Patagonia, Argentina. *Ameghiniana* **27**: 63–73.

Dickinson, H.G. and Bell, P.R. (1970). The development of the sacci during pollen formation in *Pinus banksiana*. *Grana* **10**: 101–108.

Doyle, J.A. (1987–1988). Pollen evolution in seed plants: a cladistic perspective. *J. Palynol.* **23–24**: 7–18.

Doyle, J.A. and Donoghue, M.J. (1986). Seed plant phylogeny and the origin of angiosperms: an experimental cladistic approach. *Bot. Rev.* **52**: 321–431.

Doyle, J.A. and Hotton, C.L. (1991). Diversification of early angiosperm pollen in a cladistic context. In: S. Blackmore and S.H. Barnes (editors). Pollen and Spores: Patterns of Diversification, pp. 169–195, Clarendon Press, Oxford.

Doyle, J.A., Van Campo, M. and Lugardon, B. (1975). Observations on exine structure of *Eucommiidites* and Lower Cretaceous angiosperm pollen. *Pollen & Spores* **17**: 429–486.

Erdtman, G. (1969). Handbook of Palynology. 486 pp. Munksgaard, Copenhagen.

Faegri, K. and Iversen, J. (1989). Textbook of Pollen Analysis. 4th edition, by Faegri, K., Kaland, P.E. and Krzywinski, K. 328 pp. John Wiley, Chichester.

Foster, C.B. (1979). Permian plant microfossils of the Blair Athol Coal Measures, Baralaba Coal Measures, and Basal Rewan Formation of Queensland. *Publ. Geol. Surv. Queensland* **372** *Palaeontol. Pap.* **45**: 1–244.

Foster, C.B. and Balme, B.E. (1994). Ultrastructure of *Teichertospora torquata* (Higgs) from the Late Devonian: oldest saccate palynomorph. In: M.H. Kurmann and J.A. Doyle (editors). Ultrastructure of fossil spores and pollen, pp. 87–97, Royal Botanic Gardens, Kew.

Foster, C.B. and Price, P.L. (1981). Exine infrastructure of *Praecolpatites sinuosus* (Balme & Hennelly) Bharadwaj & Srivastava, 1969, and *Marsupipollenites triradiatus* Balme & Hennelly, 1956. *Palaeobotanist* **28–29**: 177–187.

Hill, C.R. (1990). Ultrastructure of *in situ* fossil cycad pollen from the English Jurassic, with a description of the male cone *Androstrobus balmei* sp. nov. *Rev. Palaeobot. Palynol.* **65**: 165–173.

Kedves, M. (1985). Structural modification of degraded fossil sporomorphs. *Micropaleontology* **31**: 175–180.

Kedves, M. and Kincsek, I. (1989). Effect of the high temperature on the morphological characteristic features of the sporomorphs: I. *Acta Biol. (Szeged)* **35**: 233–235.

Kedves, M. and Párdutz, Á. (1973). Ultrastructure examination of fossil Pteridophyta spores and Gymnospermatophyta pollens. *Acta Bot. Acad. Sci. Hung.* **18**: 307–313.

Kedves, M. and Párdutz, Á. (1974). Ultrastructural studies on Mesozoic inaperturate Gymnospermatophyta pollen grains. *Acta Biol. (Szeged)* **20**: 81–88.

Krassilov, V.A. (1977). Contributions to the knowledge of the Caytoniales. *Rev. Palaeobot. Palynol.* **24**: 155–178.

Kurmann, M.H. (1990). Exine ontogeny in conifers. In: S. Blackmore and R.B. Knox (editors). Microspores: Evolution and Ontogeny, pp. 157–172, Academic Press, London.

Kurmann, M.H. (1991). Pollen ultrastructure in *Elatides williamsonii* (Taxodiaceae) from the Jurassic of North Yorkshire. *Rev. Palaeobot. Palynol.* **69**: 291–298.

Kurmann, M.H. (1992). Exine stratification in extant gymnosperms: a review of published transmission electron micrographs. *Kew Bull.* **47**: 25–39.

Kurmann, M.H. and Taylor, T.N. (1984). The ultrastructure of *Boulaya fertilis* (Medullosales) pollen. *Pollen & Spores* **26**: 109–116.

Médus, J. (1977a). The ultrastructure of some Circumpolles. *Grana* **16**: 23–28.

Médus, J. (1977b). Indications de l'ultrastructure d'une espèce de forme du Crétacé supérieur, *Oculopollis maximus* W. Krutzsch. *Rev. Palaeobot. Palynol.* **24**: 209–215.

Médus, J., Gajardo, R. and Woltz, P. (1989). Exine ultrastructure of *Dacrydium fonkii*, *Saxegothaea conspicua* and *Stachycarpus andina* (Podocarpaceae) from southern South America. *Grana* **28**: 19–23.

Meyen, S.V. (1987). Fundamentals of Palaeobotany. 432 pp. Chapman and Hall, London.

Meyer-Berthaud, B. (1989). First gymnosperm fructifications with trilete prepollen. *Palaeontographica, Abt. B, Paläophytol.* **211**: 87–112.

Meyer-Berthaud, B. and Galtier, J. (1986). Studies on a Lower Carboniferous flora from Kingswood near Pettycur, Scotland: II. *Phacelotheca*, a new synangiate fructification of pteridospermous affinities. *Rev. Palaeobot. Palynol.* **48**: 181–198.

Millay, M.A. and Taylor, T.N. (1974). Morphological studies of Paleozoic saccate pollen. *Palaeontographica, Abt. B, Paläophytol.* **147**: 75–99.

Millay, M.A. and Taylor, T.N. (1976). Evolutionary trends in fossil gymnosperm pollen. *Rev. Palaeobot. Palynol.* **21**: 65–91.
Millay, M.A., Eggert, D.A. and Dennis, R.L. (1978). Morphology and ultrastructure of four Pennsylvanian prepollen types. *Micropaleontology* **24**: 303–315.
Niklas, K.J. (1980). Exinite chemodiagenesis and spore wall compressional failure. Int. Palynol. Conf., 4th, Lucknow, 1976–77. Vol. 2, pp. 597–609.
Osborn, J.M. (1991). Comparative ultrastructure of fossil gymnosperm pollen and implications regarding the origin of angiosperms. 212 pp. Unpublished Ph.D. Thesis, The Ohio State University, Columbus.
Osborn, J.M. and Taylor, T.N. (1993) Pollen morphology and ultrastructure of the Corystospermales: permineralized *in situ* grains from the Triassic of Antarctica. *Rev. Palaeobot. Palynol.* **79**: 205–219.
Osborn, J.M., Taylor, T.N. and Crane, P.R. (1991). The ultrastructure of *Sahnia* pollen (Pentoxylales). *Amer. J. Bot.* **78**: 1560–1569.
Osborn, J.M., Taylor, T.N. and de Lima, M.R. (1993). The ultrastructure of fossil ephedroid pollen with gnetalean affinities from the Lower Cretaceous of Brazil. *Rev. Palaeobot. Palynol.* **77**: 171–184.
Pedersen, K. R. and Friis, E.M. (1986). *Caytonanthus* pollen from the Lower and Middle Jurassic. In: J.T. Moller (editor). 25 Years of Geology in Aarhus, pp. 255–267, Geologisk Institut Aarhus Universitet (Geoskrifter no. 24).
Pedersen, K.R., Crane, P.R. and Friis, E.M. (1989). Pollen organs and seeds with *Eucommiidites* pollen. *Grana* **28**: 279–294.
Pettitt, J.M. (1966). Exine structures in some fossil and Recent spores and pollen as revealed by light and electron microscopy. *Bull. Brit. Mus.* (*Nat. Hist.*), *Geol.* **13**: 15–257.
Pettitt, J.M. and Chaloner, W.G. (1964). The ultrastructure of the Mesozoic pollen *Classopollis*. *Pollen & Spores* **6**: 611–620.
Pocock, S.A.J. and Vasanthy, G. (1988). *Cornetipollis reticulata,* a new pollen with angiospermid features from Upper Triassic (Carnian) sediments of Arizona (U.S.A.), with notes on *Equisetosporites*. *Rev. Palaeobot. Palynol.* **55**: 337–356.
Pocknall, D.T. (1981). Pollen morphology of the New Zealand species of *Dacrydium* Solander, *Podocarpus* L'Heritier, and *Dacrycarpus* Endlicher (Podocarpaceae). *New Zealand J. Bot.* **19**: 67–95.
Rothwell, G.W. (1972). Evidence of pollen tubes in Paleozoic pteridosperms. *Science* **175**: 772–774.
Rothwell, G.W. and Basinger, J.F. (1979). *Metasequoia milleri* n. sp., anatomically preserved pollen cones from the Middle Eocene (Allenby Formation) of British Columbia. *Canad. J. Bot.* **57**: 958–970.
Rowley, J. R. and Srivastava, S.K. (1986). Fine structure of *Classopollis* exines. *Canad. J. Bot.* **64**: 3059–3074.
Rowley, J.R., Rowley, J.S. and Skvarla, J.J. (1990). Corroded exines from Havinga's leaf mold experiment. *Palynology* **14**: 53–79.
Scheuring, B. W. (1974). "Protosaccate" Strukturen, ein weitverbreitetes Pollenmerkmal zur frühen und mittleren Gymnospermenzeit. *Geol. Paläontol. Mitt. Innsbruck* **4**: 3–30.
Scheuring, B.W. (1976). Proximal exine filaments, a widespread feature among Triassic Protosaccites and Circumpolles to secure the dispersal of entire tetrads. *Pollen & Spores* **18**: 611–639.
Sengupta, S. (1977). A comparative study of the gradual degradation of exines, resulting from the effects of temperature. *Rev. Palaeobot. Palynol.* **24**: 239–246.
Serbet, R. and Stockey, R.A. (1991). Taxodiaceous pollen cones from the Upper Cretaceous (Horseshoe Canyon Formation) of Drumheller, Alberta, Canada. *Rev. Palaeobot. Palynol.* **70**: 67–76.
Stidd, B.M. (1978). An anatomically preserved *Potoniea* with *in situ* spores from the Pennsylvanian of Illinois. *Amer. J. Bot.* **65**: 677–683.

Stidd, B.M., Rischbieter, M.O. and Phillips, T.L. (1985). A new lyginopterid pollen organ with alveolate pollen exines. *Amer. J. Bot.* **72**: 501–508.

Taylor, T.N. (1970). *Lasiostrobus* gen. n., a staminate strobilus of gymnospermous affinity from the Pennsylvanian of North America. *Amer. J. Bot.* **57**: 670–690.

Taylor, T.N. (1971). *Halletheca reticulatus* gen. et sp. n.: a synangiate Pennsylvanian pteridosperm pollen organ. *Amer. J. Bot.* **58**: 300–308.

Taylor, T.N. (1973). A consideration of the morphology, ultrastructure, and multicellular microgametophyte of *Cycadeoidea dacotensis* pollen. *Rev. Palaeobot. Palynol.* **16**: 157–164.

Taylor, T.N. (1976a). The ultrastructure of *Schopfipollenites:* orbicules and tapetal membranes. *Amer. J. Bot.* **63**: 857–862.

Taylor, T.N. (1976b). Fossil ubisch bodies. *Trans. Amer. Microscop. Soc.* **95**: 133–136.

Taylor, T.N. (1978). The ultrastructure and reproductive significance of *Monoletes* (Pteridospermales) pollen. *Canad. J. Bot.* **56**: 3105–3118.

Taylor, T.N. (1980). Ultrastructural studies of pteridosperm pollen: *Nanoxanthiopollenites* Clendening and Nygreen. *Rev. Palaeobot. Palynol.* **29**: 15–21.

Taylor, T.N. (1982). Ultrastructural studies of Paleozoic seed fern pollen: sporoderm development. *Rev. Palaeobot. Palynol.* **37**: 29–53.

Taylor, T.N. (1988). Pollen and pollen organs of fossil gymnosperms: phylogeny and reproductive biology. In: C. B. Beck (editor). Origin and Evolution of Gymnosperms, pp. 177–217, Columbia University Press, New York.

Taylor, T.N. (1990). Microsporogenesis in fossil plants. In: S. Blackmore and R.B. Knox (editors). Microspores: Evolution and Ontogeny, pp. 121–145, Academic Press, London.

Taylor, T.N. and Alvin, K.L. (1984). Ultrastructure and development of Mesozoic pollen: *Classopollis. Amer. J. Bot.* **71**: 575–587.

Taylor, T.N. and Daghlian, C.P. (1980). The morphology and ultrastructure of *Gothania* (Cordaitales) pollen. *Rev. Palaeobot. Palynol.* **29**: 1–14.

Taylor, T.N. and Millay, M.A. (1977). The ultrastructure and reproductive significance of *Lasiostrobus* microspores. *Rev. Palaeobot. Palynol.* **23**: 129–137.

Taylor, T.N. and Rothwell, G.W. (1982). Studies of seed fern pollen: development of the exine in *Monoletes* (Medullosales). *Amer. J. Bot.* **69**: 570–578.

Taylor, T.N. and Taylor, E.L. (1987). The ultrastructure of fossil gymnosperm pollen. *Bull. Soc. Bot. France, Actual. Bot.* **134**: 121–140.

Taylor, T.N. and Taylor, E.L. (1993). The Biology and Evolution of Fossil Plants. 982 pp. Prentice-Hall, New Jersey.

Taylor, T.N. and Zavada, M.S. (1986). Development and functional aspects of fossil pollen. In: S. Blackmore and I. K. Ferguson (editors). Pollen and Spores: Form and Function, pp. 165–178, Academic Press, London.

Taylor, T.N., Cichan, M.A. and Baldoni, A.M. (1984). The ultrastructure of Mesozoic pollen: *Pteruchus dubius* (Thomas) Townrow. *Rev. Palaeobot. Palynol.* **41**: 319–327.

Taylor, T.N., Delevoryas, T. and Hope, R.C. (1987). Pollen cones from the Late Triassic of North America and implications on conifer evolution. *Rev. Palaeobot. Palynol.* **53**: 141–149.

Taylor, T.N., Zavada, M.S. and Archangelsky, S. (1987). The ultrastructure of *Cyclusphaera psilata* from the Cretaceous of Argentina. *Grana* **26**: 74–80.

Taylor, T.N., Osborn, J.M. and Taylor, E.L. (in press). The importance of *in situ* pollen and spores in understanding the biology and evolution of fossil plants. In: J. Jansonius and D.C. McGregor (editors). Stratigraphy and Palynology, Amer. Assoc. Stratigr. Palynologists Publ.

Thomas, B.A. (1987). The use of *in-situ* spores for defining species of dispersed spores. *Rev. Palaeobot. Palynol.* **51**: 227–233.

Traverse, A. (1988). Paleopalynology. 600 pp. Unwin Hyman, Boston.

Trevisan, L. (1980). Ultrastructural notes and considerations on *Ephedripites, Eucommiidites* and *Monosulcites* pollen grains from Lower Cretaceous sediments of southern Tuscany (Italy). *Pollen & Spores* **22**: 85–132.

Van Campo, M. (1971). Précisions nouvelles sur les structures comparées des pollens de Gymnospermes et d'Angiospermes. *Compt. Rend. Hebd. Séances Acad. Sci. Sér. D.* **272**: 2071–2074.

Van Konijnenburg-van Cittert, J.H.A. (1971). *In situ* gymnosperm pollen from the Middle Jurassic of Yorkshire. *Acta Bot. Neerl.* **20**: 1–97.

Ward, J.V., Doyle, J.A. and Hotton, C.L. (1989). Probable granular magnoliid angiosperm pollen from the Early Cretaceous. *Pollen & Spores* **31**: 113–132.

Willard, D.A. (1989). *Lycospora* from Carboniferous *Lepidostrobus* compressions. *Amer. J. Bot.* **76**: 1429–1440.

Willemse, M.T.M. (1971). Morphological and fluorescence microscopical investigation on sporopollenin formation in *Pinus sylvestris* and *Gasteria verrucosa*. In: J. Brooks, P.R. Grant, M.D. Muir, P. van Gijzel and G. Shaw (editors). Sporopollenin, pp. 68–108, Academic Press, London.

Wodehouse, R.P. (1928). The phylogenetic value of pollen grain characters. *Ann. Bot.* (*Oxford*) **42**: 891–934.

Zavada, M.S. (1984). Angiosperm origins and evolution based on dispersed fossil gymnosperm pollen ultrastructure. *Ann. Missouri Bot. Gard.* **71**: 444–463.

Zavada, M.S. (1987). The occurrence of *Cyclusphaera* sp. in southern Africa. Actas Simp. Argent. Paleobot. Palin. 7th, Buenos Aires, 1987 pp. 101–105.

Zavada, M.S. (1990). The ultrastructure of three monosulcate pollen grains from the Triassic Chinle Formation, western United States. *Palynology* **14**: 41–51

Zavada, M.S. (1991a). The ultrastructure of pollen found in dispersed sporangia of *Arberiella* (Glossopteridaceae). *Bot. Gaz.* **152**: 248–255.

Zavada, M.S. (1991b). Determining character polarities in pollen. In: S. Blackmore and S.H. Barnes (editors). Pollen and Spores: Patterns of Diversification, pp. 239–256, Clarendon Press, Oxford.

Zavada, M.S. (1992). The wall ultrastructure of fossil discoid pollen. *Bull. Torrey Bot. Club* **119**: 44–49.

Zavada, M.S. and Benson, J.M. (1987). First fossil evidence for the primitive angiosperm family Lactoridaceae. *Amer. J. Bot.* **74**: 1590–1594.

Zavada, M.S. and Crepet, W.L. (1985). Pollen wall ultrastructure of the type material of *Pteruchus africanus, P. dubius,* and *P. papillatus. Pollen & Spores* **27**: 271–276.

Zavada, M.S. and Crepet, W.L. (1986). Pollen wall ultrastructure of *Caytonanthus arberi. Pl. Syst. Evol.* **153**: 259–264.

Zavada, M.S. and Dilcher, D.L. (1988). Pollen wall ultrastructure of selected dispersed monosulcate pollen from the Cenomanian, Dakota Formation, of central USA. *Amer. J. Bot.* **75**: 669–679.

Zavada, M.S. and Gabarayeva, N. (1991). Comparative pollen wall development of *Welwitschia mirabilis* and selected primitive angiosperms. *Bull. Torrey Bot. Club* **118**: 292–302.

Zavada, M.S. and Taylor, T.N. (1986a). Pollen morphology of Lactoridaceae. *Pl. Syst. Evol.* **154**: 31–39.

Zavada, M.S. and Taylor, T.N. (1986b). The role of self-incompatibility and sexual selection in the gymnosperm-angiosperm transition: a hypothesis. *Amer. Naturalist* **128**: 538–550.

Kurmann, M.H. and Zavada, M.S. (1994). Pollen morphological diversity in extant and fossil gymnosperms. In: M.H. Kurmann and J.A. Doyle (editors). Ultrastructure of fossil spores and pollen, pp. 123–137, Royal Botanic Gardens, Kew.

8. POLLEN MORPHOLOGICAL DIVERSITY IN EXTANT AND FOSSIL GYMNOSPERMS

MARIE H. KURMANN[1] AND MICHAEL S. ZAVADA[2]

[1]Palynology Unit, Herbarium, Royal Botanic Gardens, Kew, Richmond, Surrey, TW9 3AE, U.K.
[2]Department of Biology, University of Southwestern Louisiana, Lafayette, LA 70504, U.S.A.

Abstract

Pollen morphology in gymnosperms generally falls within a narrow range of variation relative to angiosperms. This may be related to the limited number of evolutionary paradigms that are successful with the gymnospermous condition and wind pollination. Gymnospermous pollen grains from the Upper Paleozoic and Mesozoic exhibit more diversity than extant ones, and some display features more indicative of the insect pollinated angiosperms than the extant, mostly wind-pollinated gymnosperms. The occurrence of angiosperm-like features in some probable Mesozoic gymnosperms suggests that some groups might have tended more toward ambophily or entomophily, rather than anemophily.

Introduction

It has long been recognized that pollen morphological characters are useful in generating hypotheses of relationships between groups of plants. In recent years, pollen morphological studies at the ultrastructural level of extant and fossil gymnosperms have been accumulating. Several publications have reviewed and summarised pollen morphological characters in extant gymnosperms (e.g., Audran and Masure, 1977, 1978; Deghan and Deghan, 1988; Xi and Wang, 1989; Kurmann, 1992). The ultrastructure of fossil gymnosperm pollen has been reviewed and discussed in detail by Taylor and Taylor (1987) and Osborn and Taylor (1994). This contribution attempts to assess the pollen morphological diversity in extant and fossil gymnosperms and to discuss the variation with regard to pollination and reproductive biology of this major group of seed plants.

Morphology

A. Pollen Unit

The majority of extant gymnosperm pollen is shed as monads (Figs. 1–4). This is not surprising, given that many extant gymnosperms are known or presumed to be wind pollinated. Studies of fossil pollen, however, have shown the occurrence of dyads and tetrads. Trevisan (1971) described a pollen type from the Lower Cretaceous, *Dicheiropollis*, in which mature pollen grains are connected in pairs by two bundles of delicate strings. The pollen grains of the genus *Classopollis* (Triassic–Cretaceous) are frequently recovered in tetrads (e.g., Pettitt and Chaloner, 1964; Taylor and Alvin, 1984; Rowley and

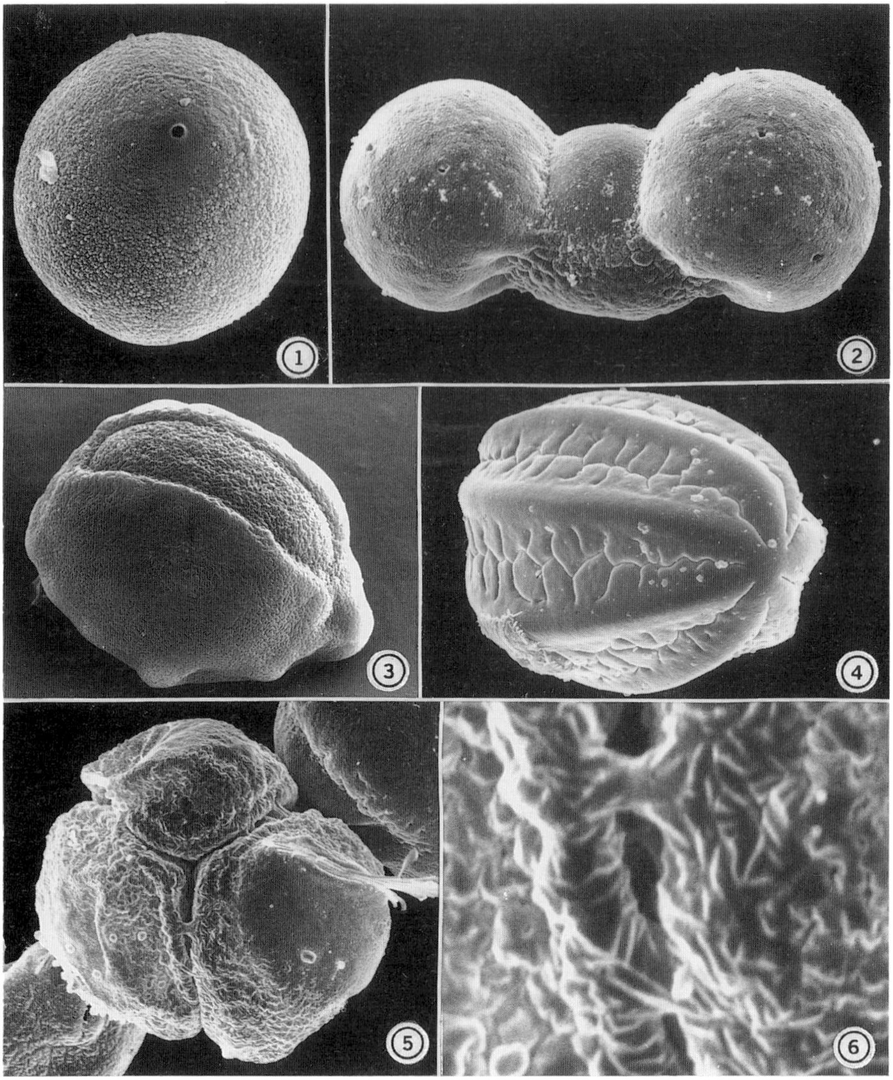

FIGS. 1–6. Pollen units and pollen shape. **1**: *Calocedrus decurrens*, SEM, × 2,000. **2**: *Podocarpus micropedunculatus*, SEM, × 1,200. **3**: *Ceratozamia mexicana*, SEM, × 1,600. **4**: *Ephedra helvetica*, SEM, × 2,000. **5**: *Ginkgo biloba*, mature pollen held in a tetrad by pollen connecting filaments, SEM, × 2,000. **6**: High magnification of pollen connecting filaments, SEM, × 10,000.

Srivastava, 1986; Pocock et al., 1990). Scheuring (1976) reported on the common occurrence of proximal exine filaments in the dispersed pollen genus *Triadispora*, which allows for adhesion of individual pollen grains in a tetrad. Such pollen connecting filaments are sometimes present in *Ginkgo biloba* (Figs. 5, 6).

Compound pollen grains held together by pollen connecting threads are also found in angiosperms (Hesse, 1986). Despite the apparent diversity of pollen units, the gymnosperms never achieve the variety of pollen units observed in angiosperms. Knox and McConchie (1986) listed 56 angiosperm families with species producing compound pollen. The occurrence of multiple pollen units is generally associated with insect pollination.

B. Pollen Aperture

The most common type in extant gymnosperms is a single, distal aperture which is either circular (Figs. 1, 7, 8) or elongate (Figs. 2, 3, 9, 10) in outline. Circular apertures are often referred to as germinal areas. In the germinal area, a thinning of the exine occurs through reduction of the ektexine. Germinal areas occur predominantly in the Araucariaceae, Cupressaceae, Taxodiaceae and Cephalotaxaceae. In the Taxodiaceae and the Cupressaceae, the centre of the germinal area is modified to form either a papilla or a pore (Figs. 1, 8, 9). The elongate or monosulcate aperture type occurs in the Cycadales, Ginkgoaceae, Pinaceae, Podocarpaceae and Welwitschiaceae (Figs. 2, 3, 9–13). Pollen grains of the Taxaceae, Ephedraceae and Gnetaceae appear to be inaperturate. A trilete mark is sometimes visible on the proximal surface of extant gymnosperm pollen indicating the position relative to the tetrad (Kurmann, 1989b).

In fossil gymnosperm pollen, the single, distal aperture also predominates. However, there are a number of exceptions. *Monoletes*, the pollen of the Paleozoic pteridosperm group Medullosaceae, appears to be triaperturate, possessing a proximal monolete mark and two equidistant grooves on the distal face (e.g., Kurmann and Taylor, 1984). In the area of the proximal monolete mark, the ektexine is reduced and the endexine thickened while in the area of the distal grooves only the ektexine is modified. This indicates a proximal germination, and grains with this type of germination are referred to as pre-pollen (Chaloner, 1970).

Foster and Price (1981) described the ultrastructure of *Praecolpatites sinuosus* from the Permian of Australia. These grains commonly have five longitudinal furrows, characterised by a thinning of the ektexine. Each of these furrows may have functioned as a potential exit for the pollen content. Pollen macerated from a Paleozoic cordaitalean sporangium, *Gothania*, has a proximal triradiate mark that probably functioned like the apertural mechanism in many spore bearing plants (Taylor and Daghlian, 1980). The occurrence of the proximal triradiate mark is also found among the Paleozoic seed bearing Lyginopteridaceae. Some Permian bisaccate pollen has one to many proximal furrows that may have released the pollen content (Fig. 12, 24).

A number of dispersed pollen taxa with seed plant affinities from the Permian have both a proximal triradiate mark and a distal sulcus-like aperture (e.g. *Lueckisporites, Jugasporites, Triadispora* and *Ovalipollis*; Scheuring, 1974). Among Mesozoic dispersed pollen, the Circumpolles group has attracted considerable attention. Pollen of this group has a proximal, triradiate mark and a distal pore-like structure (Pettitt and Chaloner, 1964; Kedves, 1976; Taylor and Alvin, 1984; Pocock et al., 1990).

The pollen of *Cyclusphaera*, a presumed gymnosperm, may have had two opposing porate apertures (Taylor et al. 1987; Zavada, 1992). The position of these apertures in relation to the proximal and distal poles is unknown. Cornet (1989) reported a number of multi-aperturate dispersed pollen types from the Triassic Newark Basin that he compared with angiosperms but we consider of unknown taxonomic affinity (Doyle and Hotton, 1991).

C. Pollen Shape and Size

Unlike the numerous shape classes in angiosperms, extant gymnosperm pollen is relatively constant with regard to shape and usually falls within three groups. Non-saccate grains are either spheroidal (Fig. 1) or elliptic (Figs. 3, 4). Spheroidal, non-saccate grains usually have a distal germinal area or are inaperturate, while elliptical, non-saccate pollen is monosulcate or inaperturate. In these grains, the long axis of the pollen is parallel to the long axis of the sulcus. The shape of saccate pollen varies according to the number and the shape of sacci (Figs. 9, 12, 13). In monosaccate pollen, the corpus is

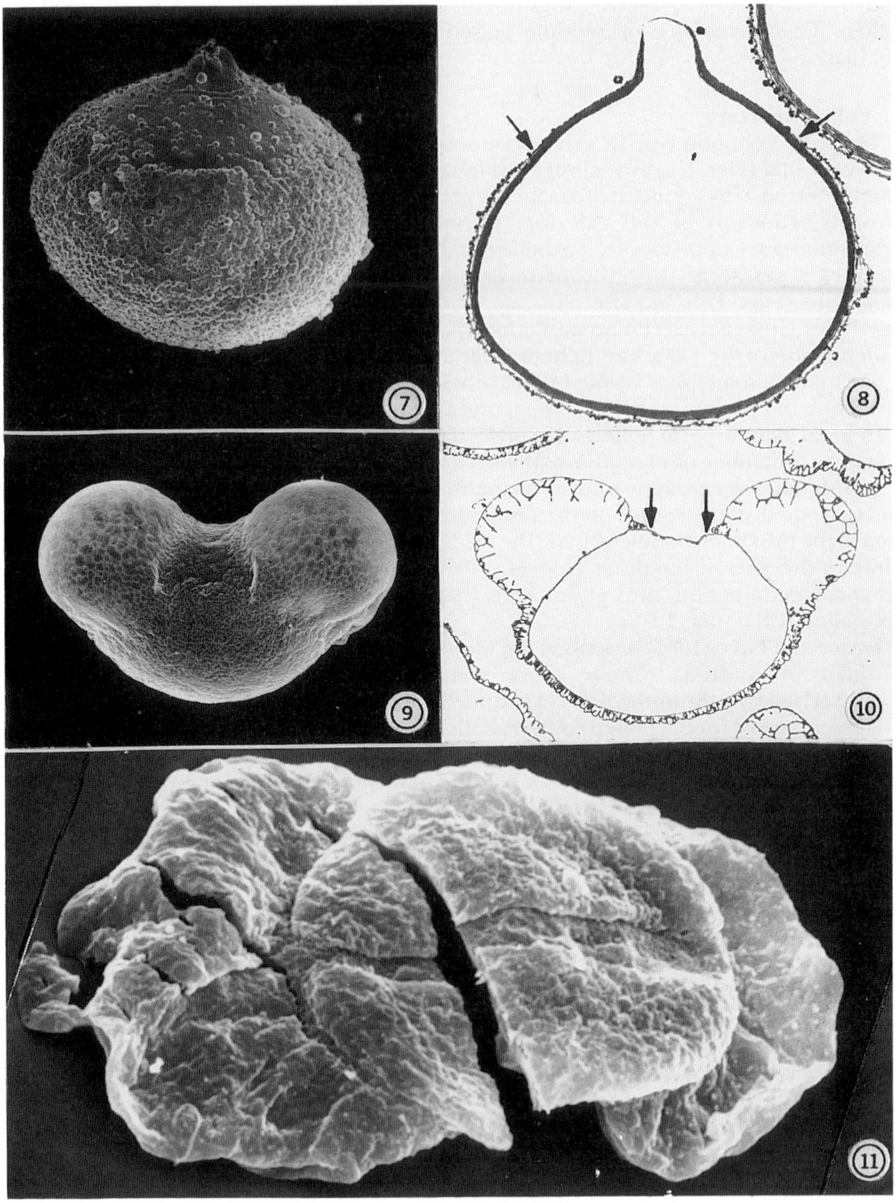

FIGS. 7-11. Pollen apertures. **7**: *Taxodium mucronatum*, germinal area with central papilla, SEM, × 2,200. **8**: *Taxodium mucronatum*, germinal area (between arrows) with central papilla, TEM, × 2,200. **9**: *Pinus armandii*, distal sulcus, SEM, × 650. **10**: *Pinus armandii*, distal sulcus (between arrows), TEM, × 650. **11**: Bulgarian fossil pine pollen with proximal sulcus, SEM, × 1,250.

usually spheroidal, while in bisaccate grains it is elliptical. In contrast to non-saccate, monosulcate pollen, the long axis of the pollen grain is perpendicular to the long axis of the sulcus. The presence of sacci generally increases the dimension of the long axis of the pollen grain.

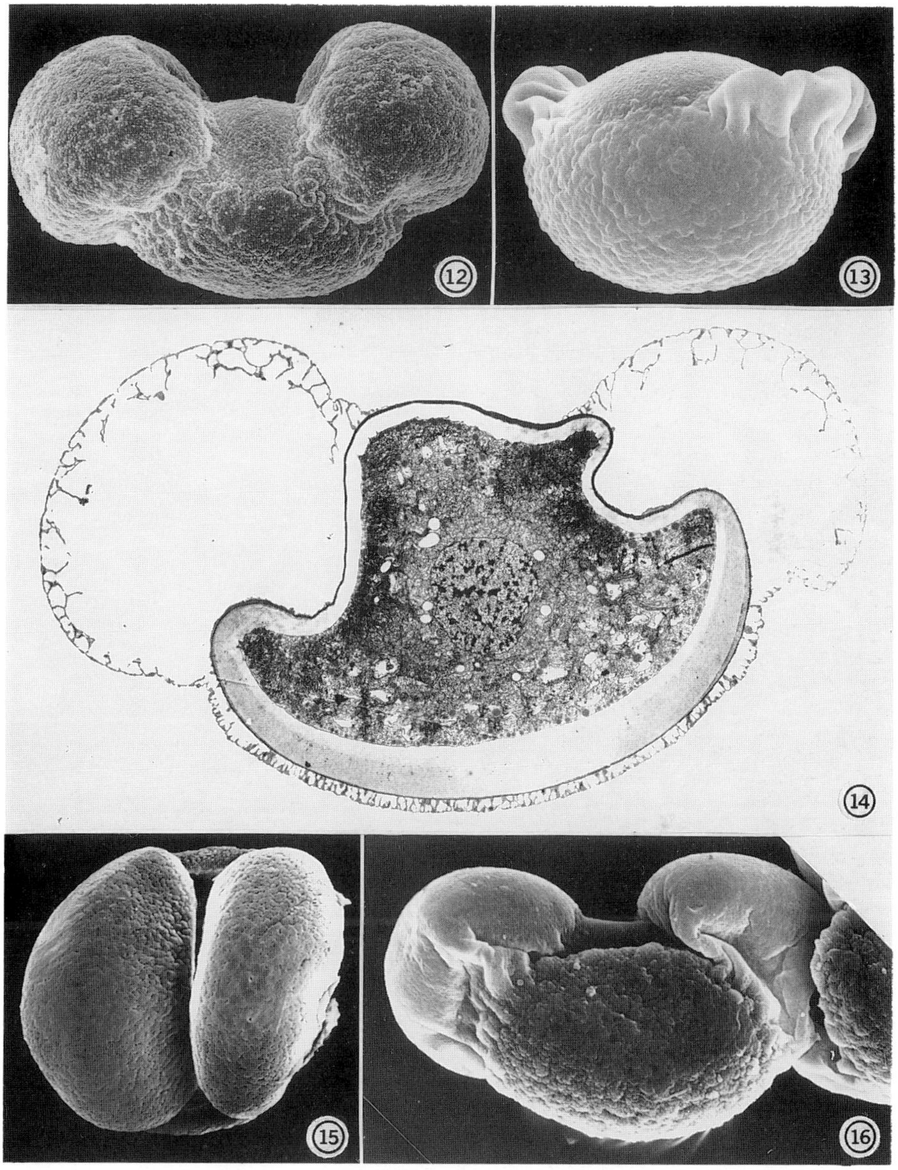

Figs. 12–16. Pollen shapes. **12**: *Nageia wallichiana* with two large lateral sacci, SEM, × 1,800. **13**: *Phyllocladus glaucus* with two small lateral sacci, SEM, × 1,800. **14**: *Pinus ponderosa* showing protoplasm restricted to corpus area, TEM, × 1,5000. **15**: *Pinus radiata*, dehydrated pollen grain with infolded sacci, SEM, × 1,200. **16**: *Pinus radiata*, rehydrated pollen grain with expanded sacci, SEM, × 1,200.

Gymnosperm pollen grains fall into two broad size categories. The longest axis in non-saccate grains is generally shorter than 45 μm, while in saccate pollen it exceeds 45 μm. The shape and size of non-saccate gymnosperm pollen falls within the range of

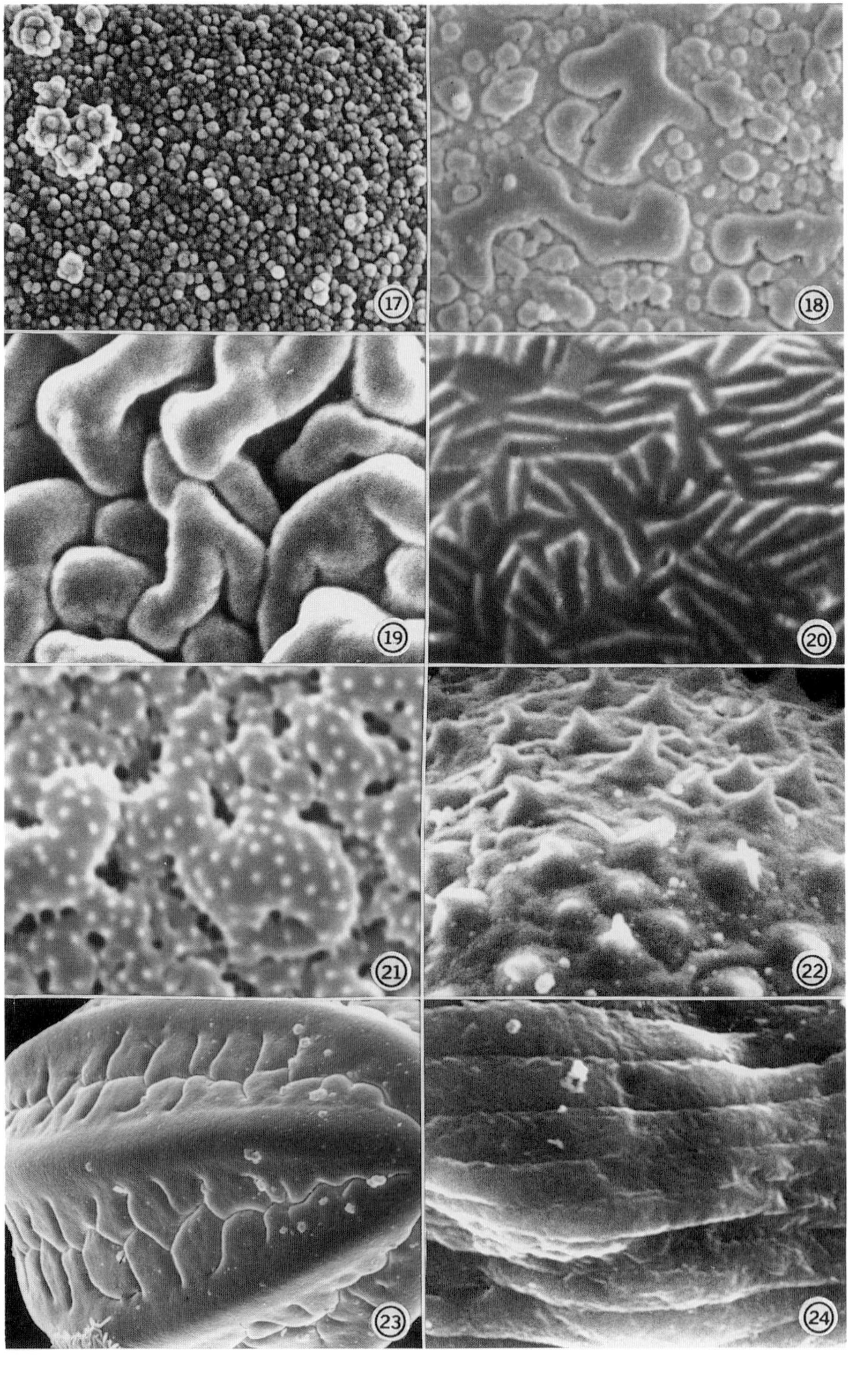
17
18
19
20
21
22
23
24

shape and size reported for most wind pollinated taxa (e.g., Faegri and van der Pijl, 1979; Whitehead, 1983; Niklas, 1985). Although saccate pollen often exceeds the size range thought to be ideal for wind pollination, sacci do not increase the weight significantly. Rather, they lower the pollen density and thus facilitate transport through the air (Whitehead, 1983). In saccate pollen, the protoplasm is confined to the corpus (Fig. 14), and many saccate grains are spheroidal in outline when released from the pollen sacs or dehydrated (Fig. 15). They are only extended to their full measure after hydration in the micropyle (Fig. 16).

Variation of shape in fossil gymnosperm pollen is similar to that present in extant ones. Pollen size, however, is much more variable. *Monoletes* grains, for example, are known to exceed 300 μm in length (Taylor, 1978). It is difficult to imagine how pollen grains of such a size could be successfully dispersed by wind. Some other mode of pollination might have been present (perhaps insect). It has been shown that gymnosperms, as early as the Paleozoic, interacted with insects (Taylor, 1982; Taylor and Scott, 1983; Taylor and Taylor, 1993). Whether this included insect pollination is currently unknown.

D. Pollen Sculpture

The sculpturing patterns in extant gymnosperm pollen range from psilate to echinate, but they are rarely very elaborate (Figs. 17–24). In saccate pollen, the surface sculpture varies from granulate to rugulate (Figs. 17, 19). In addition to these ornamentation types, striate, verrucate and echinate surface patterns occur in non-saccate pollen grains. Striate exine sculpturing is observed in *Ginkgo* (Fig. 20), *Ephedra* (Fig. 23) and *Welwitschia*. *Sciadopitys* (Fig. 21) has a verrucate-foveolate sculpture pattern, and in *Gnetum* (Fig. 22) the surface is echinate.

Fossil dispersed taeniate pollen from Upper Palaeozoic — Lower Mesozoic sediments of the Southern Hemisphere are thought to have affinities with the glossopterids. This saccate pollen has proximal taeniae (or broad striae) parallel to the long axis of the pollen grain (e.g., Foster, 1979; Zavada, 1991; Fig. 24).

The fossil pollen *Nanoxanthiopollenites* has a foveolate-echinate ornamentation. As in *Sciadopitys*, these perforations do not extend into the infratectal layer (Taylor, 1980). Pocock and Vasanthy (1988) and Cornet (1989) described a number of pre-Cretaceous dispersed pollen grains of unknown affinity with distinct tectal perforations. Until the taxonomic position of these grains is known the occurrence of perforated tecta in gymnosperms remains equivocal.

The sculpture types observed in fossil and extant gymnosperms are somewhat reminiscent of the sculpture types observed in wind pollinated Hamamelidae (Zavada and Dilcher, 1986). The lack of a wide range of exine sculpture types in the gymnosperms may be due to the preponderance of wind pollination among the gymnosperms.

E. Exine Structure

Exine stratification in extant gymnosperms was first characterized by Van Campo (1971) and Van Campo and Lugardon (1973), and has recently been reviewed by Kurmann (1992). The exine consists of an inner, lamellated endexine and an outer ektexine, which is either alveolar or granular. The alveolar type is best exemplified in the cycads (Figs. 25, 26). The ektexine of cycadaceous pollen has an outer tectum that is underlain by an infratectal layer comprised of vertically oriented rods or partitions in median section. In tangential sections of the pollen wall the infra-structural layer appears honeycomb-like (Fig. 26). The alveolar pollen walls in the Pinaceae and Podocarpaceae are somewhat

FIGS. **17–24**. Pollen sculpture. **17**: *Athrotaxis laxifolia*, granulate, SEM, × 10,000. **18**: *Saxegothaea conspicua*, granulate-rugulate, SEM, × 10,000. **19**: *Falcatifolium falcatiformae*, rugulate, SEM, × 10,000. **20**: *Ginkgo biloba*, striate, SEM, × 10,000. **21**: *Sciadopitys verticillata*, foveolate, SEM, × 10,000. **22**: *Gnetum gnemon*, echinate, SEM, × 10,000. **23**: *Ephedra helvetica*, striate, SEM, × 2,500. **24**: Proximal striation (taeniae) in pollen of *Arberiella*, SEM, × 2,500.

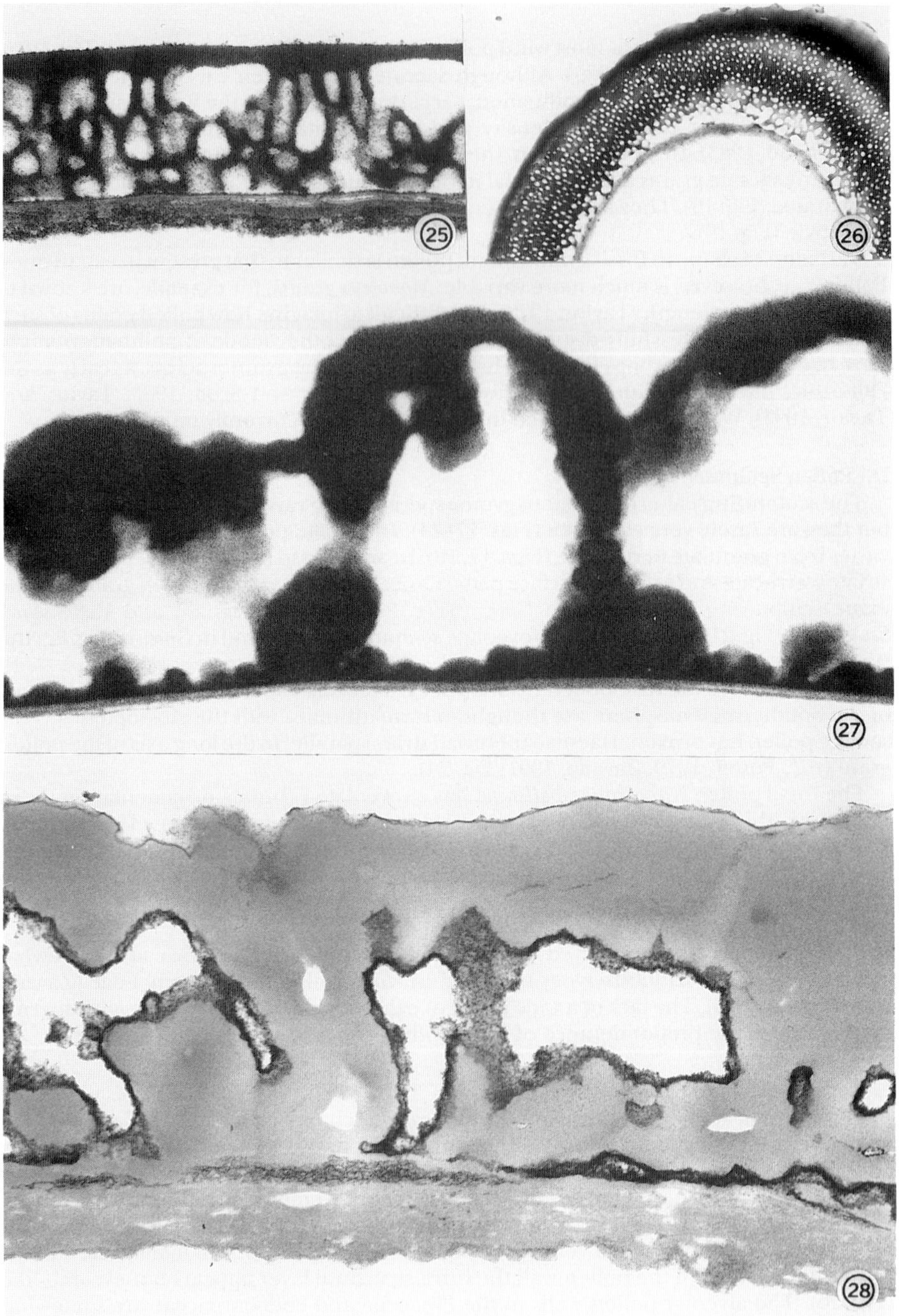

Figs. 25-28. Alveolar exine structure. **25**: *Cycas siamensis*, TEM, × 30,000. **26**: *Zamia floridana*, TEM, × 7,100. **27**: *Pinus armandii*, TEM, × 30,000. **28**: *Dacrydium beccarii*, TEM, × 30,000.

different. The ektexine of pinaceous pollen has an outer tectum underlain by an infratectal layer that appears as a network of anastomosing rods or partitions in median section (Fig. 27). In tangential section the infratectal layer is somewhat more irregular in appearance than that of the cycads. In podocarpaceous pollen, the infratectal rods or partitions are wider and sparser, giving the appearance in median section of short, stout, irregular shaped rods.

Granular ektexine structure is the most common type among extant gymnosperms. In the Coniferales, it consists mostly of spherical to irregular-shaped granules of various sizes, closely appressed to each other but not forming a tectum (Figs. 29–31). In the Gnetales, the infratectal granules are covered by a solid tectum (Fig. 32). The pollen of *Tsuga* is unique among extant gymnosperms because it lacks an infratectal layer. The convoluted tectal layer rests directly on the footlayer (Kurmann, 1990).

Further from the exine structures observed in extant gymnosperms, an infratectal layer comprised of columellar-like structures is present in the fossil taxon *Classopollis* (e.g., Pettitt and Chaloner, 1964; Taylor and Alvin, 1984; Pocock et al., 1990). Exterior to the columellar-like layer is a homogeneous layer with closely spaced supratectal elements (Pettitt and Chaloner, 1964; Taylor and Alvin, 1984).

Several of the dispersed pollen grains described by Pocock and Vasanthy (1988) and Cornet (1989) from the early Mesozoic have angiosperm-like columellate infrastructure, although Doyle and Hotton (1991) argue that their endexine structure is more typically gymnospermous. If these dispersed pollen grains prove to be gymnospermous, their morphology would expand the known variation among gymnosperm pollen for aperture type, sculpturing and wall structure.

A homogeneous exine is found only in fossil gymnosperm pollen grains. This wall typs is comprised of a solid ektexine which occasionally has lacunae that give the appearance of a densely granular infratectal layer. This type has been observed in the Pentoxylales (Osborn et al., 1991), Cycadeoideaceae (Taylor, 1973) and some dispersed pollen of possible gymnosperm affinity (Zavada, 1990).

F. Sacci

Sacci are a unique and conspicuous features in many fossil and extant gymnosperm pollen grains. In extant gymnosperm pollen the saccus is generally formed by a separation within the ektexine: the tectum and infratectum forming the saccus roof and the footlayer the saccus floor. Sacci increase the size of the pollen grain without adding much weight.

Scheuring (1974) distinguished two types of sacci in fossil gymnosperm pollen. In the protosaccate type, the infratectal elements are structurally continuous with the footlayer and retain the alveolar infrastructure (Fig. 34). In the eusaccate type, the infratectal elements are restricted to the saccus roof and large cavities are present (Fig. 33). Meyen (1987) noted that the term protosaccate might imply a pleisiomorphic condition, and therefore advocated the use of the term quasisaccate. Saccus types are also discussed in detail by Osborn and Taylor (1994). If sacci are sectioned tangetially, it is difficult to differentiate between the protosaccate (Fig. 35) and the eusaccate (Fig. 36) form. It is also interesting to note that in extant gymnosperms there is a difference in size between protosaccate and eusaccate pollen. Protosaccate pollen grains are considerably smaller than eusaccate ones, and this might indicate a density related phenomenon.

Sacci are generally considered a feature restricted to gymnosperms, however, sacci-like structures are known from a few angiosperm families (Zavada and Taylor, 1986a; Zavada and Dilcher, 1986). The number of sacci can vary between taxa (e.g., Taylor, 1970; Millay and Taylor, 1974; Baldoni et al., 1982; Pocknall, 1981) and within species (Kurmann, 1989a).

The function of the sacci has intrigued many researchers and a number of hypotheses have been put forward. Wodehouse (1935) suggested that the lateral sacci, in *Pinus* for example, protect the germinal furrow from dessication. Doyle (1945) proposed that sacci are devices for flotation in the micropyle, and may secondarily function to protect the germinal furrow. Zavada and Taylor (1986b) proposed that the increase in pollen size, without the increase in protoplasm, may reduce pollen competition within the finite micropylar space. In their view, this would lead to reduced competition for egg cells

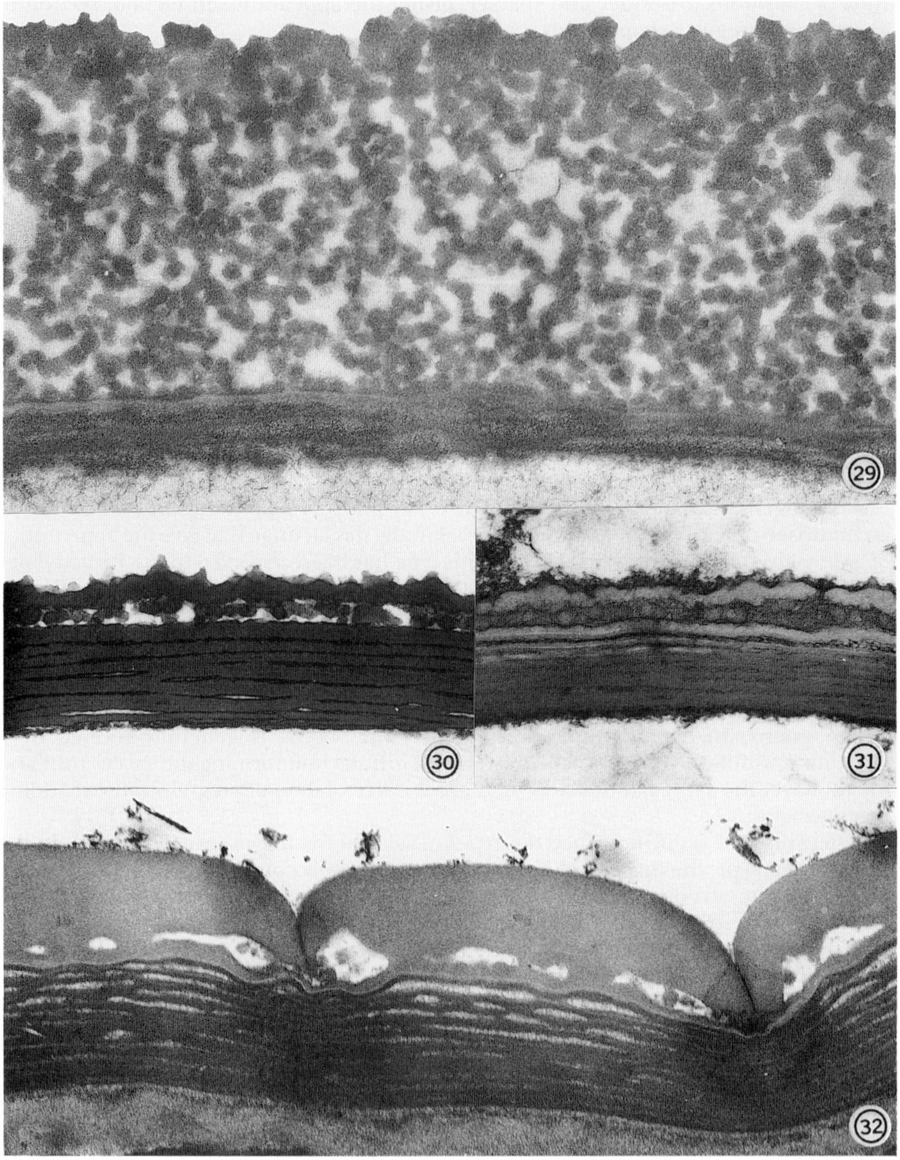

FIGS. 29–32. Granular exine structure. **29**: *Larix leptolepis*, TEM, × 30,000. **30**: *X Cypressocyparis notabilis*, acetolysed exine, TEM, × 30,000. **31**: *X C. notabilis*, non-acetolysed exine, TEM, × 30,000. **32**: *Ephedra distachya*, TEM, × 30,000.

within each gametophyte. Whatever the function of sacci may be, their widespread occurrence among fossil and extant gymnospermous plants, and their almost universal absence from angiosperms, suggests that this feature may be related to the reproductive biology of the gymnospermous condition.

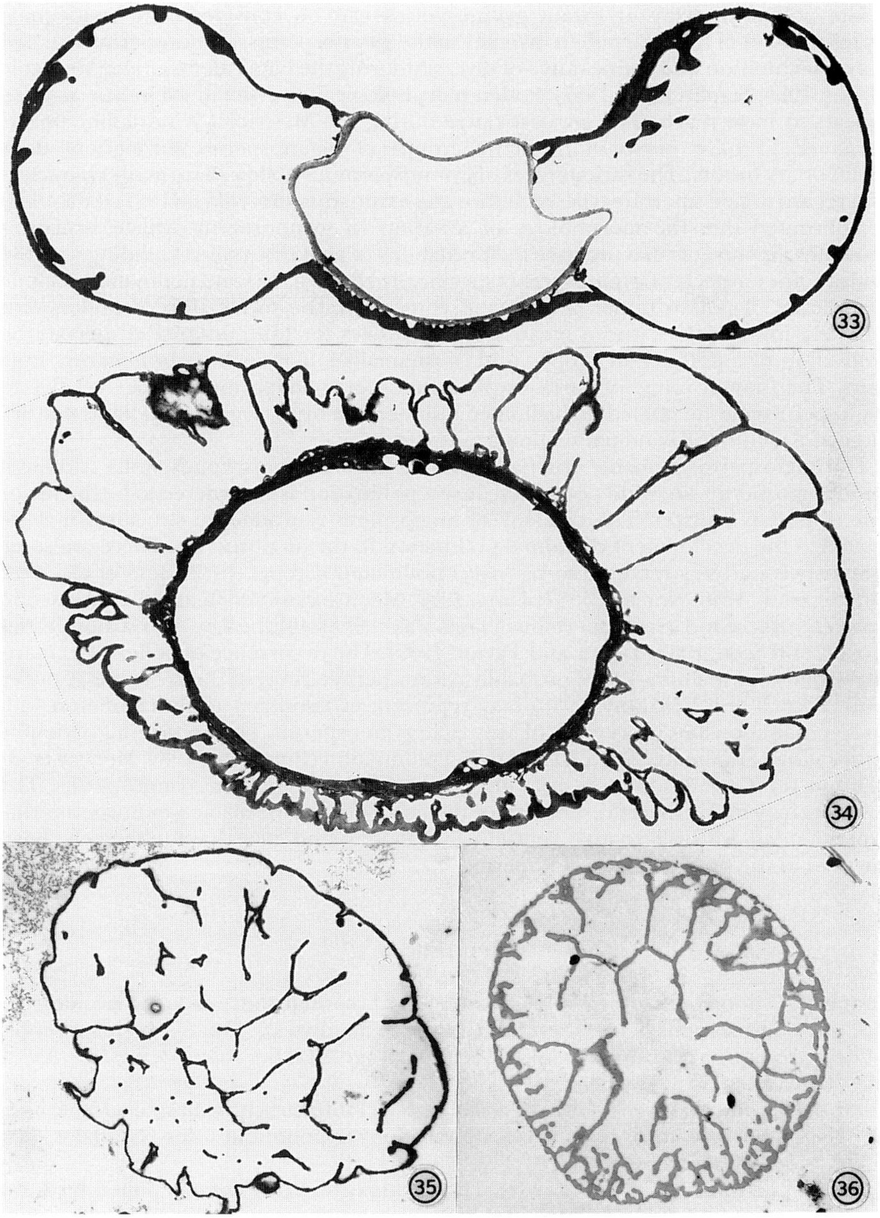

FIGS. 33–36. Saccus organisation. **33**: *Podocarpus micropedunculatus*, eusacci with large lumina, TEM, × 1,700. **34**: *Falcatifolium falcatiformae*, protosacci retaining alveolar structure, TEM, × 2,600. **35**: *F. falcatiformae*, tangential section through protosaccus, TEM, × 3,000. **36**: *Pinus radiata*, tangential section through eusaccus, TEM, × 4,000.

Discussion

The pollen diversity of extant gymnosperms can be considered an inadequate representation of the total pollen diversity in the gymnosperms. Gymnosperms reached global distribution and a wide range of diversity during the Late Paleozoic and Mesozoic (Niklas, 1983; Staplin et al., 1967). Pollen morphological diversity in gymnosperms also appears to have reached its greatest range during the Mesozoic. Wind pollination is presumed to have persisted in many groups of gymnosperms throughout their evolutionary history. The various types of gymnospermous pollen share many characters with regard to size, aperture type, sculpture and exine structure. Niklas (1981, 1983, 1985) demonstrated that the morphology of a variety of gymnosperm ovulate structures generates air currents that increase the probability of airborne pollen colliding with the ovules. Other aspects of gymnosperms specifically adapted for wind pollination include the release of pollen under active wind conditions, the production of pollen with relatively low densities and a mean size that allows for long distance dispersal, the production of a pollination droplet and of stigma-like, micropylar lobes bearing long hairs. The limited range of variation in gymnosperm pollen morphology relative to angiosperms may be related to the limited number of morphological paradigms that are successful within the wind pollination syndrome.

The angiosperms probably originated among the Mesozoic gymnosperms. Although some angiosperms are wind pollinated, insect pollination is considered to be the major type of pollen transfer. The diversity of angiosperm reproductive structures may be related to the development of animal pollination in the angiosperms. There are some gymnosperm groups reported to be insect pollinated (Crepet, 1972; Bino et al., 1984; Norstog et al., 1986; Norstog, 1987; Tang, 1987; Meeuse et al., 1990) and the interaction between insects and gymnospermous plants was well established by the Carboniferous (Taylor and Scott, 1983; Taylor and Taylor, 1993). The occurrence of pollen with many angiospermous features, but of probable gymnosperm affinity in the Mesozoic (Cornet, 1989; Pocock and Vasanthy, 1988) may represent evolutionary experimentation with insect pollination among a variety of Mesozoic gymnosperms. The occurrence of double fertilization (Friedman, 1990a, b) and insect pollination (Bino et al., 1984, Meeuse et al., 1990) in the Gnetales is more reminiscent of angiosperms than gymnosperms. The Gnetales may be extant representative of the more diverse Mesozoic gymnosperms that experimented with alternative reproductive modes and may have ultimately been involved in the origin of the angiosperms.

References

Audran, J.-C. and Masure, E. (1977). Contribution à la connaissance de la composition des sporodermes chez les Cycadales (Préspermaphytes). Étude en microscopie électronique à transmission (M.E.T.) et à balayage (M.E.B.). *Palaeontographica, Abt. B, Paläophytol.* **162**: 115–158.

Audran, J.-C. and Masure, E. (1978). La sculpture et l'infrastructure du sporoderme chez *Ginkgo biloba* comparées à celles de enveloppes polliniques des Cycadales. *Rev. Paleobot. Palynol.* **26**: 363–387.

Baldoni, A.M. and Taylor, T.N. (1982). The ultrastructure of *Trisaccites* pollen from the Cretaceous of southern Argentina. *Rev. Palaeobot. Palynol.* **38**: 23–33.

Bino, R.J., Dafni, A. and Meeuse, A.D.J. (1984). Entomophily in the dioecious gymnosperm *Ephedra aphylla* Forsk. (*E. alte* C.A. Mey.) with some notes on *E. campylopoda* C.A. Mey. I. Aspects of the entomophilous syndrome. *Proc. Kon. Ned. Akad. Wetensch. C* **87**: 1–13.

Chaloner, W.G. (1970). The evolution of miospore polarity. *Geosci. & Man* **1**: 47–56.

Cornet, B. (1989). Late Triassic angiosperm-like pollen from the Richmond Rift Basin of Virginia, U.S.A. *Palaeontographica, Abt. B., Paläophytol.* **213**: 37–87.

Crepet, W.L. (1972). Investigations of North American cycadeoids: pollination mechanism in *Cycadeoidea. Amer. J. Bot.* **59**: 1048–1056.

Dehgan, B. and Dehgan, N.B. (1988). Comparative pollen morphology and taxonomic affinities in Cycadales. *Amer. J. Bot.* **75**: 1501–1516.

Doyle, J. (1945). Developmental lines in pollination mechanisms in the Coniferales. *Sci. Proc. Roy. Dublin Soc.* **24**: 43–62.

Doyle, J.A. and Hotton, C.L. (1991). Diversification of early angiosperm pollen in a cladistic context. In: S. Blackmore and S.H. Barnes (editors). Pollen and Spores. Patterns of Diversification, pp. 169–195, Clarendon Press, Oxford.

Faegri, K. and van der Pijl, L. (1979). The principles of pollination ecology. 248 pp. Pergamon Press, Oxford.

Foster, C.B. (1979). Permian plant microfossils of the Blair Athol Coal Measures, Baralaba Coal Measures, and Basal Rewan Formation of Queensland. *Publ. Geol. Surv. Queensland* **372**, *Palaeontol. Pap.* **45**: 1–244.

Foster, C.B. and Price, P.L. (1981). Exine intrastructure of *Praecolpatites sinuosus* (Blame et Hennelly) Bharadwaj et Srivastava, 1969, and *Marsupipollenites triradiatus* Blame et Hennelly, 1956. *Palaeobotanist* **28–29**: 177–187.

Friedman, W.E. (1990a). Double fertilization in *Ephedra*, a nonflowering plant: its bearing on the origin of the angiosperms. *Science* **247**: 951–954.

Friedman, W.E. (1990b). Sexual reproduction in *Ephedra nevadensis* (Ephedraceae): further evidence of double fertilization in a nonflowering seed plant. *Amer. J. Bot.* **77**: 1581–1598.

Hesse, M. (1986). Nature, form and function of pollen-connecting threads in angiosperms. In: S. Blackmore and I.K. Ferguson (editors). Pollen and Spores: Form and Function, pp. 109–118, Academic Press, London.

Kedves, M. (1976). Scanning electron-microscope investigations on the pollen grains of the *Operculati* Venk et Gocz. 1964. *Acta Biol.* (*Szeged*) **22**: 29–36.

Knox, R.B. and McConchie, C.A. (1986). Structure and function of compound pollen. In: S. Blackmore and I.K. Ferguson (editors). Pollen and Spores: Form and Function, pp. 265–282, Academic Press, London.

Kurmann, M.H. (1989a). A reinvestigation of the pollen wall in *Pinus balfouriana* Murray. *Pollen & Spores* **21**: 23–28.

Kurmann, M.H. (1989b). Pollen wall formation in *Abies concolor* and a discussion on wall layer homologies. *Canad. J. Bot.* **67**: 2489–2504.

Kurmann, M.H. (1990). Development of the pollen wall in *Tsuga canadensis* (Pinaceae). *Nordic J. Bot.* **10**: 63–78.

Kurmann, M.H. (1992). Exine stratification in extant gymnosperms: a review of published transmission electron micrographs. *Kew Bull.* **47**: 25–39.

Kurmann, M.H. and Taylor, T.N. (1984). The ultrastructure of *Boulaya fertilis* (Medullosales) pollen. *Pollen & Spores* **26**: 109–116.

Meeuse, A.D.J., de Meijer, A.H., Mohr, O.W.P. and Wellings, S.M. (1990). Entomophily in the dioecious gymnosperm *Ephedra aphylla* Forsk. (*E. alte* C.A. Mey.) with some notes on *E. campylopoda* C.A. Mey. III. Further anthecological studies and relative importance of entomophily. *Israel J. Bot.* **39**: 113–123.

Meyen, S.V. (1987). Fundamentals of Palaeobotany. 432 pp. Chapman and Hall, London and New York.

Millay, M.A. and Taylor, T.N. (1974). Morphological studies of Paleozoic saccate pollen. *Palaeontographica, Abt. B, Paläophytol.* **147**: 75–99.

Niklas, K.J. (1981). Airflow patterns around some early seed plant ovules and cupules: implications concerning efficiency in wind pollination. *Amer. J. Bot.* **68**: 635–650.

Niklas, K.J. (1983). The influence of Paleozoic ovule and cupule morphologies on wind pollination. *Evolution* **37**: 968–986.

Niklas, K.J. (1985). The aerodynamics of wind pollination. *Bot. Rev.* **51**: 328–386.

Norstog, K. (1987). Cycads and the origin of insect pollination. *Amer. Sci.* **75**: 270–279.

Norstog, K.J., Stevenson, D.W. and Niklas, K.J. (1986). The role of beetles in the pollination of *Zamia furfuracea* L. fil (Zamiaceae). *Biotropica* **18**: 300–306.
Osborn, J.M. and Taylor, T.N. (1994). Comparative ultrastructure of fossil gymnosperm pollen and its phylogenetic implications. In: M.H. Kurmann and J.A. Doyle (editors). Ultrastructure of Fossil Spores and Pollen, pp. 99–121, Royal Botanic Gardens, Kew.
Osborn, J.M., Taylor, T.N. and Crane, P.R. (1991). The ultrastructure of *Sahnia* pollen (Pentoxylales). *Amer. J. Bot.* **78**: 1560–1569.
Pettitt, J.M. and Chaloner, W.G. (1964). The ultrastructure of the Mesozoic seed fern pollen *Classopollis. Pollen & Spores* **6**: 611–620.
Pocknall, D.T. (1981). Pollen morphology of the New Zealand species of *Dacrydium* Solander, *Podocarpus* L'Heritier, and *Dacrycarpus* Endlicher (Podocarpaceae). *New Zealand J. Bot.* **19**: 67–95.
Pocock, S.A.J. and Vasanthy, G. (1988). *Cornetipollis reticulata*, a new pollen with angiospermid features from Upper Triassic (Carnian) sediments of Arizona (U.S.A.) with notes on *Equisetosporites. Rev. Palaeobot. Palynol.* **55**: 337–356.
Pocock, S.A.J., Vasanthy, G. and Venkatachala, B.S. (1990). Pollen of Circumpolles — an enigma or morphotrends showing evolutionary adaptation. *Rev. Palaeobot. Palynol.* **65**: 179–193.
Rowley, J.R. and Srivastatva, S.K. (1986). Fine structure of *Classopollis. Canad. J. Bot.* **64**: 3059–3074.
Scheuring, B.W. (1974). "Protosaccate" Strukturen, ein weitverbreitetes Pollenmerkmal zur frühen und mittleren Gymnospermenzeit. *Geol. Paläontol. Mitt. Innsbruck* **4**: 1–30.
Scheuring, B.W. (1976). Proximal exine filaments, a widespread feature among Triassic protosaccites and Circumpolles to secure dispersal of entire tetrads. *Pollen & Spores* **18**: 611–639.
Staplin, F.L., Pocock, S.J. and Jansonius, J. (1967). Relationships among gymnospermous pollen. *Rev. Paleobot. Palynol* **3**: 297–310.
Tang, W. (1987). Insect pollination in the cycad *Zamia pumila* (Zamiaceae). *Amer. J. Bot.* **74**: 90–99.
Taylor, T.N. (1970). *Lasiostrobus* gen. n., a staminate strobilus of gymnospermous affinity from the Pennsylvanian of North America. *Amer. J. Bot.* **57**: 6700–690.
Taylor, T.N. (1973). A consideration of the morphology, ultrastructure and multicellular microgametophyte of *Cycadeoidea dacotensis* pollen. *Rev. Palaeobot. Palynol.* **16**: 157–164.
Taylor, T.N. (1978). The ultrastructure and reproductive significance of *Monoletes* (Pteridospermales) pollen. *Canad. J. Bot.* **56**: 3105–3118.
Taylor, T.N. (1980). Ultrastructural studies of pteridosperm pollen: *Nanoxanthiopollenites* Clendening and Nygreen. *Rev. Palaeobot. Palynol.* **29**: 15–21.
Taylor, T.N. (1982). Reproductive biology in early seed plants. *BioScience* **32**: 23–28.
Taylor, T.N. and Alvin, K.L. (1984). Ultrastructure and development of Mesozoic pollen: *Classopollis. Amer. J. Bot.* **71**: 575–587.
Taylor, T.N. and Daghlian, C.P. (1980). The morphology and ultrastructure of *Gothania* (Cordaitales) pollen. *Rev. Palaeobot. Palynol.* **29**: 1–14.
Taylor, T.N. and Scott, A.C. (1983). Interactions of plants and animals during the Carboniferous. *BioScience* **33**: 488–493.
Taylor, T.N. and Taylor, E.L. (1987). The ultrastructure of fossil gymnosperm pollen. *Bull. Soc. Bot. France, Actual. Bot.* **134**: 121–140.
Taylor, T.N. and Taylor, E.L. (1993). The Biology and Evolution of Fossil Plants. 982 pp. Prentice Hall, New Jersey.
Taylor, T.N., Zavada, M.S. and Archangelsky, S. (1987). The ultrastructure of *Cyclusphaera psilata* from the Cretaceous of Argentina. *Grana* **26**: 74–80.
Trevisan, L. (1971). *Dicheiropollis*, a pollen type from Lower Cretaceous sediments of southern Tuscany (Italy). *Pollen & Spores* **13**: 561–596.

Van Campo, M. (1971). Précisions nouvelles sur les structures comparées des pollens de Gymnospermes et d'Angiospermes. *Compt. Rend. Hebd. Séances Acad. Sci., Sér. D* **272**: 2071–2074.

Van Campo, M. and Lugardon, B. (1973). Structure grenue infractectale de l'ectexine des pollens de quelques Gymnospermes et Angiospermes. *Pollen & Spores* **15**: 171–187.

Whitehead, D.R. (1983). Wind pollination: some ecological and evolutionary prespectives. In: L. Real (editor). Pollination Biology, pp. 97–108, Academic Press, New York.

Wodehouse, R.P. (1935). Pollen Grains: Their Structure, Identification and Significance in Science and Medicine. 574 pp. McGraw-Hill, New York.

Xi, Y.-Z. and Wang, F.H. (1989). Pollen exine ultrastructure of extant Chinese gymnosperms. *Cathaya* **1**: 119–142.

Zavada, M.S. (1990). The ultrastructure of three monosulcate pollen grains from the Triassic Chinle Formation, Western United States. *Palynology* **14**: 41–51.

Zavada, M.S. (1991). The ultrastructure of pollen found in the dispersed sporaniga of *Arberiella* (Glossopteridaceae). *Bot. Gaz.* **152**: 248–255.

Zavada, M.S. (1992). The wall ultrastructure of fossil discoid pollen. *Bull. Torrey Bot. Club* **119**: 44–49.

Zavada, M.S. and Dilcher, D.L. (1986). Comparative pollen morphology and its relationship to phylogeny of pollen in the Hamamelidae. *Ann. Missouri Bot. Gard.* **73**: 348–381.

Zavada, M.S. and Taylor, T.N. (1986a). Pollen morphology of Lactoridaceae. *Pl. Syst. Evol.* **154**: 31–39.

Zavada, M.S. and Taylor, T.N. (1986b). The role of self-incompatibility and sexual selection in the gymnosperm-angiosperm transition: a hypothesis. *Amer. Naturalist* **128**: 538–550.

Pedersen, K.R., Friis, E. and Crane, P.R. (1994). Ultrastructure of pollen from Cretaceous angiosperm reproductive structures. In: M.H. Kurmann and J.A. Doyle (editors). Ultrastructure of fossil spores and pollen, pp. 139–159, Royal Botanic Gardens, Kew.

9. ULTRASTRUCTURE OF POLLEN FROM CRETACEOUS ANGIOSPERM REPRODUCTIVE STRUCTURES

KAJ RAUNSGAARD PEDERSEN[1], ELSE MARIE FRIIS[2] AND PETER R. CRANE[3]

[1]Department of Geology, Universitetsparken, University of Aarhus, DK-8000 Århus C, Denmark
[2]Department of Palaeobotany, Swedish Museum of Natural History, Box 50007, 104 05 Stockholm, Sweden
[3]Department of Geology, Field Museum of Natural History, Roosevelt Road at Lake Shore Drive, Chicago, IL 60605, U.S.A.

Abstract

Pollen wall ultrastructure has been studied for a number of pollen types discovered in well-preserved angiosperm reproductive organs from Cretaceous strata of Europe and North America. The study focuses mainly on pollen from Early and mid-Cretaceous forms, but pollen from a few Late Cretaceous taxa and from the non-angiospermous *Eucommiidites*-producing plant have been included for comparison. The pollen grains were observed either *in situ* within anthers or adhering to stigmatic surfaces. Early and mid-Cretaceous angiosperm pollen types examined are monoaperturate (monocolpate or trichotomocolpate) or triaperturate (tricolpate or tricolporate). They all show the same general pattern in pollen wall ultrastructure, typically with a thick, homogeneous foot layer and distinct columellae. Another striking feature is the very unevenly distributed endexine, which is typically granular and sometimes slightly layered, but never distinctly lamellated. In the monoaperturate pollen the endexine is very thin in the non-apertural regions, but strongly thickened under the apertures. The endexine in the triaperturate forms is also unevenly distributed, but particularly in Early Cretaceous forms it is much thicker in the non-apertural regions than is seen in the monoaperturate grains. Thus the differences between apertural and non-apertural endexine are not as marked in the triaperturate grains. The distribution and structure of the endexine clearly distinguish the Early Cretaceous pollen found *in situ* in angiosperm reproductive organs from pollen of other seed plants, which typically has a thick, lamellated and more or less evenly distributed endexine (as exemplified by the *Eucommiidites* pollen).

Introduction

Dispersed pollen is an invaluable source of data for understanding patterns of diversification and extinction, as well as the geographical distribution of different plant groups through time. However, the number of dispersed grains of uncertain systematic affinity is still relatively large, and knowledge about the plants that produced many distinct and stratigraphically important pollen types remains rudimentary. Some of the unassigned dispersed grains may belong to hitherto unknown extinct plant taxa, or they may belong to plant groups that are already relatively well known from macrofossil remains, but for which pollen is either unrecorded or poorly characterized. The situation is further confused by the fact that several plant groups produce rather generalized, monocolpate pollen that may be difficult to distinguish based on both light microscope and scanning electron microscope observations. Features of the pollen wall as observed with transmission electron microscopy provide additional characters of systematic importance, but so far relatively few ultrastructural studies of pollen *in situ* in Cretaceous angiosperm reproductive organs have been carried out. This has limited the extent to which dispersed pollen may be used to clarify the timing of evolutionary events and to test phylogenetic models hypothesized from studies of extant plants and macrofossil remains.

During the past few years numerous angiosperm reproductive organs have been discovered from Cretaceous strata of Europe and North America, and several of these have pollen grains either *in situ* within anthers or on the stigmatic surfaces. The pollen grains found *in situ* in several of these reproductive structures have yielded important information on the systematic relationships of a variety of pollen grains otherwise known only from the dispersed pollen record. In the present work we summarize the ultrastructural features of pollen from these previously described reproductive organs as well as from a number of new forms.

Material

Cretaceous reproductive organs with pollen *in situ* have been reported from numerous localities in Europe and North America (e.g., Friis 1985a, b; Friis et al., 1988, 1991, 1992; Friis and Crane, 1989; Endress and Friis, 1991; Crane et al., 1989, 1993; Drinnan et al., 1991; Pedersen et al., 1991, in press; Crepet et al., 1992; Friis and Pedersen, in press). Material described in this work has been collected in the Potomac Group sediments of the Atlantic Coastal Plain of the USA (Virginia and Maryland) and from the western Portuguese basin in the Estremadura and Beira Litoral regions, Portugal. The bulk of the material is of Late Cretaceous age, while Early Cretaceous reproductive structures with pollen *in situ* are more rare. In this paper we focus on angiosperm reproductive structures and their pollen from Early and mid-Cretaceous rocks, but a few relevant Late Cretaceous angiosperm taxa, as well as one non-angiospermous seed plant (the *Eucommiidites*- plant), have been included for comparison.

Eastern North America: The Potomac Group sediments of the Atlantic Coastal Plain have been intensively studied for dispersed pollen and leaves (e.g., Brenner, 1963; Doyle, 1969; Doyle and Hickey, 1976; Hickey and Doyle, 1977). A pollen zonation for the sequence was proposed by Brenner (1963) and Doyle and Robbins (1977) covering the time interval from the late Barremian or early Aptian to the mid-Cenomanian (Doyle, 1992). The Potomac Group sediments have also yielded rich assemblages of reproductive organs, including well-preserved angiosperm flowers (Friis et al., 1988; Crane et al., 1989; Drinnan et al., 1990, 1991; Pedersen et al., 1991) and a number of reproductive structures assignable to other seed plants (e.g., the *Eucommiidites*-plant and the *Brenneria*-plant: Brenner, 1967; Pedersen et al., 1989, 1993).

The material described here is from two localities in Virginia (Puddledock, Bank near Brooke) and five localities in Maryland (Kenilworth, Red Point, West Brothers, Bull Mountain, Mauldin Mountain). The localities range in age from the early or middle Albian to the earliest Cenomanian (early-middle Albian, Pollen Zone IIA–B, Kenilworth; middle-late Albian Pollen Zone IIB, Bank near Brooke, Puddledock, Red Point, West Brothers; late Albian Pollen Zone IIC, Bull Mountain; early Cenomanian, Pollen Zone III, Mauldin Mountain). In addition to the Potomac Group floras of Early and mid-Cretaceous age, material from the Late Cretaceous (Campanian) Neuse River locality of North Carolina has been included to show the diversity in the platanoid group. The non-angiospermous seed plant, *Erdtmanitheca texensis*, described from the Early Cenomanian of the Arthurs Bluff locality, Texas (Pedersen et al., 1989), was included for comparison.

Portugal: The material described here has been extracted from two mid-Early Cretaceous floras in the Western Portuguese Basin (Catefica near Torres Vedras, Estremadura, and Famalicão southeast of Leiria, Beira Litoral). Several common elements in the floras indicate that they are of approximately the same age, and although the precise age assignment is still unsettled they may be as old as Barremian and probably not younger than the Aptian. A third flora probably somewhat older than the Catefica and Famalicão floras was recovered from Torres Vedras. All three floras have yielded a large number of angiospermous reproductive organs. Most are fruits and seeds, but the material also includes some flowers and dispersed stamens. Pollen grains are only rarely preserved in the flowers, many of which are preserved as charcoal, and so far only a few pollen types have been discovered from these floras. The pollen grains are monoaperturate or more rarely tricolpate, and reticulate, with the exception of a finely striate and monocolpate pollen type that was recovered in a small epigynous flower with massive, valvate stamens. The most spectacular of the Early Cretaceous reproductive structures recovered so far in these Portuguese floras is a minute conical reproductive axis composed of densely packed anthers arranged in whorls. The structure of the axis indicates that it maybe an inflorescence consisting of unistaminate and naked male flowers, arranged in whorls along a short axis. Pollen grains observed on the reproductive structure are monoaperturate and reticulate. This taxon is represented by a single specimen only, and so far no preparations for ultrastructural studies of the pollen wall have been made.

In addition to the mid-Early Cretaceous reproductive organs a large number of floral structures with pollen *in situ* have been discovered from Late Cretaceous rocks in the Beira Litoral region. They include several rosid taxa (e.g., the combretaceous *Esgueiria*, Friis et al., 1992) and a variety of Normapolles forms, which will be described elsewhere.

Sweden: A number of well-preserved angiosperm reproductive organs have been described from Santonian-Campanian sediments at the Åsen locality north of Lake Ivösjön in Scania, southern Sweden (e.g., Friis 1985a, b, 1990; Friis et al., 1988). The floral assemblages recovered at Åsen are diverse, with fossil taxa related to the Hamamelididae and Rosidae being particularly well represented. Pollen wall ultrastructure has been reported previously from two platanoid taxa (Friis et al., 1988), from the chloranthoid fossil, *Chloranthistemon endressii* (Crane et al., 1989), and from several Normapolles-type pollen grains found *in situ* within dispersed anthers (Skarby, 1986; Skarby et al., 1990). In this work we include only one of the pollen types from the Åsen locality, which was described from platanoid inflorescences and stamens (Friis et al., 1988).

Denmark: A number of well-preserved plant fossils including seeds and vegetative fragments were discovered from Upper Berriasian-Valanginian strata of Bornholm, Denmark (C. Nielsen A/S sand pit southeast of Rønne; Friis and Pedersen, 1990). So far no fragments of angiosperms have been identified, but several different non-angiospermous seeds with pollen grains *in situ* in the micropyles have been recognized (Pedersen et al., 1989). In the present study pollen grains of the *Eucommiidites* type extracted from the seeds of *Erdtmanispermum balticum* have been included for comparison as an example of an extinct non-angiospermous seed plant.

Results

A. Ultrastructure in Cretaceous monoaperturate angiosperm pollen

Trichotomocolpate pollen from unnamed fruits – Kenilworth (*?early-middle Albian, Potomac Group, Pollen Zone IIA–B*). The pollen grains are minute, spherical, about 14 μm in diameter and have a triradiate colpus. The colpus is sharply delimited and open, with

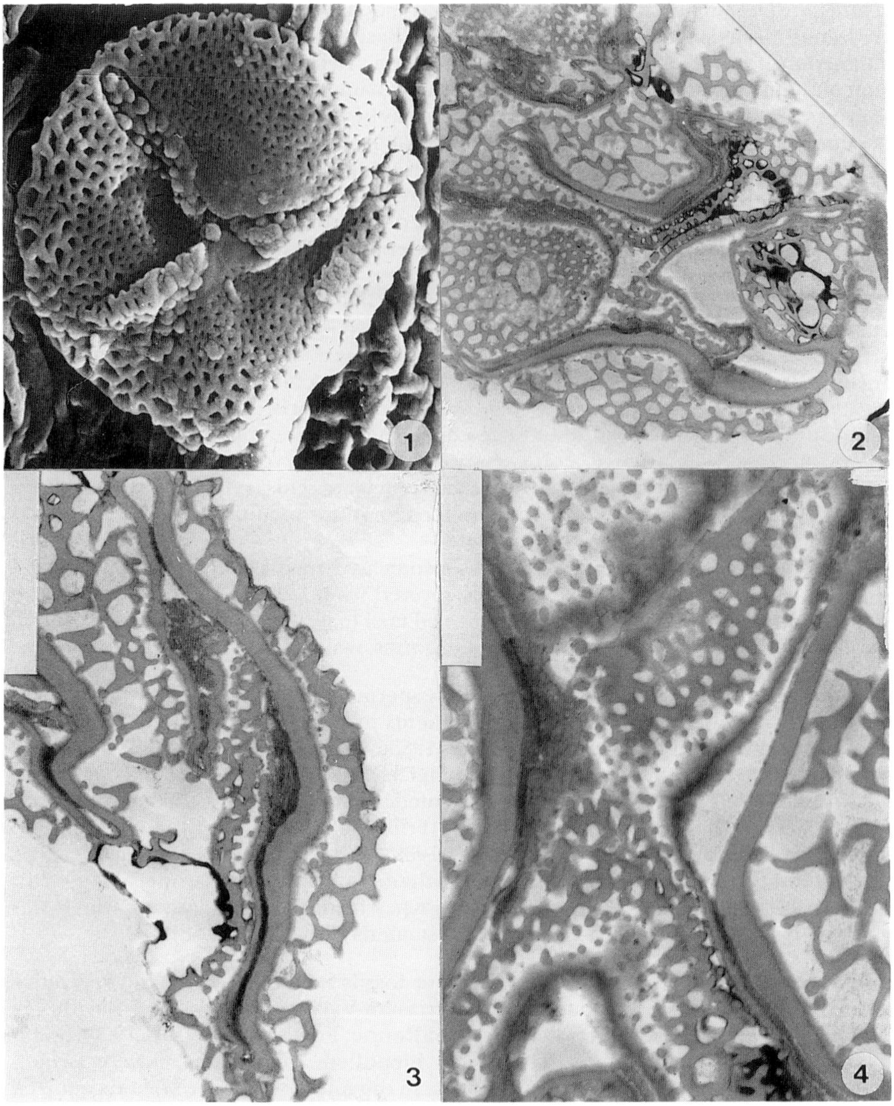

Figs. 1–4. Trichotomocolpate pollen from unnamed fruits – Kenilworth (Maryland, USA, Potomac Group – early or middle Albian). **1**: SEM-micrograph showing distal part of pollen with triradiate aperture and granular aperture membrane, × 4000. **2–4**: TEM-micrographs showing pollen wall ultrastructure with reticulate tectum, columellae, thick foot layer and darker staining endexine (**2**: × 4200; **3**: × 7000; **4**: × 10000).

arms that have rounded ends and may differ in length (Fig. 1). The colpus membrane is granular to verrucate with exine elements of varying size and shape. The tectum is coarsely reticulate to foveolate with irregular areas at the proximal pole and around the colpus, where the tectum is finely foveolate to almost non-perforate. The lumina vary from distinctly angular to rounded (Figs. 2–4) with a maximum dimension of about 1 μm. The surface of the muri is smooth and rounded.

Ultrathin sections of the pollen wall in some areas (probably apertural regions) show a moderately thick, granular endexine, about 0.5 μm, which may be weakly layered. In other areas the endexine is extremely thin (Fig. 3). The foot layer of the ektexine is thick, about 0.5 μm, and homogeneous. The columellae are long, about 0.3 μm, simple and closely aligned below the muri, which are about 0.3 μm high (Fig. 4).

Dispersed trichotomocolpate pollen is known from several mid-Cretaceous palynofloras, and dispersed grains from the upper part of Pollen Zone IIB in the Potomac Group were studied in detail by Walker and Walker (1984). These grains, referred to as *Liliaciidites* sp. 2, are much larger than the material reported here, being about 30 μm in diameter, but otherwise they are similar in colpus outline and tectum configuration. In particular, they have non-perforate areas at the proximal pole and around the arms of the aperture as in the Kenilworth material. In pollen wall ultrastructure, *Liliaciidites* sp. 2 differs in having a much thinner foot layer, short columellae and a very thick tectum with high, sharply pointed muri in section.

The trichotomocolpate pollen from Kenilworth was observed consistently occurring on the stigmatic surfaces of small fruits with a sessile and indistinct stigmatic area. The fruits are unilocular with a single anatropous seed, and the fruit wall is membranous with scattered resin bodies. They are similar in general morphology to the fruits of *Couperites mauldinensis* described from the early Cenomanian Mauldin Mountain flora (Pedersen et al. 1991, see below), but differ in details of seed wall structure and perhaps seed attachment. The systematic affinity of the Kenilworth fruits and pollen is clearly with some group of magnoliid dicotyledons or possibly basal monocotyledons, but their precise relationships are still uncertain.

Monocolpate pollen of the ***Clavatipollenites*** *type from* ***Couperites*** *fruits –Mauldin Mountain (early Cenomanian, Potomac Group, Pollen Zone III)*. The monocolpate pollen grains of the *Clavatipollenites*-type are elongated to rounded in equatorial outline, about 22–25 μm long, and tectate-reticulate. The colpus is clearly delimited and extends almost the full length of the grain (Fig. 5). The reticulum has irregular, angular to rounded lumina of variable size from about 0.5 to 1 μm. Muri are about 0.5 μm wide, rounded in cross-section and ornamented with more or less elongated micro-verrucae often arranged in two marginal rows. The colpus membrane is typically covered with verrucae or towards the ends with fully developed columellae and tectum (Fig. 6).

Ultrathin sections of the pollen wall show a very thin, about 0.005 μm, endexine in non-apertural areas, which thickens strongly towards the aperture to form a granular and weakly laminar layer, up to about 1.3 μm thick (Figs. 6–8). The foot layer of the ektexine is very thick, up to about 0.8 μm. The columellae are scattered, about 0.5 μm high, expanding in diameter towards the top. Muri are about 0.7 μm high.

Dispersed grains of the *Clavatipollenites*-type have been widely reported from Cretaceous strata of Europe and North America and are among the earliest pollen that can securely be attributed to the angiosperms (e.g., Couper, 1958; Muller, 1970; Walker and Walker, 1984; Pedersen et al., 1991).

The *Clavatipollenites* type pollen grains from the Mauldin Mountain flora were observed adhering to the stigmatic surfaces of small chloranthoid fruits described as *Couperites mauldinensis* (Pedersen et al., 1991). These fruits are unilocular and uniovulate with a membranous fruit wall containing resin bodies. Seeds are anatropous, a character that distinguishes the fossils from all living members of the Chloranthaceae (Pedersen et al., 1991).

The pollen wall ultrastructure of the *Couperites* pollen is comparable to that of living Chloranthaceae (e.g., Walker and Walker 1984) and to pollen observed *in situ* in the chloranthoid fossil *Chloranthistemon endressii* from the Santonian/Campanian of southern Sweden (Crane et al., 1989).

Monocolpate/trichotomocolpate pollen of ***Clavatipollenites*** *type from unnamed anther – Famalicão (mid-Early Cretaceous)*. The pollen grains are almost circular in equatorial outline, about 15 μm in diameter, and semitectate-reticulate. The aperture configuration varies from

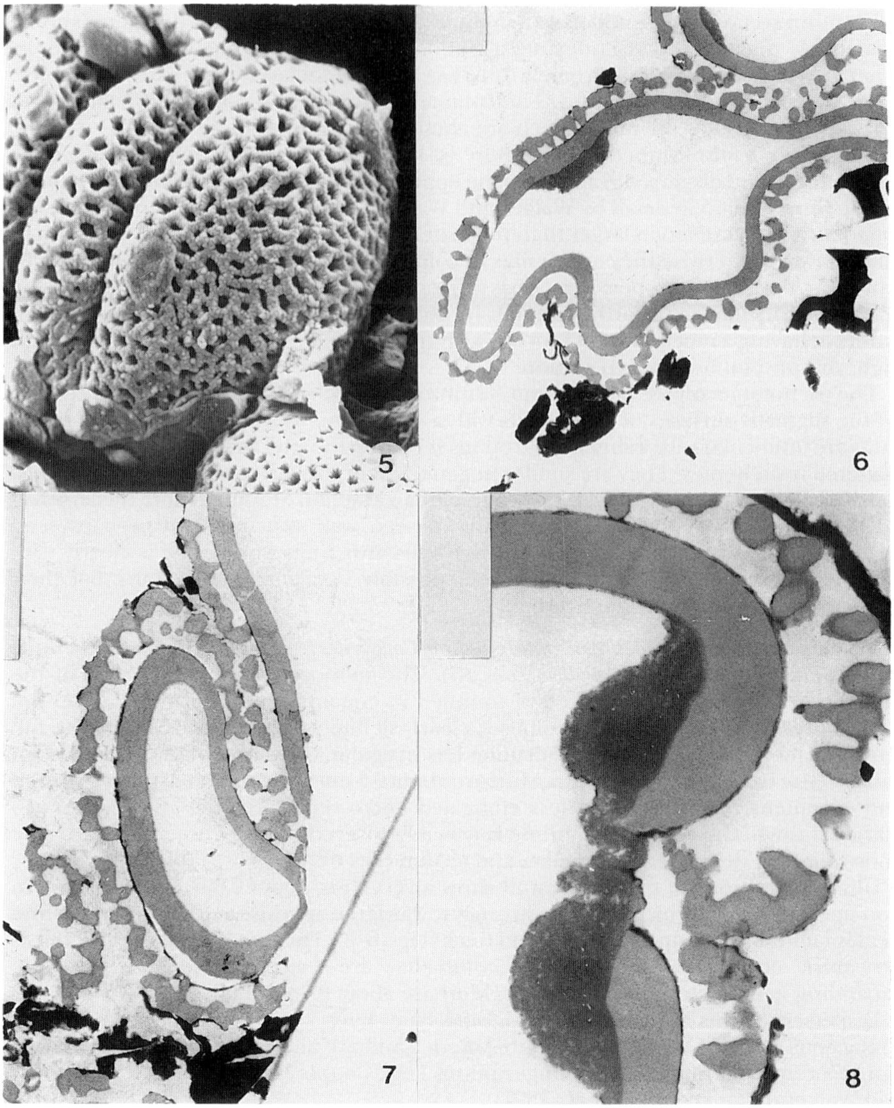

Figs. 5–8. Monocolpate pollen of the *Clavatipollenites* type from *Couperites* fruits – Mauldin Mountain (Maryland, USA, Potomac Group – early Cenomanian). **5**: SEM-micrograph of pollen showing distal part of grain with colpus area, × 2750. **6–8.** TEM-micrographs showing pollen wall ultrastructure with reticulate tectum, columellae, thick foot layer and darker staining endexine. **6**: Section in marginal part of colpus, × 2800. **7**: Longitudinal section, × 4200. **8**: Cross-section through colpus showing details of endexine, × 10000.

monocolpate to trichotomocolpate (Fig. 9). The colpus is clearly delimited and has a verrucate membrane. The reticulum is fine and has irregular angular to rounded lumina of variable size from about 0.1 to 0.6 μm in diameter. Muri are about 0.3 μm wide and rounded in cross-section with a beaded ornamentation (Fig. 10).

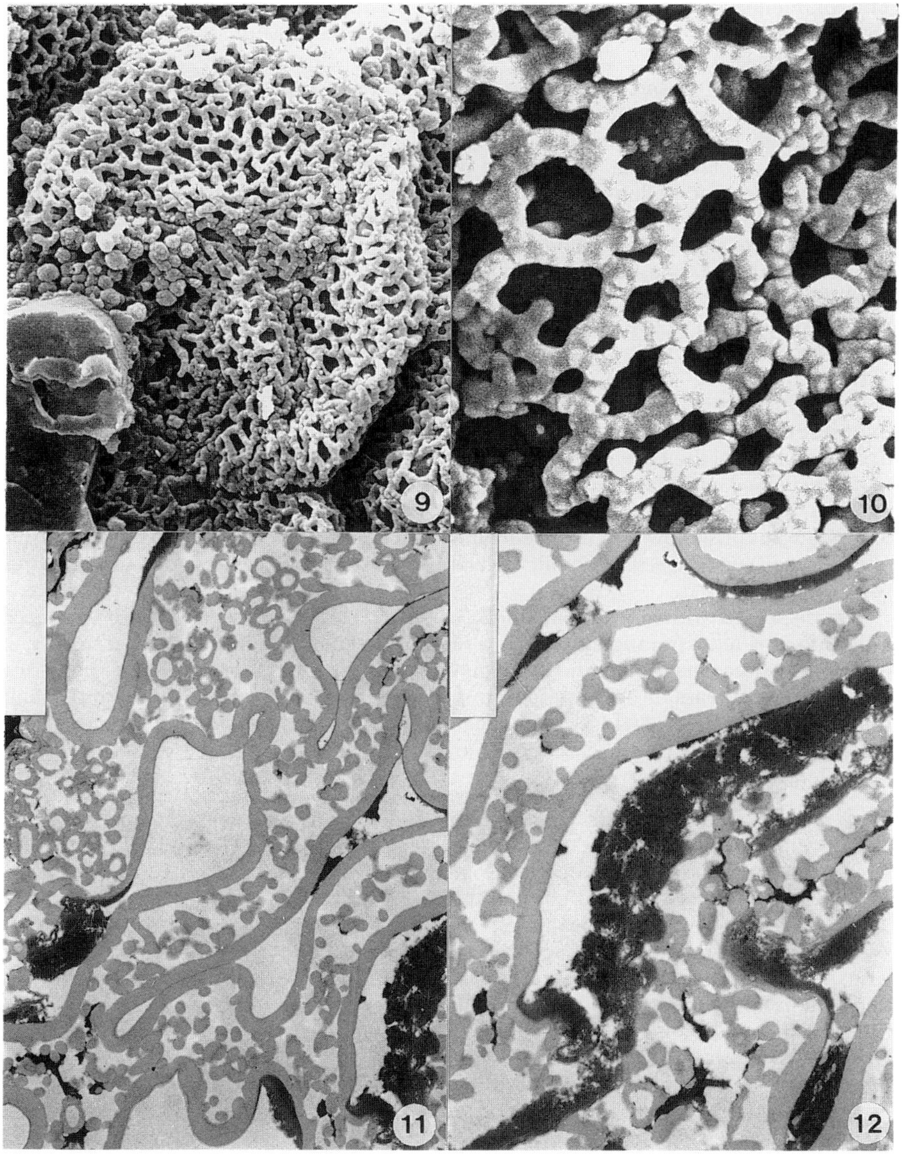

FIGS. 9–12. Monocolpate/trichtomocolpate pollen of *Clavatipollenites* type from unnamed anther – Famalicao (Beira Litoral, Portugal, mid-Early Cretaceous). **9**: SEM-micrograph of pollen showing triradiate aperture and reticulate tectum, × 4000. **10**: SEM-micrograph of pollen showing details of reticulum and surface ornamentation, × 15000. **11–12**: TEM-micrographs showing open reticulate tectum, columellae, thick foot layer and darker staining endexine. **11**: Cross-section through aperture, × 7000. **12**: Longitudinal section through aperture, × 10000.

Ultrathin sections of the pollen wall show an extremely thin endexine in non-apertural regions. The endexine thickens strongly towards the apertures forming a granular and weakly laminar layer (Figs. 11–12). The foot layer of the ektexine is thick, about 0.3 μm.

The columellae are simple and scattered, about 0.3 μm high. Muri are about 0.3 μm high. Orbiculae occur along the inner margin of the pollen sac and scattered between the pollen grains (Fig. 11).

The pollen grains from Famalicão are similar to some dispersed pollen assigned to the *Clavatipollenites* complex, which includes mono- as well as trichotomocolpate pollen with a semitectate- to tectate-reticulate pollen wall and beaded, spiny or micro-verrucate surface ornamentation (e.g., Walker and Walker, 1984). They are distinguished from the *Clavatipollenites* type pollen associated with *Couperites mauldinensis* (see above) by their smaller size, the more open and much finer reticulum, as well as the much thinner wall layers. They are also distinguished from the trichotomocolpate pollen from the Potomac Group sediments (Zone IIB) assigned to aff. *Clavatipollenites* sp. 1 (Walker and Walker, 1984) by their smaller size and much finer reticulum, but are similar to this material in ornamentation of the muri. Several dispersed pollen types assignable to the *Clavatipollenites* complex have also been reported from the Early Cretaceous of Portugal (Groot and Groot, 1962; Chapman, 1987; Trincão, 1990). *Clavatipollenites hughesii* recorded from the Valanginian to the late Albian in Portugal (Trincão, 1990) differs from the present material mainly in its larger size. A more detailed comparison of surface ornamentation of the Portuguese material is not possible based on the light micrographs of the dispersed material. *Clavatipollenites* sp.1 from the late Aptian of Portugal (Trincão, 1990) is also somewhat larger than the pollen described here from Famalicão, and SEM-micrographs show a denser reticulum with surface ornamentation more like that of *Clavatipollenites hughesii*, with micro-verrucae that are often arranged in two marginal rows.

The dispersed anther from Famalicão containing the *Clavatipollenites* type pollen is bithecate and tetrasporangiate. It has a distinct apical extension of the connective as observed in other magnoliid and hamamelidid stamens from the Cretaceous. The floral assemblages of Famalicão, Catefica and Torres Vedras comprise numerous small, unilocular and one-seeded fruits similar in general structure to those of *Couperites*. It is possible that some of these fruits belong to a *Clavatipollenites*-producing plant, but so far no pollen has been observed adhering to the stigmatic surface of the fruits.

Monocolpate pollen from unnamed anther – Catefica (mid-Early Cretaceous). The pollen grains are circular to slightly elongate in equatorial outline, about 15–20 μm long, often irregularly compressed. The colpus is broad and only weakly delimited, sometimes with a slightly branched, star-shaped configuration (Fig. 13). The pollen is tectate-reticulate. The reticulum is irregular, typically with small, angular lumina and broad muri. In some areas where the tectum is absent columellae may be fully developed or missing. In most cases the absence of the tectum appears to be a result of loose attachment of the reticulum and columellae to the foot layer, but in some areas it appears to be an original feature in the pollen wall. Over the colpus membrane the pollen wall is also tectate, but the reticulum is suppressed.

Ultrathin sections of the pollen wall show an inner darker staining and granular endexine. It is very thin in the non-aperturate areas and strongly thickened and weakly layered under the aperture (Figs. 14–16). The foot layer of the ektexine is very thick, about 0.5 μm and homogeneous. The columellae are regularly aligned and relatively long, about 0.5 μm, except in the colpus area where they are suppressed. The muri are broad with a smooth and rounded surface.

The monocolpate pollen grains from Catefica were recovered in small, dispersed stamens with tetrasporangiate and bithecate elongated anthers. Anther dehiscence was apparently longitudinal. The connective has a distinct apical extension as observed in many other Cretaceous magnoliid and hamamelidid floral structures. So far the complete floral structure producing these grains is unknown.

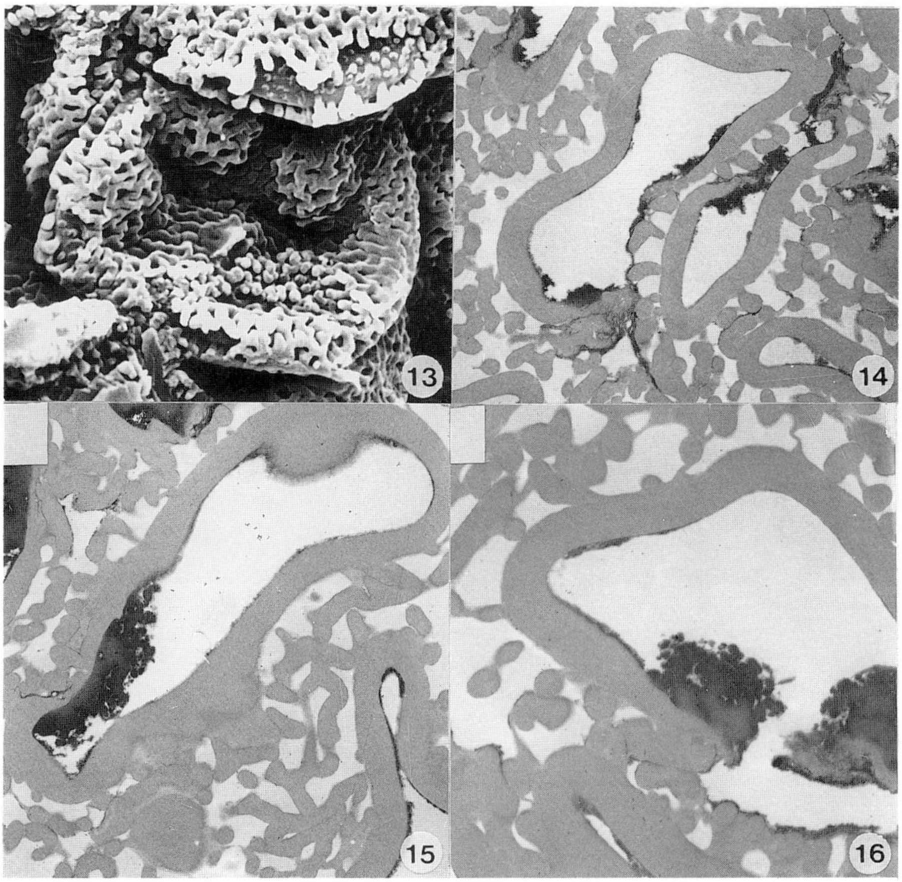

FIGS. 13–16. Monocolpate pollen from unnamed anther – Catefica (Estremadura, Portugal, mid-Early Cretaceous). **13**: SEM-micrograph of pollen showing aperture region and loosely attached reticulate tectum, × 3000. **14–16**: TEM-micrographs showing pollen wall ultrastructure with open reticulate tectum, columellae, thick foot layer and darker staining endexine. **14**. Longitudinal section through colpus, × 5600. **15**: Cross-section through colpus, × 7000. **16**: Cross-section through colpus showing details of endexine, × 10000.

B. Ultrastructure in Cretaceous triaperturate angiosperm pollen

Tricolpate pollen from ***Aquia*** *– Bank near Brooke (middle Albian, Potomac Group, Pollen Zone IIB)*. The pollen grains are tricolpate and weakly prolate, about 12 μm long. The colpi are about two thirds the length of the grains. They are broad with rounded ends and a verrucate colpus membrane. The tectum is foveo-reticulate with broad muri (Fig. 17). Lumina are often elongated and angular in the inter-apertural areas, while the pollen wall around the colpus margins is foveolate to almost imperforate.

Ultrathin sections of the pollen wall show a darker staining and moderately thick (about 0.1 μm), granular endexine which thickens to about 1 μm under the apertures and becomes distinctly layered (Figs. 18–19). The ektexine is thick with a homogeneous foot layer, about 0.5 μm. The columellae are few, simple and short, about 0.3 μm, and the perforated tectum is about 0.3 μm thick. Orbiculae occur along the inner margin of the pollen sac together with incompletely developed pollen.

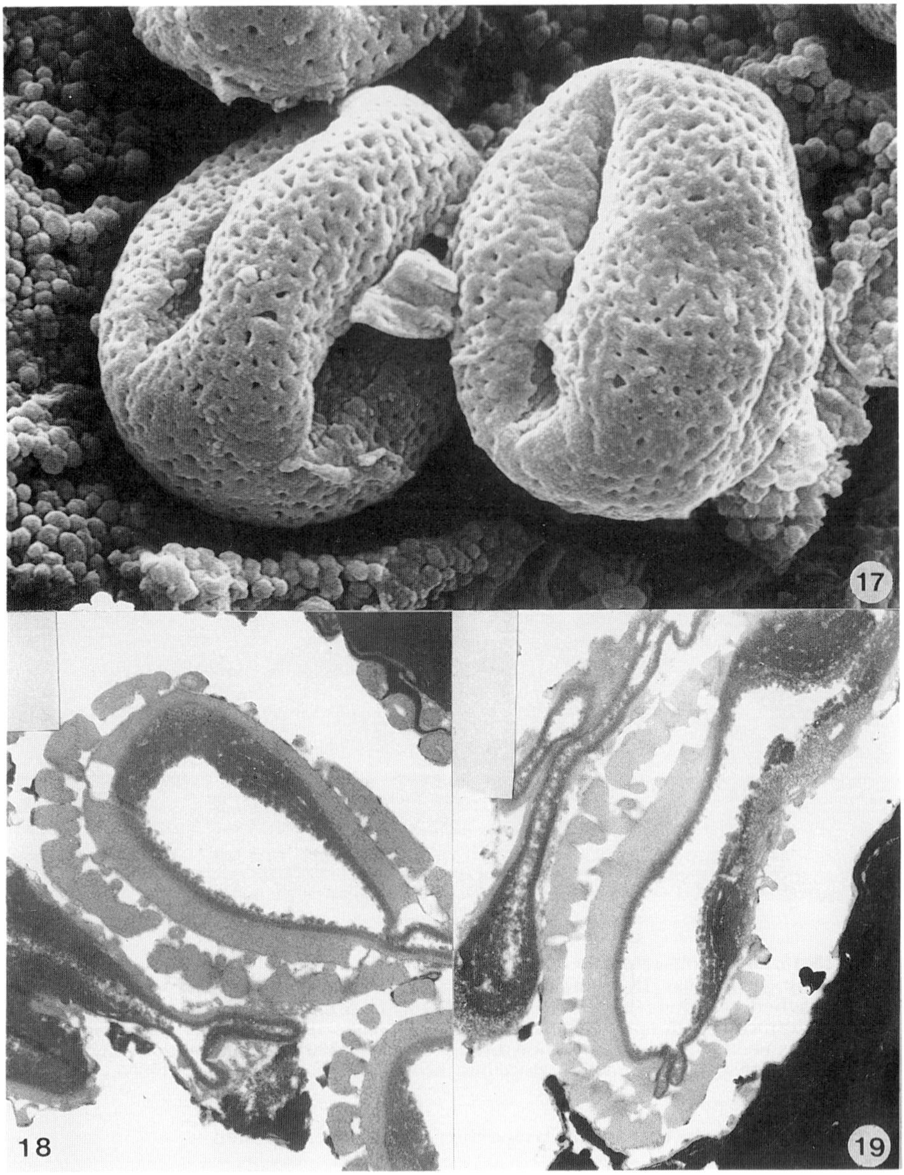

FIGS. 17–19. Tricolpate pollen from *Aquia* – Bank near Brooke (Virginia, USA, Potomac Group – middle Albian). **17**: SEM-micrograph of pollen grains showing colpi and foveolate tectum, × 5000. **18–19**: TEM-micrographs showing pollen wall ultrastructure with foveolate, thick tectum, columellae, thick foot layer and darker staining endexine. **18**: Cross-section through colpus in equatorial region, × 7000; **19**: Longitudinal section through colpus, × 7000.

The pollen grains from the Bank near Brooke locality occur in staminate inflorescences of platanaceous affinity. The flowers have a pentamerous androecium. The stamens consist of bithecate and tetrasporangiate anthers borne on long filaments. Extensive connective tissue is developed between the pollen sacs and also apically to form

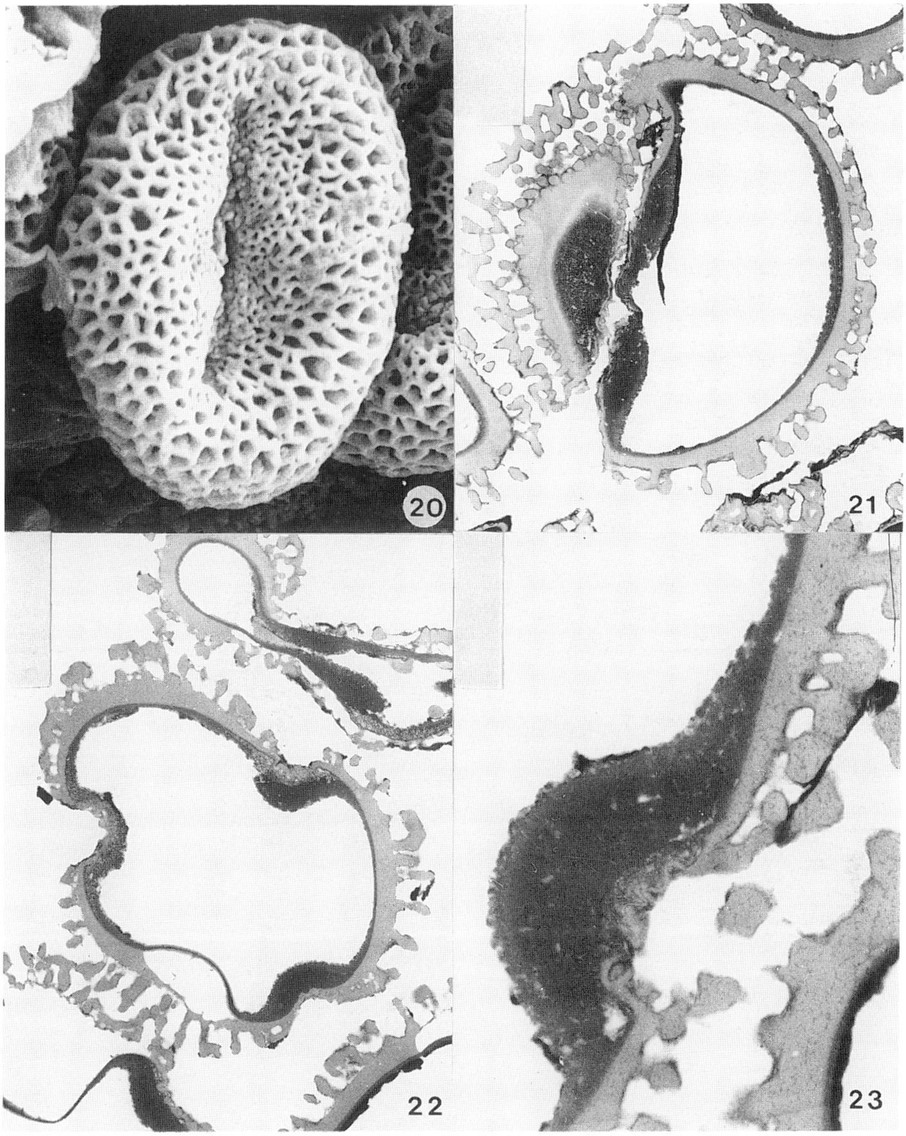

Figs. 20–23. Tricolporate pollen from *Hamatia* – Bull Mountain (Maryland, USA, Potomac Group – late Albian). **20**: SEM-micrograph of pollen in equatorial view showing colpus and reticulum, × 5000. **21–23**: TEM micrographs showing pollen wall ultrastructure with reticulate tectum, columellae, thick foot layer and darker staining endexine. **21**: Longitudinal section through colpus and endopore, × 5600. **22**: Cross-section through colpi in equatorial region, × 5600. **23**: Cross-section of colpus showing details of endexine, × 18000.

a dome-shaped extension. The staminate flowers are referred to as *Aquia brookensis* and co-occur with distinct platanoid pistillate heads described as *Platanocarpus brookensis* and leaves assigned to the extinct genus *Sapindopsis* (Crane et al., 1993). They comprise some of the earliest platanoid reproductive structures so far reported and are distinguished from other Cretaceous and Tertiary platanoid fossils by the long filament of the stamen and the foveo-reticulate tectum of the pollen (Crane et al., 1993).

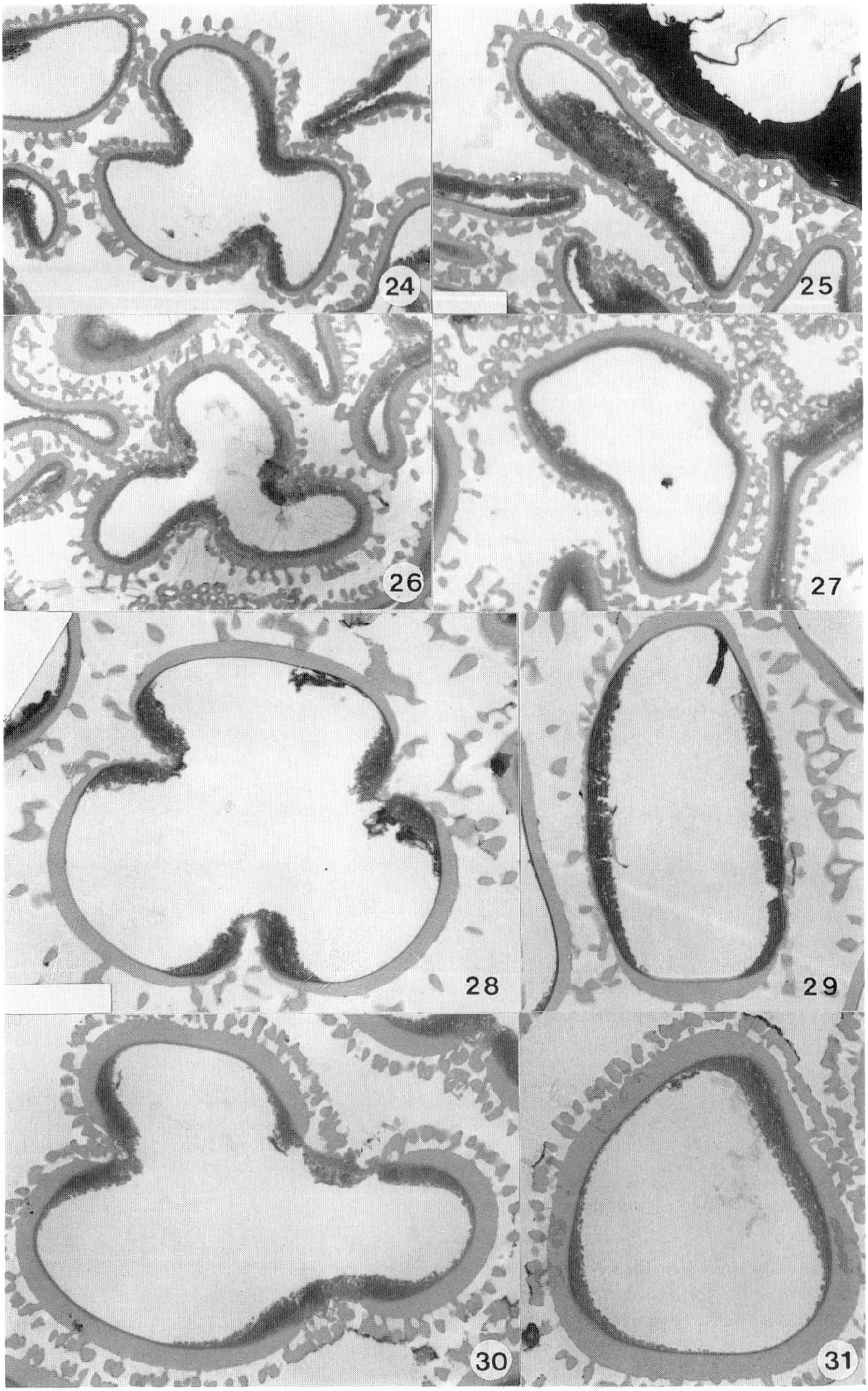
24
25
26
27
28
29
30
31

Tricolporate pollen from ***Hamatia*** *– Bull Mountain (late Albian, Potomac Group, Pollen Zone IIC)*. The pollen grains are tricolporate, tectate-reticulate and weakly prolate, about 12 μm long. The colpi are about two thirds the length of the grain and have a finely verrucate colpus membrane. The reticulum has elongated polygonal lumina decreasing in size towards the apertures (Fig. 20). The muri are narrow and parallel-sided.

Ultrathin sections of the pollen wall show a thin, about 0.1 μm, darker staining, granular endexine in non-apertural areas that thickens to about 0.7 μm under the apertures (Figs. 21–23). The foot layer of the ektexine is thick, up to 0.5 μm, and homogeneous. The columellae are high, about 0.5 μm, simple and regularly spaced below the muri (Fig. 22). The endopore is circular in outline and about 1 μm in diameter. The endopore is rarely visible in SEM since the colpi are usually infolded, but it is very distinct in light microscopy and in some of the TEM sections (Fig. 21).

The pollen grains were found *in situ* in small unisexual staminate heads assigned to *Hamatia elkneckensis*. The individual staminate flowers have a pentamerous androecium and distinct hooked apical extension of the connective. Anthers are bithecate and tetrasporangiate with valvate dehiscence. They are associated with pistillate flowers described as *Platanocarpus elkneckensis* (Pedersen et al., in press). In general organization the staminate flowers are very similar to those of *Platananthus*, and are thought to be part of the diverse Cretaceous platanoid complex. The pollen differs, however, in having tricolporate aperture configuration in contrast to the simple tricolpate apertures of *Platananthus* and *Platanus* (Pedersen et al., in press).

Tricolpate pollen from ***Platananthus*** *– Red Point, West Brothers, Neuse River, Åsen (late Albian to Campanian)*. Staminate flowers of platanaceous affinity referred to the extinct genus *Platananthus* have been recorded from numerous localities in Europe and North America in Cretaceous and Lower Tertiary strata. In the present work, pollen grains from four different species and from four different localities are described: the Red Point (*Platananthus* sp., Potomac Group, Pollen Zone IIB) and West Brothers localities in Maryland (*P. potomacensis*, Potomac Group, Pollen Zone IIB), the Neuse River locality of North Carolina (*P. hueberi*, Campanian) and the Åsen locality, Scania (*P. scanicus*, Santonian-Campanian) (e.g., Friis et al., 1988).

Platananthus pollen grains are all tricolpate, weakly prolate and tectate-reticulate. The various species differ in configuration of the reticulum. They also differ in size, with the smallest form being from the Early Cretaceous and larger forms occurring in the Late Cretaceous (*Platananthus* sp. from Red Point: 6μm; *P. potomacensis*: 10 μm; *P. hueberi* and *P. scanicus* about 15 μm).

Ultrathin sections of the pollen wall show a darker staining granular endexine with distinct thickenings under the colpi. In the two Early Cretaceous species the endexine is also relatively thick in the non-apertural regions (Figs. 24–27), while in the two Late Cretaceous species it is very thin in the non-apertural regions (Figs. 28–31). The foot layer is homogeneous and typically thick, constituting about half of the wall thickness in most of the species, but is thinner in *P. hueberi*. Columellae are relatively high and closely spaced. The tectum is thick, and muri of the Early Cretaceous forms show a rounded profile in cross-section, while muri of the Late Cretaceous species are rounded to triangular in cross-section.

Figs. 24–31. Tricolpate pollen from *Platananthus*. TEM-micrographs showing reticulate tectum, columellae, thick foot layer and darker staining endexine, × 5600. **24, 26, 28, 30**: Cross-sections through colpi in equatorial region. **25, 27, 29, 31**: Sections more or less longitudinal through colpi. **24–25**: *Platananthus* sp. from Red Point (Maryland, USA, Potomac Group — late Albian). **26–27**: *Platananthus marylandensis* from West Brothers (Maryland, USA, Potomac Group — late Albian). 28–**29**: *Platananthus hueberi* from Neuse River (North Carolina, USA, Campanian). **30–31**: *Platananthus scanicus* from Åsen (Scania, Sweden, Santonian/Campanian).

The fossil genus *Platanocarpus* was first established on the basis of petrified material from the Early Tertiary (Manchester, 1986). It has subsequently been recognized from many Cretaceous localities and several new species have been described from the Cretaceous (e.g., Friis et al., 1988). The staminate flowers of *Platananthus* are borne in small unisexual and globose heads. Individual flowers consist of five stamens. Stamens consist of a short filament and a bithecate, tetrasporangiate anther typically with a pronounced apical extension of the connective, and with valvate dehiscence. The staminate flowers mentioned here from the Cretaceous are all found in association with pistillate flowers ascribed to the extinct genus *Platanocarpus*.

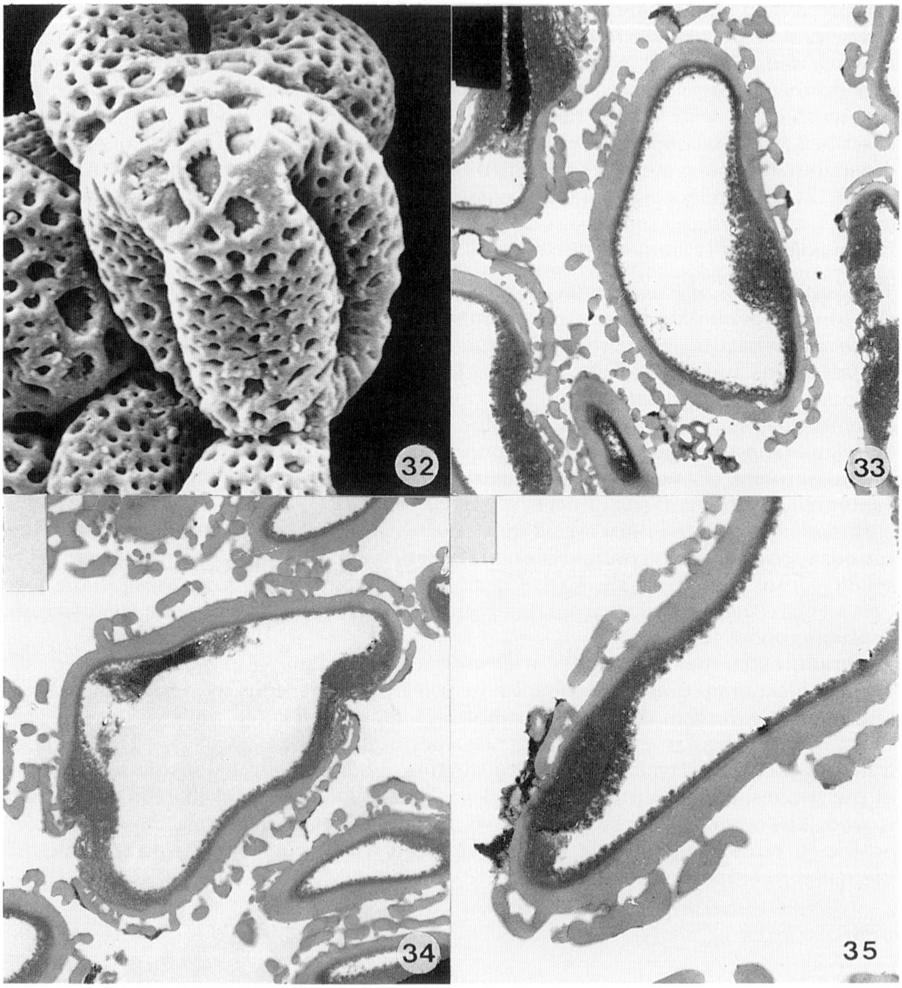

Figs. 32–35. Tricolpate pollen from unnamed stamens – Puddledock (Virginia, USA, Potomac group – middle Albian). **32**: SEM-micrograph of pollen showing larger lumina of the reticulum in the polar region, × 4000. **33–35**: TEM-micrographs showing pollen wall ultrastructure with reticulum, columellae, thick foot layer and darker staining endexine. **33**: Longitudinal section through colpus, × 5600. **34**: Cross-section through colpi near equatorial region, × 5600. **35**: Details of endexine near colpus, × 10000.

Tricolpate pollen from unnamed stamens – Puddledock (middle Albian, Potomac Group, Pollen Zone IIB). The pollen grains are tricolpate and prolate, about 12 to 15 μm long, with a characteristic, heterogeneous reticulum (Fig. 32). The colpi are moderately long, about two thirds the length of the grain. The tectum is reticulate-foveolate with larger lumina, up to about 1.5 μm in diameter, at the flat polar areas, and much smaller lumina in the equatorial regions of the grain. Areas along the colpi are foveolate. The muri are broad with small foveae at their junctions.

Ultrathin sections of the pollen wall show a darker staining granular endexine, which is about 0.1 μm thick in non-apertural regions and thickens considerably towards the apertures (Figs. 33–35). The foot layer is homogeneous and relatively thick, up to about 0.5 μm. The columellae are simple, about 0.2 μm high, and closely spaced below the muri. The tectum is thick, about 0.3 μm, with smooth and rounded surfaces of the muri.

The pollen grains were observed *in situ* in dispersed stamens. The stamens have bithecate and tetrasporangiate anthers basifixed on a slender filament, and distinctly valvate dehiscence. The apical part has a marked dome-shaped extension of the connective.

Dispersed pollen grains similar to those described here from Puddledock have been recorded from the Early Cretaceous (Albian) of Oklahoma (Hedlund and Norris, 1968) and assigned to the dispersed pollen species *Retitricolpites fragosus*. LM illustrations of this material show the same characteristic reticulation and similar colpus arrangement. Similar dispersed pollen grains were also reported by Doyle and Robbins (1977) from the Pollen Zone IIB of the Potomac Group of the Atlantic Coastal Plain and assigned to "*Retitricolpites*" *fragosus*.

Tricolpate pollen from **Spanomera** *– West Brothers (late Albian, Potomac Group, Pollen Zone IIB) and Mauldin Mountain (early Cenomanian, Potomac Group, Pollen Zone III)*. Two species of the angiosperm reproductive structure *Spanomera* have been described from Maryland: *S. marylandensis* from the West Brothers locality and *S. mauldinensis* from the Mauldin Mountain locality (Drinnan et al., 1991).

Pollen grains observed in the two species are tricolpate and prolate, about 18–20 μm, and semitectate. The colpi are clearly delimited and extend for about two thirds the length of the grains. The colpus membrane is verrucate and has a granular ektexinous layer below the verrucae. In *S. marylandensis* the reticulum consists of widely spaced, irregularly undulating muri that form a striate-rugulate sculpture pattern. In *S. mauldinensis* exine ornamentation is similar but the reticulum is more densely striate.

Ultrastructural sections of the pollen wall show a darker staining and finely granular, weakly layered endexine, which is thin (about 0.2 μm) in the non-apertural regions but thickens to about 1 μm under the apertures (Figs. 36–39). The ektexine of *S. marylandensis* is thick (about 1 μm), with a prominent, homogeneous foot layer, and thins at the aperture. The columellae are simple and short, about 0.3 μm high, and closely spaced. The muri are 1–3 μm high, parallel-sided, with a rounded profile in cross-section (Figs. 36–37). The ektexine of *S. mauldinensis* has a thick foot layer, about 0.5 μm thick, and simple, short columellae, about 0.2 μm high. The muri are about 0.8 μm high, parallel-sided, with a sharply pointed triangular apical profile in cross-section (Figs. 38–39).

Dispersed pollen grains similar to those of *Spanomera mauldinensis* have been reported from several dispersed pollen assemblages from North America ranging in age from middle Albian to early Cenomanian and have been referred to the dispersed pollen species *Striatopollis* (*Retitricolpites*) *paraneus* (e.g., Brenner, 1963; Norris, 1967; Singh, 1971; Doyle and Robbins, 1977). Dispersed pollen grains similar to those of *Spanomera marylandensis* are common in many Potomac Group samples of Albian age and are referred to the highly variable dispersed pollen species *Striatopollis* (*Retitricolpites*) *vermimurus* (e.g., Brenner, 1963; Doyle and Robbins, 1977). Dispersed pollen of *Striatopollis vermimurus* has also been reported from other North American localities of Albian age (e.g., Ward, 1986).

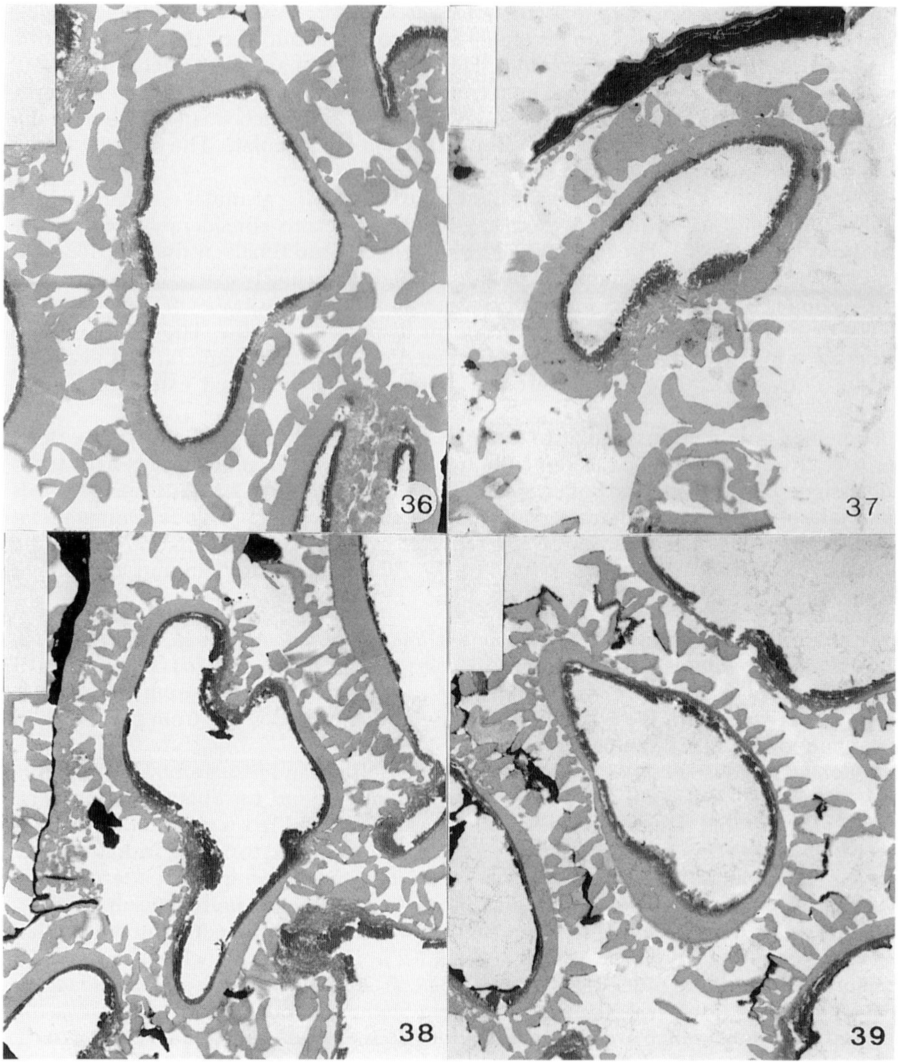

Figs. 36–39. Tricolpate pollen from *Spanomera*. TEM-micrographs showing open reticulate tectum, short columellae, very thick foot layer and darker staining endexine, × 5600. **36**, **38**: Cross-section through colpi in equatorial region. **37**, **39**: Oblique longitudinal sections through colpus area. **36–37**: *Spanomera marylandensis* – West Brothers (Maryland, USA, Potomac Group – late Albian). **38–39**: *Spanomera mauldinensis* – Mauldin Mountain (Maryland, USA, Potomac Group – early Cenomanian).

The *Spanomera* flowers are unisexual, with pentamerous (*S. mauldinensis*) or tetramerous (*S. marylandensis*) staminate flowers and bicarpellate pistillate flowers. Pistillate and staminate flowers are borne in the same inflorescence. The plants that produced these reproductive organs are thought to be closely related to modern members of the Buxaceae (Drinnan et al., 1991).

C. Ultrastructure in Cretaceous non-angiospermous seed plant pollen

*Triaperturate pollen of the **Eucommiidites** type from **Erdtmanitheca** and **Erdtmanispermum** –Arthurs Bluff and southeast of Rønne (late Berriasian–early Cenomanian).* Pollen grains have been studied from the male reproductive organ *Erdtmanitheca texensis* (Arthurs Bluff, Texas, USA - early Cenomanian) and from the seeds of *Erdtmanispermum balticum* (C.Nielsen A/S sand pit, southeast of Rønne, Bornholm, Denmark - late Berriasian-Valanginian). They are elliptical in outline and 20–25 μm long. They have a distinct distal

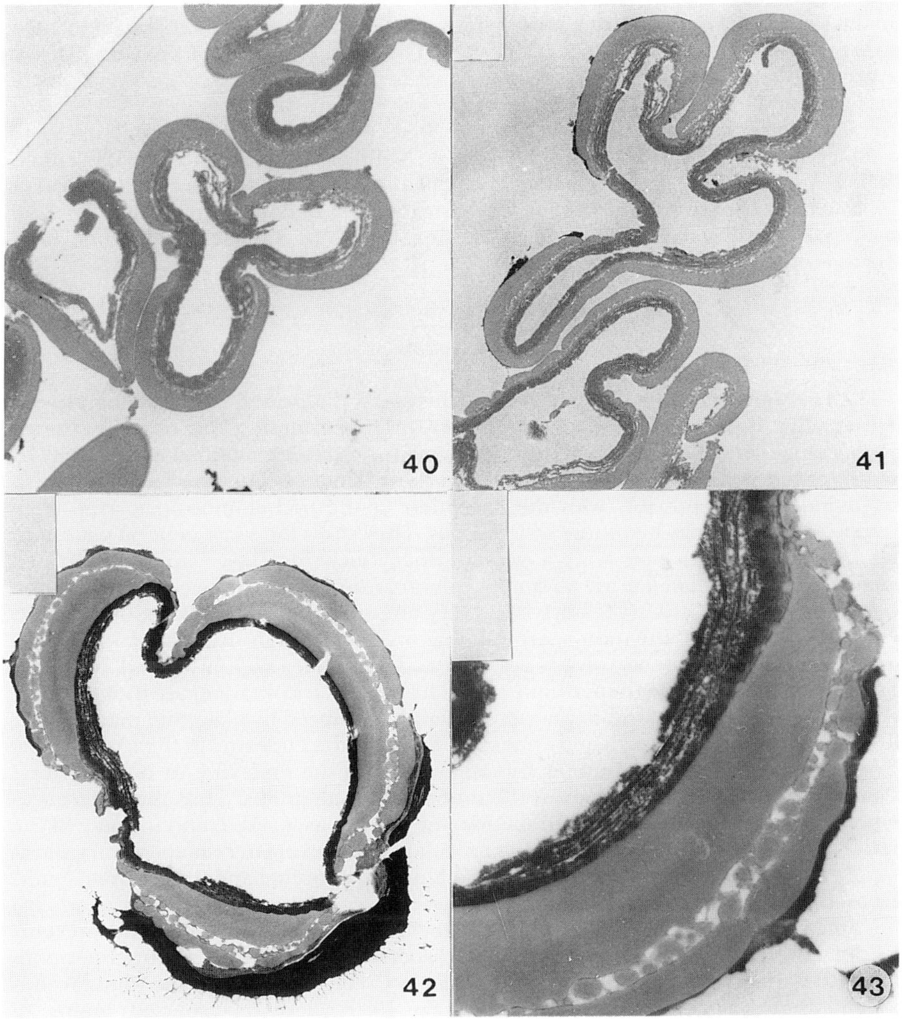

Figs. 40–43. Triaperturate pollen of the *Eucommiidites* type. **40–41**: TEM-micrographs of pollen from *Erdtmanitheca texensis* (Arthurs Bluff, Texas, USA - early Cenomanian) showing massive tectum, granular layer of ektexine and darker staining endexine; section perdendicular to colpi, × 4200. **42–43**: TEM-micrographs of pollen from *Erdtmanispermum balticum* (southeast of Rønne, Bornholm, Denmark - late Berriasian-Valginian) showing foveolate tectum, middle granular layer and inner massive layer of ektexine, and darker staining endexine; section perpendicular to colpi. **42**: × 3500. **43**: × 10000.

colpus flanked by two lateral colpi approximately in an equatorial position. The configuration of the colpi is distinct from that of tricolpate angiosperm pollen in being di-symmetrical rather than radially symmetrical, and the colpi are perpendicular to the presumed polar axis rather than parallel to it. The surface of the grains is psilate in pollen from *Erdtmanitheca texensis* or foveolate in pollen from *Erdtmanispermum balticum*.

Ultrathin sections of the pollen wall show three distinct layers in *E. texensis* (Figs. 40–41). The darker staining endexine is thick, about 0.3 μm, granular and lamellated, and almost evenly distributed in both non-apertural and apertural regions. The inner layer of the ektexine is granular, about 0.25 μm thick, while the homogeneous outer layer is about 0.5 μm thick, and the thickest of the three wall layers. Pollen grains of *E. balticum* are similar to those of *E. texensis* in having a thick, darker staining, granular and lamellated endexine, which is almost evenly distributed. They differ, however, in having a three-layered ektexine, with a thick homogeneous inner layer ("foot layer"), about 1 μm, a granular middle layer, and an outer perforated layer (Figs. 42–43).

Pollen grains of the *Eucommiidites* type have been observed *in situ* in the micropyles of small seeds from several localities in Europe and North America (Hughes, 1961; Brenner, 1967; Reymanówna, 1968; Pedersen et al., 1989) and from the microsporangiate head of *Erdtmanitheca* (Pedersen et al., 1989). The structure and organization of the seeds and male reproductive organ indicate a relationship with Gnetales and other non-angiospermous anthophytes.

Discussion

The Early and mid-Cretaceous pollen grains so far discovered *in situ* in angiosperm reproductive organs are all monoaperturate or triaperturate. The monoaperturate (monocolpate, trichotomocolpate) and reticulate forms are exemplified in this study by four different pollen types from Portugal and North America. They are thought to be of magnoliid affinity, possibly with the exception of the trichotomocolpate type from Kenilworth, which may be monocotyledonous. They show the same general pattern in pollen wall ultrastructure, with a very distinct, thick foot layer and well-defined columellae. The columellae are all simple, but vary from species to species in length and in spacing. Another striking feature is the very unevenly distributed endexine, which is weakly developed in the non-apertural regions but strongly expressed under the apertures. The endexine is typically granular and sometimes slightly layered, but never distinctly lamellated. The triaperturate grains (tricolpate, tricolporate) are thought to be of hamamelidid affinity, and many were certainly produced by platanoid plants. They also show an unevenly distributed endexine. Typically the endexine in the triaperturate forms is strongly thickened under the apertures, but the endexine in non-apertural regions is generally thicker than in the monoaperturate grains. Thus the differences between apertural and non-apertural endexine are not as marked, and in some of the Early Cretaceous triaperturate forms in particular the endexine in non-apertural regions is quite thick. With very few exceptions in both monoaperturate and triaperturate grains the foot layer is prominent.

A similar ultrastructural pattern, with thick foot layer, distinct columellae and unevenly distributed endexine, is also seen in some dispersed pollen grains of angiospermous affinity described by other authors from the Early Cretaceous, such as the winteraceous pollen *Walkeripollis gabonensis* (Doyle et al., 1990). Several dispersed grains of angiospermous affinity described by Walker and Walker (1984) from the Early Cretaceous also show the same pattern, and in those grains where this pattern cannot be observed it may result from problems of preservation or preparation technique.

The distribution and structure of the endexine clearly distinguish the Early Cretaceous pollen found *in situ* in angiosperm reproductive organs from the pollen of other seed plants, which typically have a thick, lamellated and more or less evenly distributed endexine, as exemplified by the pollen of *Eucommiidites*. An exception is possibly the pollen of the Bennettitales, but their ultrastructure is currently poorly known. Several dispersed pollen grains with angiospermid features including forms with reticulate and sometimes columellate exine have recently been described from the Triassic (e.g., *Tricrinopollis* and *Monocrinopollis*; Cornet, 1989) and from the Early Cretaceous (e.g., species of *Afropollis*; Doyle et al., 1990). They all have a thick, lamellated and evenly distributed endexine very different from the unequivocal angiosperm grains reviewed here, raising the possibility that they may have been produced by non-angiospermous seed plants. The aperture configuration in *Tricrinopollis* is comparable to that of *Eucommiidites*, with three elongated colpi in a non-radial arrangement.

Summary

Pollen grains from Early and mid-Cretaceous angiospermous reproductive organs studied here are monoaperturate (monocolpate, trichotomocolpate) or triaperturate (tricolpate, tricolporate). In the mid-Early Cretaceous floras of Portugal most *in situ* pollen grains are monoaperturate, and several monoaperturate forms have also been reported *in situ* from the Early to mid-Cretaceous floras of the Potomac Group. In younger floras triaperturate pollen types become much more common and very diverse. Many of the Cretaceous tricolpate grains were produced by platanoid plants (e.g., Friis et al., 1988; Crane et al., 1993; Pedersen et al., in press) with unisexual flowers and crowded into small heads. Three different genera, *Platananthus*, *Hamatia* and *Aquia*, have so far been recognized. A number of dispersed stamens with bithecate and tetrasporangiate anthers showing valvate dehiscence and apical expansions of the connective are also thought to belong to the platanoid complex, but have not been described formally yet in the absence of more complete specimens. One of these (from the Puddledock locality) has also been studied for ultrastructure and is included here for comparison with the younger platanoid material.

The Early to mid-Cretaceous pollen grains found *in situ* are characterized by having thick, prominent foot layer, distinct columellae and unevenly distributed endexine, which in the monoaperturate forms is very thin in the non-apertural regions and strongly thickened under the apertures. The tectum is mostly reticulate to foveolate.

Based on studies of *in situ* fossil pollen, angiosperm grains are clearly distinguished from non-angiospermous grains particularly by the distribution and ultrastructure of the endexine. However, the ultrastructure of many extinct, non-angiospermous seed plants is still poorly known and clarification of the relationships of many dispersed pollen types with angiospermid features awaits their discovery in fossil reproductive organs.

Acknowledgements

We are grateful to J.A. Doyle and P.K. Endress for valuable comments on the manuscript, to B. Larsen and B. Kunderup, University of Aarhus, for assistance in preparing the ultrastructural sections, and to G. Danscher, for permission to use TEM facilities at the Anatomical Institute, University of Aarhus. Y. Arremo and J. Sommer are thanked for help in preparing the illustrations. Support from the Swedish Natural Research Council, the Carlsberg Foundation and the National Science Foundation (BRS 90-20237) is gratefully acknowledged.

References

Brenner, G.J. (1963). The spores and pollen of the Potomac Group of Maryland. *Maryland Dept. Geol., Mines, Water Resources, Bull.* **27**: 1–215.

Brenner, G.J. (1967). The gymnospermous affinity of *Eucommiidites* Erdtman, 1948. *Rev. Palaeobot. Palynol.* **5**: 123–127.

Chapman, J.L. (1987). Comparison of Chloranthaceae pollen with the Cretaceous "*Clavatipollenites* complex". Taxonomic implications for palaeopalynology. *Pollen & Spores* **29**: 249–272.

Cornet, B. (1989). Late Triassic angiosperm-like pollen from the Richmond Rift Basin of Virginia, USA. *Palaeontographica, Abt. B, Paläophytol.* **213**: 37–87.

Couper, R.A. (1958). British Mesozoic microspores and pollen grains. *Palaeontographica, Abt. B, Paläophytol.* **103**: 75–179.

Crane, P.R., Friis, E.M. and Pedersen, K. R. (1989). Reproductive structure and function in Cretaceous Chloranthaceae. *Pl. Syst. Evol.* **165**: 211–226.

Crane, P.R., Pedersen, K.R., Friis, E.M. and Drinnan, A.N. (1993). Early Cretaceous (middle Albian) platanoid inflorescences associated with *Sapindopsis* leaves from the Potomac Group of eastern North America. *Syst. Bot.* **18**: 328–344.

Crepet, W.L., Nixon, K.C., Friis, E.M. and Freudenstein, J.V. (1992). The oldest flowers of hamamelidaceous affinity, from the Upper Cretaceous of New Jersey, North America. *Proc. Natl. Acad. Sci. U.S.A.* **89**: 8986–8989.

Doyle, J.A. (1969). Cretaceous angiosperm pollen of the Atlantic Coastal Plain and its evolutionary significance. *J. Arnold Arbor.* **50**: 1–35.

Doyle, J.A. (1992). Revised palynological correlations of the lower Potomac Group (USA) and the Cocobeach sequence of Gabon (Barremian-Aptian). *Cretaceous Res.* **13**: 337–349.

Doyle, J.A. and Hickey, L.J. (1976). Pollen and leaves from the mid-Cretaceous Potomac Group and their bearing on early angiosperm evolution. In: C.B. Beck (editor). Origin and early evolution of angiosperms, pp. 139–206, Columbia Univ. Press, New York.

Doyle, J.A. and Robbins, E.I. (1977). Angiosperm pollen zonation of the continental Cretaceous of the Atlantic Coastal Plain and its application to deep wells in the Salisbury embayment. *Palynology* **1**: 43–78.

Doyle, J.A., Hotton, C.L. and Ward, J.V. (1990). Early Cretaceous tetrads, zonosulculate pollen, and Winteraceae. I. Taxonomy, morphology, and ultrastructure. *Amer. J. Bot.* **77**: 1544–1557.

Drinnan, A.N., Crane, P.R., Friis, E.M. and Pedersen, K.R. (1990). Lauraceous flowers from the Potomac Group (mid-Cretaceous) of eastern North America. *Bot. Gaz.* **151**: 370–384.

Drinnan, A.N., Crane, P.R., Pedersen, K.R. and Friis, E.M. (1991). Angiosperm flowers and tricolpate pollen of buxaceous affinity from the Potomac Group (mid-Cretaceous) of eastern North America. *Amer. J. Bot.* **78**: 153–176.

Endress, P.K. and Friis, E.M. (1991). *Archamamelis,* hamamelidalean flowers from the Upper Cretaceous of Sweden. *Pl. Syst. Evol.* **175**: 101–114.

Friis, E.M. (1985a). Structure and function in Late Cretaceous angiosperm flowers. Niels Bohr Volume. *Biol. Skr.* **25**: 1–37.

Friis, E.M. (1985b). *Actinocalyx* gen. nov., sympetalous angiosperm flowers from the Upper Cretaceous of Sweden. *Rev. Palaeobot. Palynol.* **45**: 171–183.

Friis, E.M. (1990). *Silvianthemum suecicum* gen. et sp. nov., a new saxifragalean flower from the Late Cretaceous of Sweden. *Biol. Skr.* **36**: 1–35.

Friis, E.M. and Crane, P.R. (1989). Reproductive structures of Cretaceous Hamamelidae. In: P.R. Crane and S. Blackmore (editors). Evolution, systematics, and fossil history of the Hamamelidae, pp. 155–174, Clarendon Press, Oxford.

Friis, E.M. and Pedersen, K.R. (1990). Structure of the Lower Cretaceous fern *Onychiopsis psilotoides* from Bornholm, Denmark. *Rev. Palaeobot. Palynol.* **66**: 47–63.

Friis, E.M. and Pedersen, K.R. (in press). Angiosperm pollen *in situ* in Cretaceous reproductive organs. In: J. Jansonius and D.C. McGregor (editors). Palynology and Stratigraphy, Amer. Assoc. Stratigr. Palynologists Publ.

Friis, E.M., Crane, P.R. and Pedersen, K.R. (1988). Reproductive structure of Cretaceous Platanaceae. *Biol. Skr.* **31**: 1–56.

Friis, E.M., Crane, P.R. and Pedersen, K.R. (1991). Stamen diversity and *in situ* pollen of Cretaceous angiosperms. In: S. Blackmore and S.H. Barnes (editors). Pollen and spores: Patterns of diversification, pp. 197–224, Clarendon Press, Oxford.

Friis, E.M., Pedersen, K.R. and Crane, P.R. (1992). *Esgueiria*, new fossil flowers from the Late Cretaceous of Portugal with combretaceous features. *Biol. Skr.* **41**: 1–45.

Groot, J.J. and Groot, C.R. (1962). Plant microfossils from Aptian, Albian and Cenomanian deposits of Portugal. *Comun. Serv. Geol. Portugal* **46**: 133–176.

Hedlund, R.W. and Norris, G. (1968). Spores and pollen grains from Fredericksburgian (Albian) strata, Marshall County, Oklahoma. *Pollen & Spores* **10**: 129–159.

Hickey, L.J. and Doyle, J.A. (1977). Early Cretaceous fossil evidence for angiosperm evolution. *Bot. Rev.* **43**: 2–104.

Hughes, N.F. (1961). Further interpretation of *Eucommiidites* Erdtman 1948. *Palaeontology* **4**: 292–299.

Manchester, S.R. (1986). Vegetative and reproductive morphology of an extinct plane tree (Platanaceae) from the Eocene of Western North America. *Bot. Gaz.* **147**: 200–226.

Muller, J. (1970). Palynological evidence on early differentiation of angiosperms. *Biol. Rev. Cambridge Philos. Soc.* **45**: 417–450.

Norris, G. (1967). Spores and pollen from the Lower Colorado Group (Albian-?Cenomanian) of central Alberta. *Palaeontographica, Abt. B, Paläophytol.* **120**: 72–115.

Pedersen, K.R., Crane, P.R. and Friis, E.M. (1989). Pollen organs and seeds with *Eucommiidites* pollen. *Grana* **28**: 279–294.

Pedersen, K.R., Crane, P.R., Drinnan, A.N. and Friis, E.M. (1991). Fruits from the mid-Cretaceous of North America with pollen grains of the *Clavatipollenites* type. *Grana* **30**: 577–590.

Pedersen, K.R., Friis, E.M. and Crane, P.R. (1993). Pollen organs and seeds with *Decussosporites* Brenner from Lower Cretaceous Potomac Group of eastern USA. *Grana* **32**: 273–289.

Pedersen, K.R., Friis, E.M., Crane, P.R. and Drinnan, A.N. (in press). Reproductive structures of an extinct platanoid from the Early Cretaceous (latest Albian) of eastern North America. *Rev. Palaeobot. Palynol.*

Reymanówna, M. (1968). On seeds containing *Eucommiidites troedssonii* pollen from the Jurassic of Grojec, Poland. *Bot. J. Linn. Soc.* **61**: 147–152.

Singh, C. (1971). Lower Cretaceous microfloras of the Peace River area, northwestern Alberta. *Bull. Res. Council Alberta* **28**: 1–542.

Skarby, A. (1986). Normapolles anthers from the Upper Cretaceous of southern Sweden. *Rev. Palaeobot. Palynol.* **46**: 234–256.

Skarby, A., Rowley, J.R. and Nilsson, L. (1990). Exine structure of upper Cretaceous Normapolles grains from anthers (Northeastern Scania, Sweden). *Palynology* **14**: 145–173.

Trincão, P.R.P. (1990). Esporos e pólenes do Cretácico inferior (Berriasiana-Aptiana) de Portugal: Paleontologia e biostratigrafia. 312 pp. + 47 plates. Unpublished dissertation, Universidade de Aveiro, Lisbon.

Walker, J.W. and Walker, A.G. (1984). Ultrastructure of Lower Cretaceous angiosperm pollen and the origin and early evolution of flowering plants. *Ann. Missouri Bot. Gard.* **71**: 464–521.

Ward, J.V. (1986). Early Cretaceous angiosperm pollen from the Cheyenne and Kiowa Formations (Albian) of Kansas, USA. *Palaeontographica, Abt. B, Paläophytol.* **202**: 1–81.

Ward, J.V. and Doyle, J.A. (1994). Ultrastructure and relationships of mid-Cretaceous polyforate and triporate pollen from Northern Gondwana. In: M.H. Kurmann and J.A. Doyle (editors). Ultrastructure of fossil spores and pollen, pp. 161–172, Royal Botanic Gardens, Kew.

10. ULTRASTRUCTURE AND RELATIONSHIPS OF MID-CRETACEOUS POLYFORATE AND TRIPORATE POLLEN FROM NORTHERN GONDWANA

JEROME V. WARD[1] AND JAMES A. DOYLE[2]

[1]**Department of Geology, Auburn University, Auburn, Alabama 36849**
[2]**Section of Botany, University of California, Davis, California 95616**

Abstract

Polyforates (*Cretacaeiporites*), appearing in the middle Albian, and triangular triporates with simple pores (*Triorites africaensis*), appearing in the late Cenomanian, are among many distinctive angiosperm pollen types endemic to Northern Gondwana in the mid-Cretaceous. LM, SEM, and TEM observations on grains of *C. scabratus* and *T. africaensis* from a sample of probable late Cenomanian (Zone C-XIII) age allow evaluation of proposed botanical affinities of these forms. *Cretacaeiporites scabratus* has spinules continuous with large columellae, a tectum with small perforations and much finer microchannels, and a thick footlayer composed of fused granules. This structure is very different from that of superficially similar pollen of Trimeniaceae, Alismataceae, and Caryophyllidae. However, similar features do occur in polyforate and tricolpate Ranunculaceae, although Ranunculaceae differ in having both a footlayer and endexine. *Triorites africaensis* has a scabrate, finely perforated tectum, a granular-columellar infratectal layer, a thick footlayer, and a thin endexine. The exine thins gradually toward the pores, by reduction of the nexine and infratectal layer. Similar exine and pore features occur in putatively primitive members of the Proteaceae, although not in combination. Exines of triporate Olacaceae are less similar. These observations are consistent with the view that *T. africaensis* represents a stem relative of Proteaceae, and with the view that Proteaceae are derived from tricolpate (lower hamamelid) rather than tricolporate (rosid) ancestors.

Introduction

The Cretaceous Northern Gondwana province of Brenner (1976), or the ASA province of Herngreen and Chlonova (1981), which covered most of Africa and South America except for their southern portions, is known for its high diversity of early angiosperm pollen, including endemic taxa and members of pollen classes that appear later elsewhere. This has led to the view that Northern Gondwana was the most important center of early diversification of angiosperms, if not their area of origin (Brenner, 1976; Doyle et al., 1977, 1982; Crane and Lidgard, 1989).

Two pollen groups that exemplify this pattern are the oldest known polyforates (periporates) and some of the oldest triangular triporates. These do not show any obvious relationship to members of these pollen classes in other provinces, such as polyforates described by Pacltová (1971) from the Cenomanian of Bohemia or the triporate Normapolles that dominate the Late Cretaceous of Laurasia. Both were first described from the Cenomanian-Turonian of Senegal by Jardiné and Magloire (1965), who named the polyforates as three species of *Multiporopollenites* and the triporates as *Triorites*

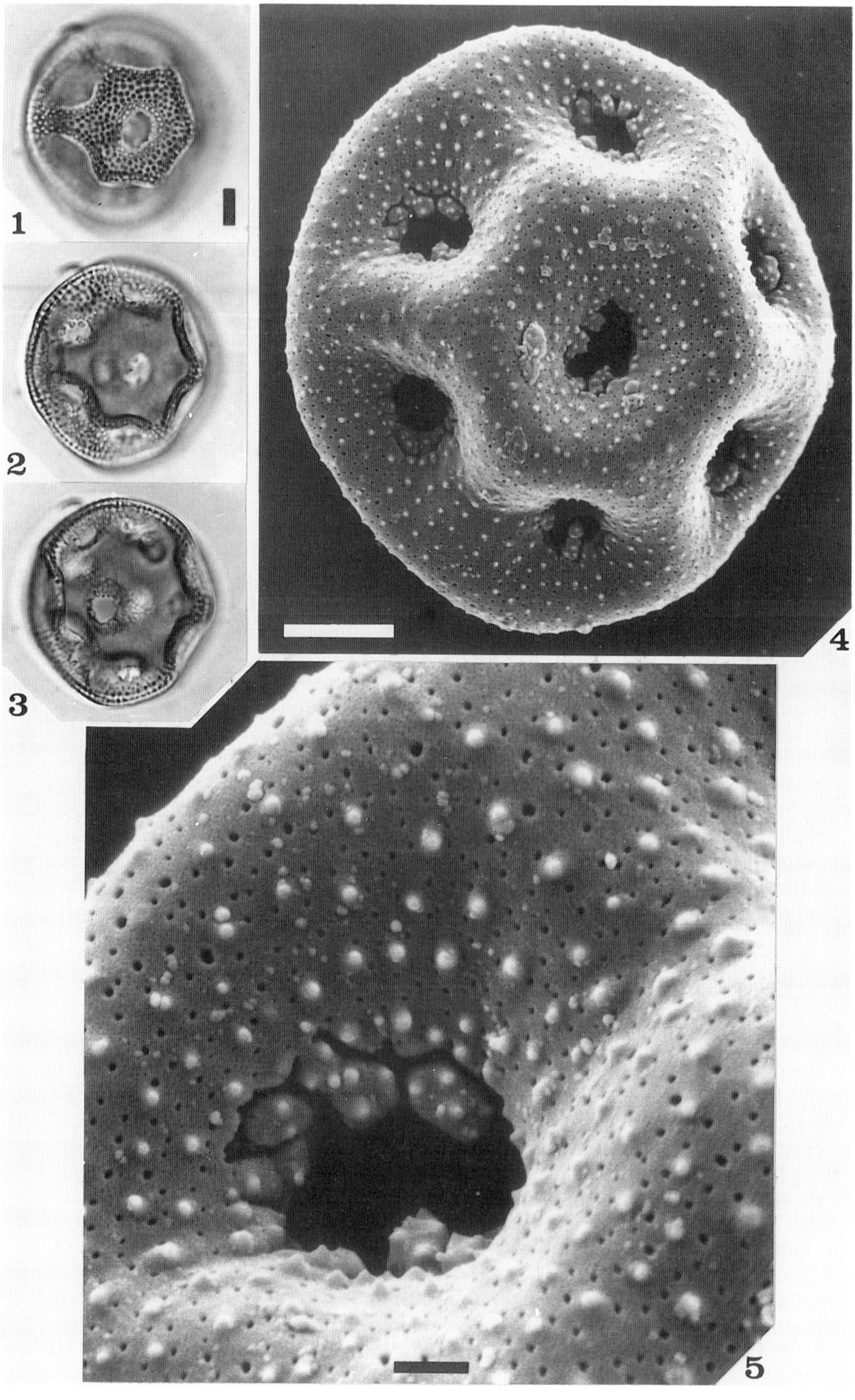
1
2
3
4
5

africaensis. The polyforates were later transferred to the new genus *Cretacaeiporites* by Herngreen (1973), who formally described the species treated in this paper, *Multiporopollenites* sp. S. CI. 105 of Jardiné and Magloire (1965), as *C. scabratus*. Among the characters distinguishing this species from other members of the genus are its large, well-spaced columellae. Subsequent work has dated the appearance of *Cretacaeiporites* as a whole as mid-Albian and the appearance of *C. scabratus* as late Albian (Zone II of Brazil: Herngreen, 1973). The appearance of *Triorites africaensis* (Zone III in Brazil, Zone C-XIII in Gabon) is dated as late Cenomanian (Herngreen, 1973; Doyle et al., 1982). The purpose of this article is to present new data on the ultrastructure of *C. scabratus* and *T. africaensis*, and to discuss the botanical affinities of these forms by comparing them with published ultrastructural data on potentially related modern taxa.

Materials and Methods

Both grains studied are from the same sample vial, labeled 4968, kindly provided by S. Jardiné at Société National Elf-Aquitaine (Production) in a series of samples said to be from Zone C-IX (Aptian) of Gabon. Since the sample is clearly much younger than Zone C-IX, we assumed that it had been mistakenly included in this series. However, subsequent checking by J.-F. Raynaud (Elf-Aquitaine, pers. comm.) showed that the original collection and prepared slides bearing the number 4968 are indeed from Zone C-IX. Apparently the residue was placed in the wrong vial after preparation of the slides. Efforts to determine what sample it was confused with have been unsuccessful. We hesitate to report on such undocumented material, but in view of its botanical interest we have decided to do so, for several reasons. First, we have no doubt that the material is of Zone C-XIII age: it contains not only fairly common *Cretacaeiporites scabratus* and abundant *Triorites africaensis*, but also the tricolpodiorate species *Hexaporotricolpites emelianovii*, *H. coronatus*, and *H. potoniei* (*chordatus*), which appear in the late Albian or Cenomanian (upper Zone C-XII) (Jardiné et al., 1972; Doyle et al., 1982), and frequent *Classopollis*, which declines abruptly above Zone C-XIII (Jardiné et al., 1974). Second, these are not new forms being described for the first time, but rather already well-known, common taxa in the Northern Gondwana province. Third, our purpose is strictly morphological and systematic, and we do not attempt to draw any stratigraphic conclusions from the material.

Individual fossil grains were picked from residue in glycerine and studied with light microscopy, SEM, and TEM using single-grain techniques (Ward et al., 1989). For TEM, grains were embedded in Epon-Araldite and sectioned, picked up on formvar-coated single-slot grids, and poststained with uranyl acetate and lead citrate.

Results and Discussion

A. *Cretacaeiporites scabratus*

Light micrographs of *Cretacaeiporites scabratus* (Figs. 1–3) show the foramina and conspicuous dots in surface view, which are seen in optical section to be columellae. SEM (Figs. 4, 5) shows verrucate sculpture on the foramina, small tectal perforations, and conspicuous spinules. The most distinctive features are in the exine structure as seen under TEM (Figs. 6–8). First, the columellae are relatively large and continuous with the spinules; the spinules appear to be the heads of columellae that "pass through" the tectum. Second, the tectum is relatively thick and perforated by channels of two sizes:

FIGS. 1–5. *Cretacaeiporites scabratus* Herngreen, single-grain preparation 4968-1. 1–3: light micrographs, three focal levels, × 1000. 4: SEM of same grain, × 3000. 5: SEM, detail of foramen and exine sculpture, × 10,000. Scale bar = 5 μm for 1–4, 1 μm for 5.

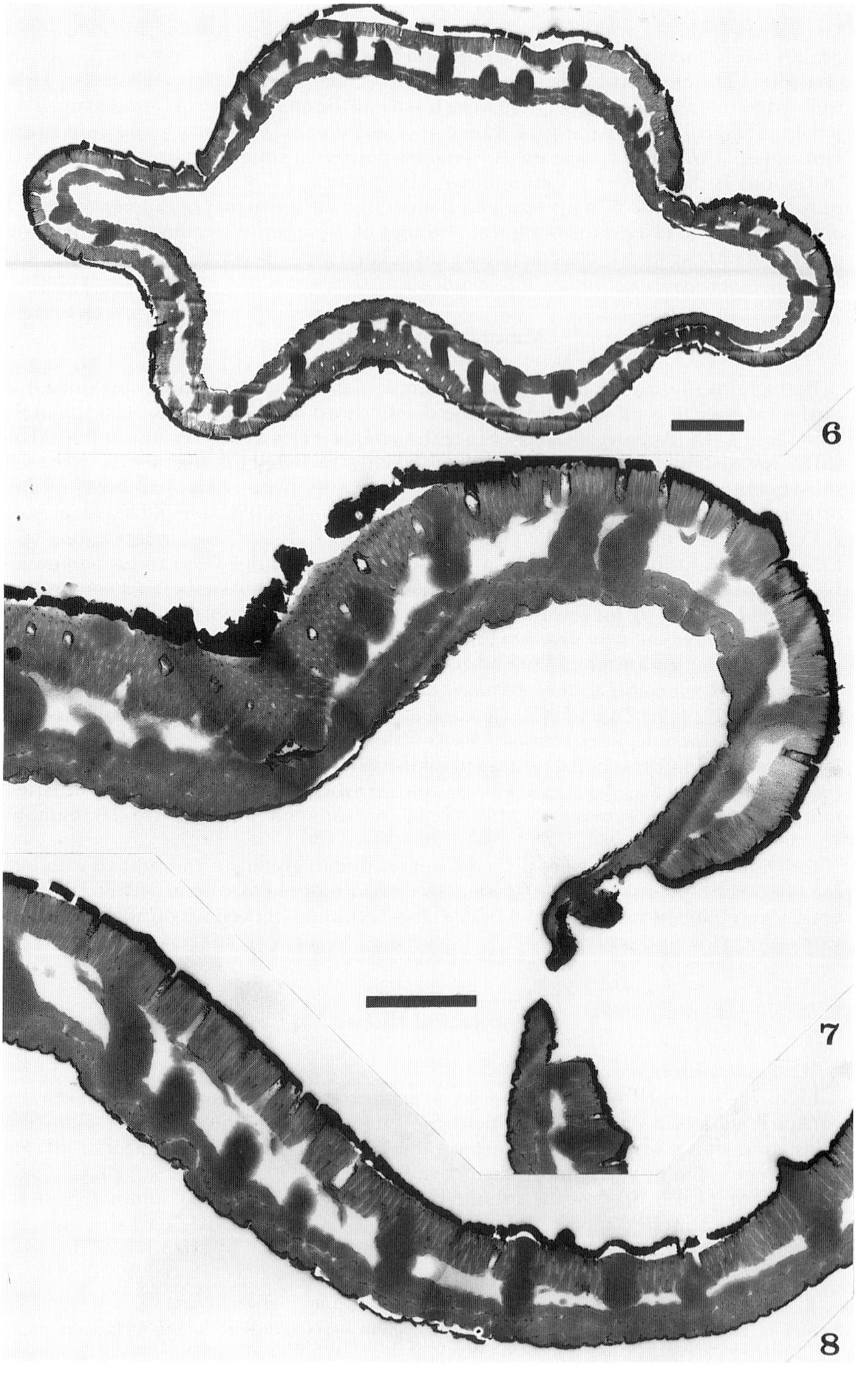
6
7
8

larger ones corresponding to the perforations seen under SEM, and much finer microchannels. Finally, the nexine appears to be entirely ectexinous (footlayer), but it is made up of two layers of fused granules.

In terms of angiosperm pollen evolution, polyforates are usually considered a rather advanced type, and it might therefore seem surprising to encounter them in the mid-Cretaceous. However, they are actually not so anomalous, since polyforate pollen occurs today in groups that are basically monosulcate (monocots, magnoliid dicots) and basically tricolpate ("non-magnoliid dicots", or eudicots of Doyle and Hotton, 1991), and both of these pollen classes were well diversified by the time *Cretacaeiporites* appears. In his original description, Herngreen (1973) remarked that *C. scabratus* shows some resemblance to *Thalictrum*, in the relatively primitive eudicot family Ranunculaceae, and two members of basically monosulcate groups, *Alisma*, a monocot, and *Trimenia*, formerly placed in Monimiaceae but now considered one of two genera in the Australasian family Trimeniaceae, in the magnoliid order Laurales.

In his critical review of pollen records of extant angiosperm taxa, Muller (1981) more specifically favored the possibility that *Cretacaeiporites scabratus* represents *Trimenia* pollen. According to cladistic analyses of extant primitive angiosperms (Donoghue and Doyle, 1989), Trimeniaceae are most closely related to Chloranthaceae (as suggested by Endress, 1987), next most closely to Amborellaceae and "higher" Laurales, and next to Austrobaileyaceae and/or Calycanthaceae. Monosulcate pollen occurs in Chloranthaceae, Amborellaceae, and Austrobaileyaceae. The presence of Trimeniaceae in the mid-Cretaceous would not be surprising, because pollen comparable to that of their sister group, Chloranthaceae, namely *Clavatipollenites* and related forms, is common at this time (Muller, 1981; Walker and Walker, 1984; Doyle and Hotton, 1991; Pedersen et al., 1991).

Muller's suggestion was supported by Dettmann and Jarzen (1990), in a discussion of *Trimenia*-like pollen from the Late Cretaceous and Tertiary of Australia (*Australopollis obscurus*, *Polyporina fragilis*). If *Cretacaeiporites* is related to Trimeniaceae, the family may have originated in Northern Gondwana, which lay in the equatorial belt in the mid-Cretaceous, and later moved into the more temperate regions of Southern Gondwana. This would parallel the scenario proposed for another magnoliid group, Winteraceae, which have probable relatives in the Early Cretaceous of Northern Gondwana but are now austral (Doyle et al., 1990). More recently Specht et al. (1992) related *A. obscurus* to Callitrichaceae rather than Trimeniaceae, but this would not invalidate the southward migration scenario based on Cretaceous and Recent forms.

Pollen ultrastructure of Trimeniaceae was studied by Sampson and Endress (1984). The family varies in aperture condition: *Piptocalyx* is disulculate, whereas *Trimenia* ranges from disulculate (*T. neocaledonica*) to inaperturate (*T. papuana*) to polyforate (*T. papuana*, *T. weinmanniifolia*). In terms of ultrastructure, a direct relation of *C. scabratus* to polyforate *Trimenia* is doubtful: both have verrucate foramina, spinules, and a thick tectum, but otherwise they have little in common. First, the columellae are smaller and show no consistent relation to the spinules, which seem strictly supratectal. Second, the tectum has nothing comparable to the abundant microchannels of the fossils. Third, the nexine consists not of a granular footlayer but rather of a laminated endexine. If more comparable structure occurred in other Trimeniaceae, it might be suggested that *C. scabratus* represents a sister lineage of polyforate *Trimenia*, but this is not the case. The disulculate grains of *T. neocaledonica* and *Piptocalyx* are like the fossils in having sculptured apertures, but again there is no relation between the columellae and the supratectal spinules, and the nexine is either laminated or solid.

Figs. 6–8. *Cretacaeiporites scabratus*, single-grain preparation 4968-1, TEM. **6**: view of whole grain, × 5000. **7, 8**: detail of exine structure and foramen, × 15,000. Solid black layer is palladium-gold coating used for SEM. Scale bar = 2 μm for 6, 1 μm for 7, 8.

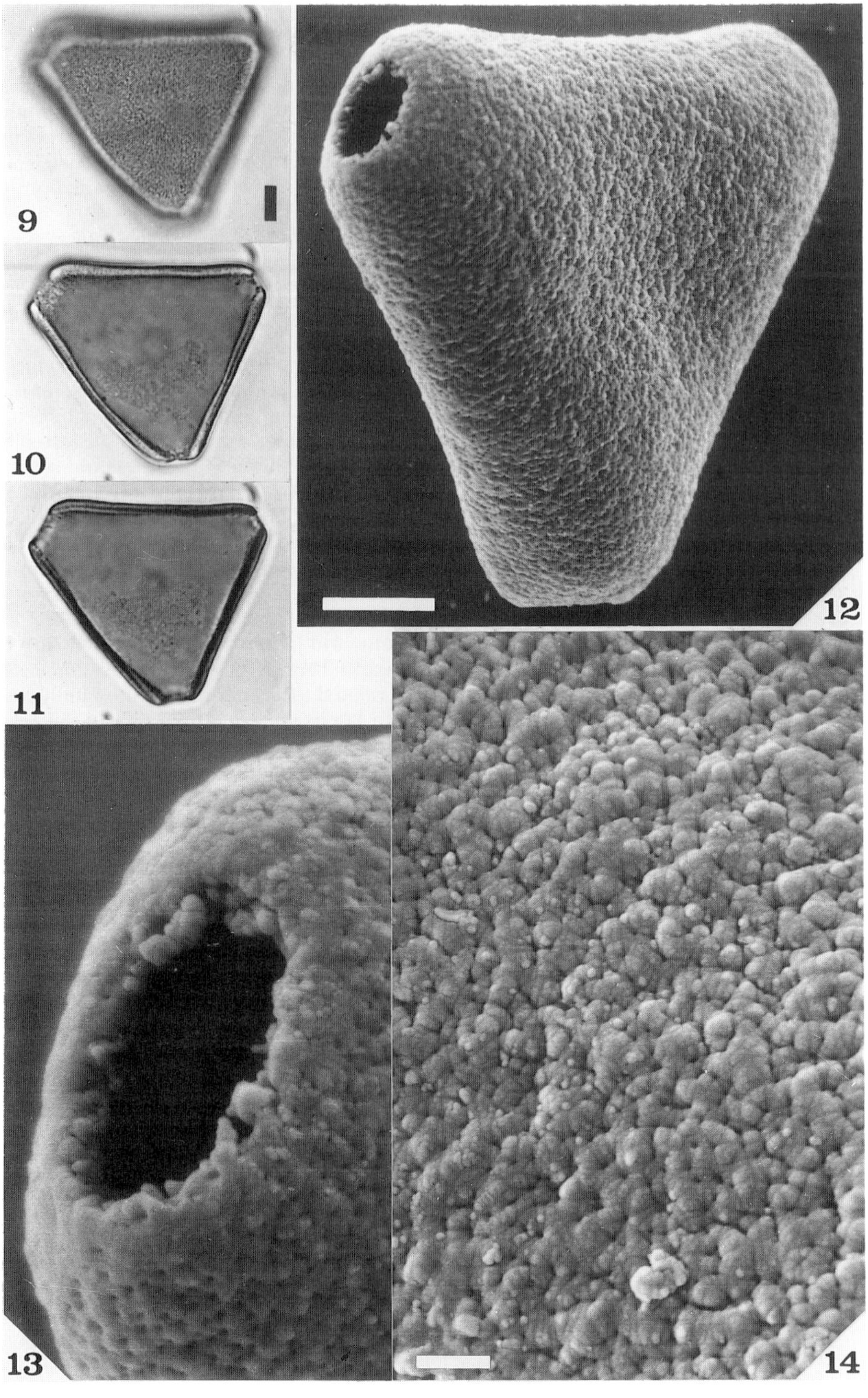
9
10
11
12
13
14

Polyforate pollen of the aquatic monocot family Alismataceae was studied with SEM and TEM by Argue (1974, 1976). Members of Argue's *Alisma* type are strikingly similar to *C. scabratus* when seen under SEM. However, TEM shows that they are very different in having a thin nexine and small, irregular columellae that have no relation to the much larger supratectal spinules.

Among basically tricolpate taxa, perhaps the most familiar group with polyforate pollen is the Caryophyllidae, which vary from tricolpate to rugate to polyforate (Nowicke and Skvarla, 1977, 1979). Some of the latter are very similar to *C. scabratus* under SEM, with sculptured foramina, spinules, and fine perforations. Examples are *Anredera* (Basellaceae) and *Gymnocarpos* (Caryophyllaceae) figured in Nowicke and Skvarla (1977) and *Sclerocephalus* (Caryophyllaceae), *Anabasis*, and *Nitrophila* (Chenopodiaceae) in Nowicke and Skvarla (1979). However, the ultrastructure of the same taxa fails to show any closer similarities. Again, the columellae are more slender than in *C. scabratus*, and they show no special relation to the spinules. In addition, the tectum has larger perforations but no microchannels, and the nexine is not granular but consists of a solid footlayer and a homogeneous endexine.

The only modern polyforate group we know that does share most of the unusual exine features of *C. scabratus* is Ranunculaceae (Nowicke and Skvarla 1979, 1981). Like caryophyllids, Ranunculaceae range from tricolpate to polyforate, and both tricolpates and polyforates have sculptured apertures and spinules. However, the ultrastructure of such polyforate taxa as *Anemone*, *Coptis* (Nowicke and Skvarla, 1979), and *Hepatica* (Nowicke and Skvarla, 1981) is more like that of *C. scabratus*: the columellae are large and continuous with the spinules, and the tectum is traversed by both larger perforations and finer microchannels. This structure, termed "typical Ranunculaceae structure" by Nowicke and Skvarla, occurs not only in the polyforates but also in some of the tricolpate taxa, such as *Batrachium* and *Hamadryas* (Nowicke and Skvarla, 1981). It does not occur in other families of the Ranunculales, such as Berberidaceae, Lardizabalaceae, and Menispermaceae (Nowicke and Skvarla, 1979, 1981; Nowicke, pers. comm.). The main differences between *C. scabratus* and Ranunculaceae are the higher density of microchannels in *C. scabratus* and the fact that the nexine of Ranunculaceae consists of both a footlayer and a well-developed endexine (Nowicke and Skvarla, 1981), rather than a two-layered, granular footlayer.

In Ranunculaceae too, it appears that the spinules are morphologically the heads of columellae that penetrate the tectum: not only are the spinules aligned with the underlying columellae, but also no tectal channels are present in the areas where the columellae fuse with the tectum (Nowicke and Skvarla, 1980; 1981; Nowicke, pers. comm.). Both criteria also hold for *C. scabratus*; microchannels are visible above some of the columellae toward the top of Fig. 7, but this seems to be because the section is oblique at this point.

We therefore conclude that the most likely modern relatives of *Cretacaeiporites scabratus* are Ranunculaceae, and that it is tricolpate- rather than monosulcate-derived. Actually, in order for the polyforate condition to be homologous in the two groups, the fossils would have to be nested within (belong to) Ranunculaceae, since the basic condition in the family is presumably tricolpate. These conclusions have interesting biogeographic implications, since modern Ranunculaceae are predominantly Laurasian. Their history might therefore parallel that of eudicots as a whole: from the late Barremian through the Aptian, tricolpate pollen is almost entirely restricted to Northern Gondwana (with rare exceptions in England and the Potomac Group: Hughes and McDougall, 1990; Doyle, 1992), but the most primitive living tricolpate groups (Ranunculales, Trochodendrales,

FIGS. 9–14. *Triorites africaensis* Jardiné & Magloire, single-grain preparation 4968-5. **9–11**: light micrographs, three focal levels, × 1000. **12**: SEM of same grain, × 3000. **13**: SEM, detail of pore, × 10,000. **14**: SEM, detail of exine sculpture on opposite side of grain, × 10,000. Scale bar = 5 μm for 9–12, 1 μm for 13, 14.

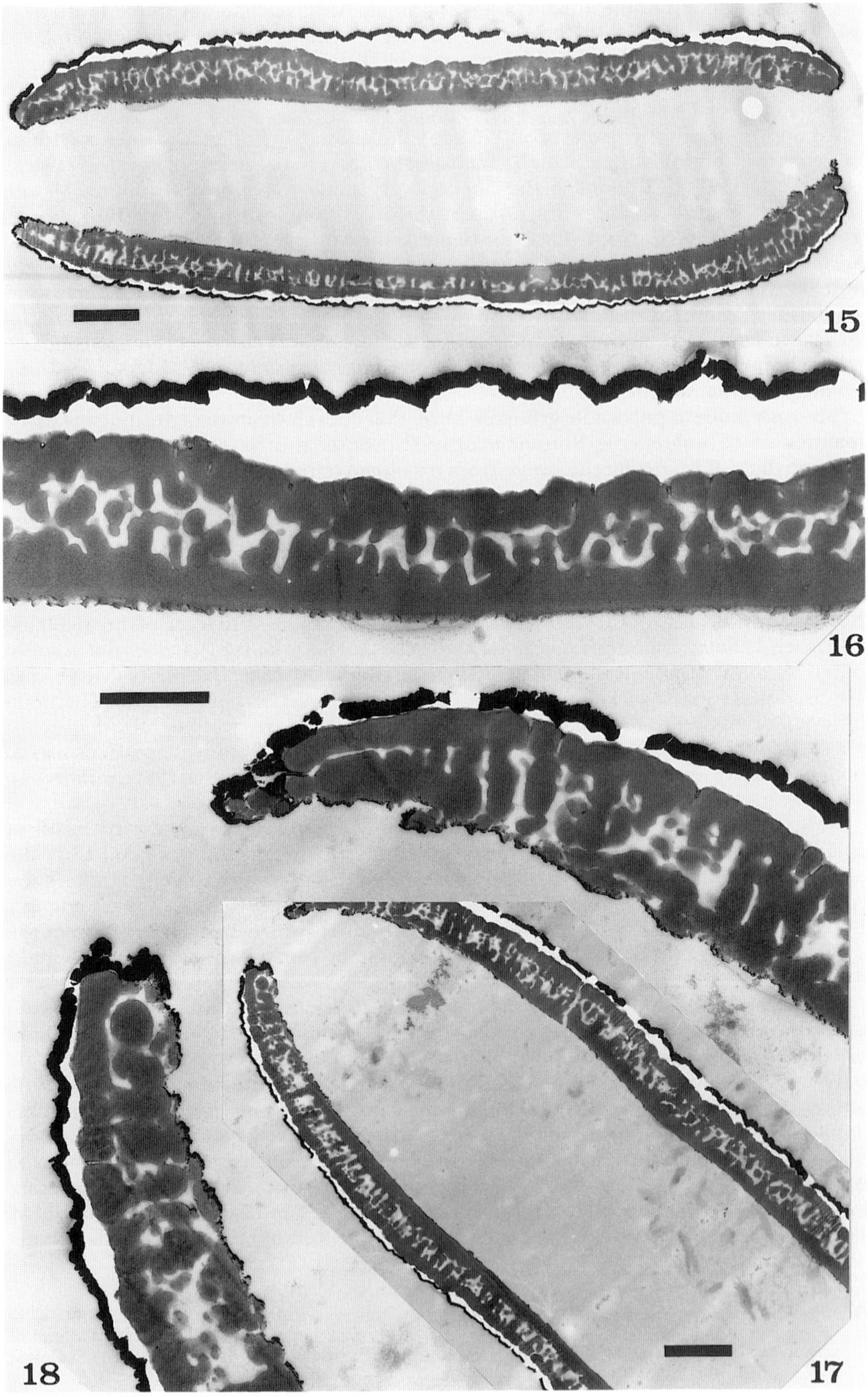
15
16
18
17

Hamamelidales) are largely Laurasian. We emphasize, however, that this proposed relationship is only a "best guess"; *C. scabratus* could represent an unrelated extinct group that simply converged with Ranunculaceae. The most serious caveat is the difference in nexine structure. This may not be an insuperable obstacle; the lack of endexine in *C. scabratus* could be an autapomorphy, resulting from either loss or transformation of the endexine into the inner part of the footlayer. The resulting two-layered, granular footlayer does not occur in Ranunculaceae, but neither have we seen it in any other group. Of course, additional data are needed in order to say whether these conclusions apply to other species of *Cretacaeiporites*.

B. *Triorites africaensis*

Under light microscopy (Figs. 9–11), *Triorites africaensis* has simple-looking pores and a finely scabrate exine surface. SEM (Figs. 12–14) shows that the sculpture consists of densely packed, irregularly sized, globular elements. TEM sections (Figs. 15–18) show an irregular tectum with fine perforations, an infratectal layer apparently intermediate between granular and columellar, with many elements resembling irregularly shaped columellae, and a fairly thick nexine, made up of a thicker, solid footlayer and a thinner, structureless endexine. The pores (Fig. 18) have a very simple structure at the TEM level as well: the exine thins gradually toward the aperture as a result of reduction of the nexine and the infratectal layer, so there is no endannulus or atrium of the sort seen in Normapolles or extant Amentiferae.

Discussion of the botanical affinities of *T. africaensis* has centered on the Proteaceae, which occur today in the former Gondwana continents and were diverse in the later Cretaceous and Tertiary of the same areas (e.g., Dettmann and Jarzen, 1991). Both Muller (1981) and Dettmann and Jarzen (1991) suggested that *T. africaensis* might be a precursor of Proteaceae.

Exines of Proteaceae show considerable variation in surface texture and infratectal structure, but *T. africaensis* falls within this range. The most interesting variations concern endexine and pore structure. According to the systematic survey of Johnson and Briggs (1975) and the palynological studies of Feuer (1990), the most primitive modern Proteaceae are in the subfamily Persoonioideae, possibly the Australian rainforest genus *Placospermum*. Many putatively more advanced taxa, such as the Embothrieae (Grevilleoideae) illustrated by Feuer (1989), have complex pores, with atriate endoapertures lined with thickened endexine. A less advanced version of this condition is seen in *Carnarvonia* and *Sphalmium*, placed in their own tribes between Persoonioideae and Grevilleoideae (Feuer, 1990). However, the putatively basal groups have pollen more like *T. africaensis*. Under LM, *Placospermum* has similar simple triporate pollen, and TEM sections (Feuer, 1990) show that the exine (which is somewhat more columellar than that of *T. africaensis*) thins gradually toward the pores. However, its nexine, including both the footlayer and the thinner endexine, is different in having a peculiar spongy structure. While interpreting *Placospermum* as primitive in most respects, Feuer (1990) suggested that its spongy nexine is not primitive, but rather an autapomorphy. The nexine is solid in other groups, such as Embothrieae, *Carnarvonia*, and another putatively primitive genus of Persoonioideae, *Bellendena* (Feuer, 1990). *Bellendena* also resembles *Placospermum* and *T. africaensis* in that the exine thins gradually toward the pore, but it has no endexine. If the common ancestor of the Proteaceae combined features of *Placospermum* and *Bellendena* —simple aperture structure, solid footlayer, and thin endexine — it would be essentially like *T. africaensis*.

FIGS. **15–18**. *Triorites africaensis*, single-grain preparation 4968-5, TEM. **15**: view of whole grain sectioned through two pores, × 4500. **16**: detail of same section, showing exine structure, × 15,000. **17**: part of another section passing through one pore, × 4500. **18**: detail of same section, showing pore structure, × 15,000. Solid black layer is palladium-gold coating used for SEM. Scale bar = 2 μm for 15, 17, 1 μm for 16, 18.

These comparisons are consistent with the hypothesis that *T. africaensis* is related to Proteaceae, possibly as a more primitive "stem relative" on the line leading to Proteaceae, rather than a member of the family in the modern sense (the "crown group": cf. Doyle and Donoghue, 1993). This would imply that the most notable exception to the triporate condition in the family, the tricolpate pollen of *Beauprea* and related fossil forms (known back to the Campanian: Dettmann and Jarzen, 1991), is secondarily derived. This agrees with Johnson and Briggs's (1975) placement of *Beauprea* within the advanced subfamily Proteoideae.

It might be premature to rule out relationships with other taxa with superficially similar triporate pollen, such as certain Sapindaceae (Muller and Leenhouts, 1976) or Olacaceae. However, TEM studies by Lobreau-Callen (1980) show that triporate Olacaceae differ from *T. africaensis* in having a finely granular infratectal structure and no endexine except at the pores. Both Sapindaceae and Olacaceae differ from Proteaceae in that their triporate members appear to be derived rather than primitive: they are highly diverse palynologically, and the most likely basic type is tricolporate. This in itself might make them less plausible relatives for Cenomanian fossils.

If *Triorites africaensis* is related to Proteaceae, it could shed light on the position of Proteaceae within the angiosperms, a highly controversial topic (Johnson and Briggs, 1975). Proteaceae have usually been associated with one or another group of Rosidae, such as Elaeagnaceae or Santalales (including Olacaceae), which seem to be basically tricolporate. However, Johnson and Briggs (1975) argue that they are considerably more primitive, possibly derived from a common ancestor with Rosidae. This could place them among so-called lower Hamamelidae, which they resemble in being apetalous (A.W. Douglas, pers. comm.). This view is supported by recent analyses of DNA sequences of the plastid gene *rbc*L, which place Proteaceae near the base of the eudicot clade (Chase et al., 1993). Since lower hamamelids have simple, tricolpate pollen, this could mean that the triporate pollen of Proteaceae was derived directly from tricolpate, rather than by way of tricolporate, as triporate pollen is in most cases (Walker and Doyle, 1975). This scenario would be consistent with the simplicity of the pores in *T. africaensis* and putatively primitive Proteaceae, which show no sign of a compound-aperturate ancestry.

Acknowledgements

We wish to thank Serge Jardiné and the Société Nationale Elf-Aquitaine (Production) for the material used in this study, Carol Hotton and especially Joan Nowicke for comments on ranunculaceous pollen, and Mark Chase and Andrew Douglas for use of unpublished results. This work was supported in part by NSF grant BSR-8415772.

References

Argue, C.L. (1974). Pollen studies in the Alismataceae (Alismaceae). *Bot. Gaz.* **135**: 338–344.

Argue, C.L. (1976). Pollen studies in the Alismataceae with special reference to taxonomy. *Pollen & Spores* **18**: 161–201.

Brenner, G.J. (1976). Middle Cretaceous floral provinces and early migrations of angiosperms. In: C. B. Beck (editor). Origin and Early Evolution of Angiosperms, pp. 23–47, Columbia Univ. Press, New York.

Chase, M.W. *and others* (1993). Phylogenetics of seed plants: an analysis of nucleotide sequences from the plastid gene *rbc*L. *Ann. Missouri Bot. Gard.* **80**: 528–580.

Crane, P.R. and Lidgard, S. (1989). Angiosperm diversification and paleolatitudinal gradients in Cretaceous floristic diversity. *Science* **246**: 675–678.

Dettmann, M.E. and Jarzen, D.M. (1990). The Antarctic/Australian rift valley: Late Cretaceous cradle of northeastern Australasian relicts? *Rev. Palaeobot. Palynol.* **65**: 131–144.

Dettmann, M.E. and Jarzen, D.M. (1991). Pollen evidence for Late Cretaceous differentiation of Proteaceae in southern polar forests. *Canad. J. Bot.* **69**: 901–906.

Donoghue, M.J. and Doyle, J.A. (1989). Phylogenetic analysis of angiosperms and the relationships of Hamamelidae. In: P.R. Crane and S. Blackmore (editors). Evolution, Systematics, and Fossil History of the Hamamelidae, vol. 1, pp. 17–45, Clarendon Press, Oxford.

Doyle, J.A. (1992). Revised palynological correlations of the lower Potomac Group (USA) and the Cocobeach sequence of Gabon (Barremian-Aptian). *Cretaceous Res.* **13**: 337–349.

Doyle, J.A. and Donoghue, M.J. (1993). Phylogenies and angiosperm diversification. *Paleobiology* **19**: 141–167.

Doyle, J.A. and Hotton, C.L. (1991). Diversification of early angiosperm pollen in a cladistic context. In: S. Blackmore and S.H. Barnes (editors). Pollen and Spores: Patterns of Diversification, pp. 169–195, Clarendon Press, Oxford.

Doyle, J.A., Biens, P., Doerenkamp, A. and Jardiné, S. (1977). Angiosperm pollen from the pre-Albian Cretaceous of Equatorial Africa. *Bull. Centres Rech. Explor.-Prod. Elf-Aquitaine* **1**: 451–473.

Doyle, J.A., Jardiné, S. and Doerenkamp, A. (1982). *Afropollis*, a new genus of early angiosperm pollen, with notes on the Cretaceous palynostratigraphy and paleoenvironments of Northern Gondwana. *Bull. Centres Rech. Explor.-Prod. Elf-Aquitaine* **6**: 39–117.

Doyle, J.A., Hotton, C.L. and Ward, J.V. (1990). Early Cretaceous tetrads, zonasulculate pollen, and Winteraceae. II. Cladistic analysis and implications. *Amer. J. Bot.* **77**: 1558–1568.

Endress, P.K. (1987). The Chloranthaceae: reproductive structures and phylogenetic position. *Bot. Jahrb. Syst.* **109**: 153–226.

Feuer, S. (1989). Pollen morphology of Embothrieae (Proteaceae) I. Buckinghamiinae, Stenocarpinae and Lomatiinae. *Grana* **28**: 225–242.

Feuer, S. (1990). Pollen aperture evolution among the subfamilies Persoonioideae, Sphalmioideae, and Carnarvonioideae (Proteaceae). *Amer. J. Bot.* **77**: 783–794.

Herngreen, G.F.W. (1973). Palynology of Albian-Cenomanian strata in borehole 1-QS-1-MA, State of Maranhão, Brazil. *Pollen & Spores* **15**: 515–555.

Herngreen, G.F.W. and Chlonova, A.F. (1981). Cretaceous microfloral provinces. *Pollen & Spores* **23**: 441–555.

Hughes, N.F. and McDougall, A.B. (1990). Barremian-Aptian angiospermid pollen records from southern England. *Rev. Palaeobot. Palynol.* **65**: 145–151.

Jardiné, S. and Magloire, L. (1965). Palynologie et stratigraphie du Crétacé des bassins du Sénégal et de Côte d'Ivoire. *Mém. Bur. Rech. Géol. Minières* **32**: 187–245.

Jardiné, S., Doerenkamp, A. and Legoux, O. (1972). Le genre *Hexaporotricolpites* Boltenhagen 1967. Morphologie, systématique, stratigraphie et extension géographique. Actes 4e Colloq. Africain Micropaléontol., Abidjan, 1970, pp. 175–191.

Jardiné, S., Kieser, G. and Reyre, Y. (1974). L'individualisation progressive du continent africain vue à travers les données palynologiques de l'ère secondaire. *Sci. Géol. Bull. (Strasbourg)* **27**: 69–85.

Johnson, L.A.S. and Briggs, B.G. (1975). On the Proteaceae — the evolution and classification of a southern family. *Bot. J. Linn. Soc.* **70**: 83–182.

Lobreau-Callen, D. (1980). Caractères comparés du pollen des Icacinaceae et des Olacaceae. *Adansonia*, sér. 2, **20**: 29–89.

Muller, J. (1981). Fossil pollen records of extant angiosperms. *Bot. Rev.* **47**: 1–142.

Muller, J. and Leenhouts, P.W. (1976). A general survey of pollen types in Sapindaceae in relation to taxonomy. In: I.K. Ferguson and J. Muller (editors). The Evolutionary Significance of the Exine, pp. 407–445, Academic Press, London.

Nowicke, J.W. and Skvarla, J.J. (1977). Pollen morphology and the relationship of the Plumbaginaceae, Polygonaceae, and Primulaceae to the order Centrospermae. *Smithsonian Contr. Bot.* **37**: 1–64.

Nowicke, J.W. and Skvarla, J.J. (1979). Pollen morphology: the potential influence in higher order systematics. *Ann. Missouri Bot. Gard.* **66**: 633–700.

Nowicke, J.W. and Skvarla, J.J. (1981). Pollen morphology and phylogenetic relationships of the Berberidaceae. *Smithsonian Contr. Bot.* **50**: 1–83.

Pacltová, B. (1971). Palynological study of Angiospermae from the Peruc Formation (?Albian-Lower Cenomanian) of Bohemia. *Sborn. Geol. Vĕd, řada P, Paleontol.* **13**: 105–141.

Pedersen, K.R., Crane, P.R., Drinnan, A.N. and Friis, E.M. (1991). Fruits from the mid-Cretaceous of North America with pollen grains of the *Clavatipollenites* type. *Grana* **30**: 577–590.

Sampson, F.B. and Endress, P.K. (1984). Pollen morphology in the Trimeniaceae. *Grana* **23**: 129–137.

Specht, R.L., Dettmann, M.E. and Jarzen, D.M. (1992). Community associations and structure in the Late Cretaceous vegetation of southeast Australasia and Antarctica. *Palaeogeogr. Palaeoclimatol. Palaeoecol.* **94**: 283–309.

Walker, J.W. and Doyle, J.A. (1975). The bases of angiosperm phylogeny: palynology. *Ann. Missouri Bot. Gard.* **62**: 664–723.

Walker, J.W. and Walker, A.G. (1984). Ultrastructure of Lower Cretaceous angiosperm pollen and the origin and early evolution of flowering plants. *Ann. Missouri Bot. Gard.* **71**: 464–521.

Ward, J.V., Doyle, J.A. and Hotton, C.L. (1989). Probable granular magnoliid angiosperm pollen from the Early Cretaceous. *Pollen & Spores* **33**: 101–120.

Hotton, C.L., Leffingwell, H.A. and Skvarla, J.J. (1994). Pollen ultrastructure of Pandanaceae and the fossil genus *Pandaniidites*. In: M.H. Kurmann and J.A. Doyle (editors). Ultrastructure of fossil spores and pollen, pp. 173–191, Royal Botanic Gardens, Kew.

11. POLLEN ULTRASTRUCTURE OF PANDANACEAE AND THE FOSSIL GENUS *PANDANIIDITES*

CAROL L. HOTTON[1], HARRY A. LEFFINGWELL[2,3] AND JOHN J. SKVARLA[4]

[1]**Department of Paleobiology, MRC NHB 121, Smithsonian Institution, Washington, DC 20560, U.S.A.**
[2]**Unocal Energy Resources Division, Unocal Corporation, 376 S. Valencia Ave., Brea, CA 92621, U.S.A.**
[3]Current address: **3180 Bonn Drive, Laguna Beach, CA 92651, U.S.A.**
[4]**Department of Botany and Microbiology and Oklahoma Biological Survey, University of Oklahoma, Norman, OK 73109, U.S.A.**

Abstract

Pandaniidites is a common Late Cretaceous-Paleogene, echinate monoporate pollen genus of North America, important as a potential paleoclimate and facies indicator. To determine affinities, its ultrastructure was compared to 24 species in three genera (*Pandanus, Freycinetia, Sararanga*) of the Pandanaceae and two species of *Lemna* (Lemnaceae), using light, scanning and transmission electron microscopy. *Pandaniidites* ultrastructurally resembles Pandanaceae, rather than Lemnaceae; combining characters of both *Pandanus* and *Freycinetia. Lemna* possesses a very thin tectum and foot layer, and relatively tall columellae, as well as tall, narrow spines which are hollow and often constricted at the base. In contrast, *Pandanus* and *Pandaniidites* both possess spines with thick bulbous bases and thickened exine beneath the spine. However, most species of *Pandanus* have a single-layered exine and lack an annulus, while *Pandaniidites* possesses a thin, granular-columellate infratectal layer and imperforate tectum, as well as an annulus. The ultrastructure of *Pandaniidites* more closely resembles that of *Freycinetia*, which differs primarily in possessing a smooth tectum. The most primitive pollen type in Pandanaceae, by outgroup comparison, occurs in the basal members of *Pandanus* (subgenus *Martellidendron*), which have smooth, tectate-perforate, columellate, annulate pollen, and in *Sararanga*, which possesses coarsely reticulate-columellate pollen. The relatively advanced nature of *Pandaniidites*, as well as the generally primitive floral morphology of living pandanaceous genera, suggest that origin of the family occurred before the appearance of *Pandaniidites* in the Maastrichtian, perhaps as early as the Early Cretaceous.

Introduction

Pandaniidites Elsik, which encompasses echinate, monoporate, roughly spheroidal pollen, is a common element in Late Cretaceous and Paleogene sediments of North and South America. Pollen of this type is ascribed by many authors to *Pandanus*, a member of the screw-pine family Pandanaceae, which it closely resembles. The combination of echinate sculpture and monoporate aperture that occurs in both *Pandaniidites* and modern species of *Pandanus* is unusual among angiosperms, and spine morphology is similar in both genera (Jarzen, 1983). However, discrepancies between the modern genus and the fossil form exist. *Pandaniidites* possesses an annulate aperture, unlike *Pandanus* (Leffingwell, 1970; Muller, 1981); and fossil forms are reported from the New World, which is at variance with the strictly paleotropical distribution of modern Pandanaceae (Stone, 1976). Furthermore, echinate, monoporate pollen is by no means restricted to *Pandanus*: similar pollen morphology occurs among palm genera (Thanikaimoni, 1970a; Ferguson et al., 1987), and in the aquatic genus *Lemna* (Erdtman, 1952; Sweet, 1986). Comparative ultrastructural analysis of *Pandaniidites* and similar modern forms is necessary to clarify the nature of its affinities.

If *Pandaniidites* represents *Pandanus* or Pandanaceae, it is a potentially valuable source of information on the timing and place of origin, and past distribution of the modern family. Furthermore, since the architecture and physiology of extant members of the family restrict them to frost-free climatic zones, *Pandaniidites* is potentially important as a paleoclimatic indicator. The fossil genus is also a useful facies indicator consistently associated with pond or marsh environments (Farley, 1989, 1990; Quattrocchio and Volkheimer, 1990), whose interpretation would be assisted by knowledge of its affinities.

We began this study with a specific goal in mind: to compare the pollen morphology of *Pandaniidites* with that of *Pandanus* and *Lemna*. It became apparent that a broad survey of species within *Pandanus* was necessary to assess morphological variability in this species-rich genus, and evaluation of related genera was necessary in order to evaluate phylogenetic significance of characters in *Pandaniidites*. Consequently, this paper has two aims: first, to present a summary of results of a systematic survey of Pandanaceae, focusing on possible affinities of *Pandaniidites*, and second, to apply evidence from pollen morphology to questions of relationships within the Pandanaceae, and among that family and possible outgroups, in order to evaluate the phylogenetic placement of *Pandaniidites*. A more complete systematic treatment of the modern species will be presented elsewhere.

Relevant studies of pollen ultrastructure are few. Huynh (1980) surveyed 16 species of one subgenus of *Pandanus* with scanning electron microscopy (SEM), and pollen of a few other species has been illustrated by Huynh (1981, 1982, 1983), Huynh and Stone (1981), Stone and Huynh (1982) and Jarzen (1983). Tarasevich (1990) has illustrated pollen ultrastructure of *Lemna* and *Spirodela* (Lemnaceae), using both scanning and transmission electron microscopy. To our knowledge, no transmission electron microscopy (TEM) has been presented on fossil pandanaceous pollen, or pollen of the modern genera of Pandanaceae.

Systematics and Sampling of the Study Groups

The Pandanaceae are an arecid monocotyledonous family widespread today in the Paleotropics. The family consists of three genera: *Sararanga* with two species, *Freycinetia* with > 200 species, and *Pandanus* with 600–700 species (Stone, 1990a). Members of this family are arborescent or lianous, dioecious, and possess small flowers aggregated into a complex bracteate inflorescence, sword-shaped leaves with toothed margins and midribs, and an unusual phyllotaxis in which the leaves are borne in tristichous ranks on a spirally twisted axis (Cronquist, 1981; Dahlgren et al., 1985). Pollination is by wind, insects, or in *Freycinetia*, bats and birds (Cox, 1984).

Stone (1974) divided *Pandanus* into four groups of eight subgenera and approximately 80 sections, based on variations in floral morphology, leaf anatomy and morphology, and habit. No formal phylogenetic analysis of the genus has been attempted to date. The subgenus *Martellidendron*, a small monophyletic group endemic to Madagascar and allied with the African subgenus *Vinsonia*, is probably the most primitive based on the consistent occurrence of pistillodia in the staminate flowers (Huynh, 1981; Stone, 1990b). Of the other subgenera, *Rykia*, *Acrostigma* and *Coronata* are likely to be monophyletic, while the others probably are not (Stone, pers. comm., 1993). We have examined 36 species of *Pandanus* with light microscopy (LM), including representatives of all seven subgenera for which staminate flowers are known. Of these, 20 species from seven subgenera have been examined with SEM and TEM (Table 1).

Freycinetia species are characterized by a number of features that represent adaptations to vertebrate pollination, such as brightly colored, fleshy, nutritious bracts, abundant pollen and partially bisexual inflorescences (Cox, 1982, 1984). Stone (1968) has divided *Freycinetia* into 17 sections, which have not been grouped into subgenera. We have examined 13 species representing nine sections with LM; of these, five species from three sections have been studied with SEM and TEM (Table 1).

In addition, use was made of data kindly provided by Stone and Merican from an unpublished manuscript on pollen morphology in *Pandanus* and *Freycinetia*.

The Lemnaceae is another arecid family, closely allied with the Araceae, consisting of six cosmopolitan genera and about 30 species. Members of this family are minute, floating thalloid aquatics, with greatly reduced inflorescences and roots and conducting tissue reduced or absent (Cronquist, 1981). We have examined two species of *Lemna* (Table 1).

Materials and Methods

The modern pollen examined in this study was obtained from herbarium material loaned from the Herbarium of the University of California, Berkeley (UC), the United States National Herbarium (US), the Laboratoire de Phanérogamie, Paris (P), the Royal Botanic Gardens, Kew (K), the Papua New Guinea National Herbarium (LAE) and the Canadian Museum of Nature, Ottawa (CANM). The species that were prepared for SEM and TEM, their classification, and sources, are displayed in Table 1. Specimens were first briefly acetolyzed, then mounted and coated for SEM, and embedded in resin, stained and sectioned for TEM. Details of preparation techniques may be found in Leffingwell and Hodgkin (1971), Skvarla and Kelly (1968) and Ward et al. (1989). Measurements of whole grains reported are based on counts of 10 to 20 specimens mounted in glycerine jelly under LM. SEM was carried out on a Cambridge Stereoscan 2 (Micrographics, Inc.), an Amray 1000 at Unocal and a Hitachi S570 electron microscope at the National Museum of Natural History. TEM was carried out on a Phillips 200 and JEOL 2000 at the University of Oklahoma and a JEOL S100 at the University of California at Davis.

Two species of *Pandaniidites* were examined in this study: *Pandaniidites typicus*, from Late Cretaceous-early Paleocene sediments of eastern central Montana, and *P. radicus* from the type Late Cretaceous Lance Formation of Wyoming, USA. The specimens of *P. typicus* that were examined with SEM and TEM were prepared from thin lignites directly overlying an iridium-enriched boundary layer (samples RPC83-5, IH81-422-424, Hotton, 1988), where they were particularly abundant. Specimens of *Pandaniidites radicus* examined with SEM and TEM were prepared from a gray siltstone approximately 700 feet below the Cretaceous-Tertiary boundary (sample RC 1636, Leffingwell, 1970). Individual grains were picked from residues and prepared for light and electron microscopy.

Results

A. *Sararanga*

Pollen of *Sararanga sinuosa* is small, averaging 18 × 17 μm. Shape is usually ovoid, with one end slightly narrower than the other, and with one or sometimes two slightly flattened faces and a broad convex face, or sometimes subspheroidal. Aspect ratio (ratio of major to minor axis) is about 1.21. Viewed from one end, grains are roughly triangular to almost spheroidal. The small, simple pore, often obscure, is usually located near the midpoint of the long axis of the grain and on the convex face. The sculpture is relatively coarse and irregularly reticulate, with smooth, flattened muri (Fig. 1). TEM shows that the tectum and foot layer are relatively thin and roughly equal in thickness, each about 0.2–0.4 μm thick, and that the columellae are tall and narrow, roughly 0.8–1.0 μm high

TABLE 1. Taxonomy and Sources of Species Studied with SEM and TEM

Pandanaceae	
Freycinetia	
Section *Filiformicarpae*	
F. reineckei Warb.	Cox 120 [UC]
F. ponapensis Mart.	Johns & Stasuno 50470 [LAE]
Section *Freycinetia*	
F. arborea Gaud.	Katik 46753 [LAE]
F. wilderi Mart.	Katik 46754 [LAE]
Section *Taiwanella*	
F. formosana Hemsl.	Faurie 534 [P]
Pandanus	
Subgenus *Acrostigma*	
Section *Acrostigma*	
P. copelandii Merr.	McGregor 23186 [US]
P. thwaitesii Mart.	Saldanha & Ghandi 2155 [US]
Subgenus *Kurzia*	
Section *Kurzia*	
P. hollrungii Schumann & Warb.	Stone 53326 [LAE]
Subgenus *Lophostigma*	
Section *Barrotia*	
cf. *P. macrocarpus* Vieill.	McKee 40275 [PH]
Subgenus *Martellidendron*	
Section *Martellidendron*	
P. hermaphroditus Mart.	de la Bâthie 11010 [P]
P. androcephalanthus Mart.	de la Bâthie 10936 [P]
Subgenus *Rykia*	
Section *Roussinia*	
P. leram Jones	Stone 11102 [US]
Section *Rykia*	
Subsection *Multispina*	
P. labyrinthicus Kurz	Ridley UC 266916
Subsection *Rykia*	
P. furcatus Roxb.	UC 235235
Section *Solmsia*	
P. helicopus Kurz	Corner 28161 [P]
Subgenus *Pandanus*	
Section *Excavata*	
P. antaresensis St. John	Stone 26 [UC]
Section *Pandanus*	
Subsection *Intraobtutus*	
P. papuanus Solms	UC 235003
Subsection *Pandanus*	
P. fragrans Gaud.	Rodin UC 789274
P. kirkii Rendle	Frazier 1146 [US]
P. odoratissimus L.	UC 360521
P. tectorius Park. ex Z.	J & M S Clemens 4054 [UC]
P. unipapillatus Denst.	Suresh & Manilal 1920 [US]

Table 1 continued

Subgenus *Vinsonia*	
Section *Heterostigma*	
P. ambongensis Mart.	de la Bâthie 11906 [P]
Section *Stephanostigma*	
P. analamazaotrensis Mart.	de la Bâthie 4607 [P]
Section *Vinsonia*	
P. utilis Bory	DeCary 5309 [US]
Sararanga sinuousa Hemsl.	Cox 613 [UC]
Cyclanthaceae	
Subfamily *Carludoviceae*	
Asplundia rhodea Schultes	Idrobo & Schultes 1052 [US]
Lemnaceae	
Lemna aequinoctialis Welw.	Massey & Nighswonger 1764 [OKL]
Lemna gibba L.	Palmer 11398 [US]

(Fig. 2). A very thin, lightly staining layer that we interpret as endexine appears to be continuous around the grain. The aperture is marked by an abrupt thinning of the ektexine, with a pore membrane formed by elements of the foot layer and endexine (Fig. 3).

B. *Freycinetia*

Pollen morphology in *Freycinetia* is quite uniform. Size class is small to medium, ranging from about 12 to 30 μm (largest diameter). Shape is generally ovoid to ellipsoidal or reniform, with an aspect ratio of about 1.2–1.45 (average 1.32). The aperture consists of a large and conspicuous, round or slightly elliptical pore, generally located near the midpoint of the long axis of the grain (Fig. 4). The pore is usually surrounded by a conspicuous annulus, although in a few species it is simple and unornamented. Sculpture in all species examined is psilate to granulate. Under SEM, the tectum is seen to range from smooth and almost entirely imperforate (Fig. 5) to microfoveolate (Fig. 6). TEM reveals that the exine is three-layered, with a continuous tectum and foot layer of about equal thickness (each about 0.2–0.5 μm) and a very narrow infratectal layer, ranging from 0.05–0.2 μm thick (Figs. 7–10). The infratectal layer usually consists of short, irregular granular columellae (Fig. 10), but may sometimes display distinct columellae (Fig. 9). The aperture is surrounded by a markedly thickened foot layer, and the pore membrane itself is composed of both endexinal and foot layer material (Fig. 8).

C. *Pandanus*

Most of the species surveyed display a remarkable degree of uniformity in size, shape, sculpture and exine structure. Size class is small to medium, ranging from about 15 to 30 μm (largest diameter). Shape is characteristically elongate and ovoid with distinctly convex and flattened faces, but it may vary from reniform to almost spheroidal. The most divergent pollen occurred in the subgenus *Martellidendron*, which will be discussed separately below. Aspect ratio in the subgenera *Lophostigma* and *Kurzia* is about 1.4,

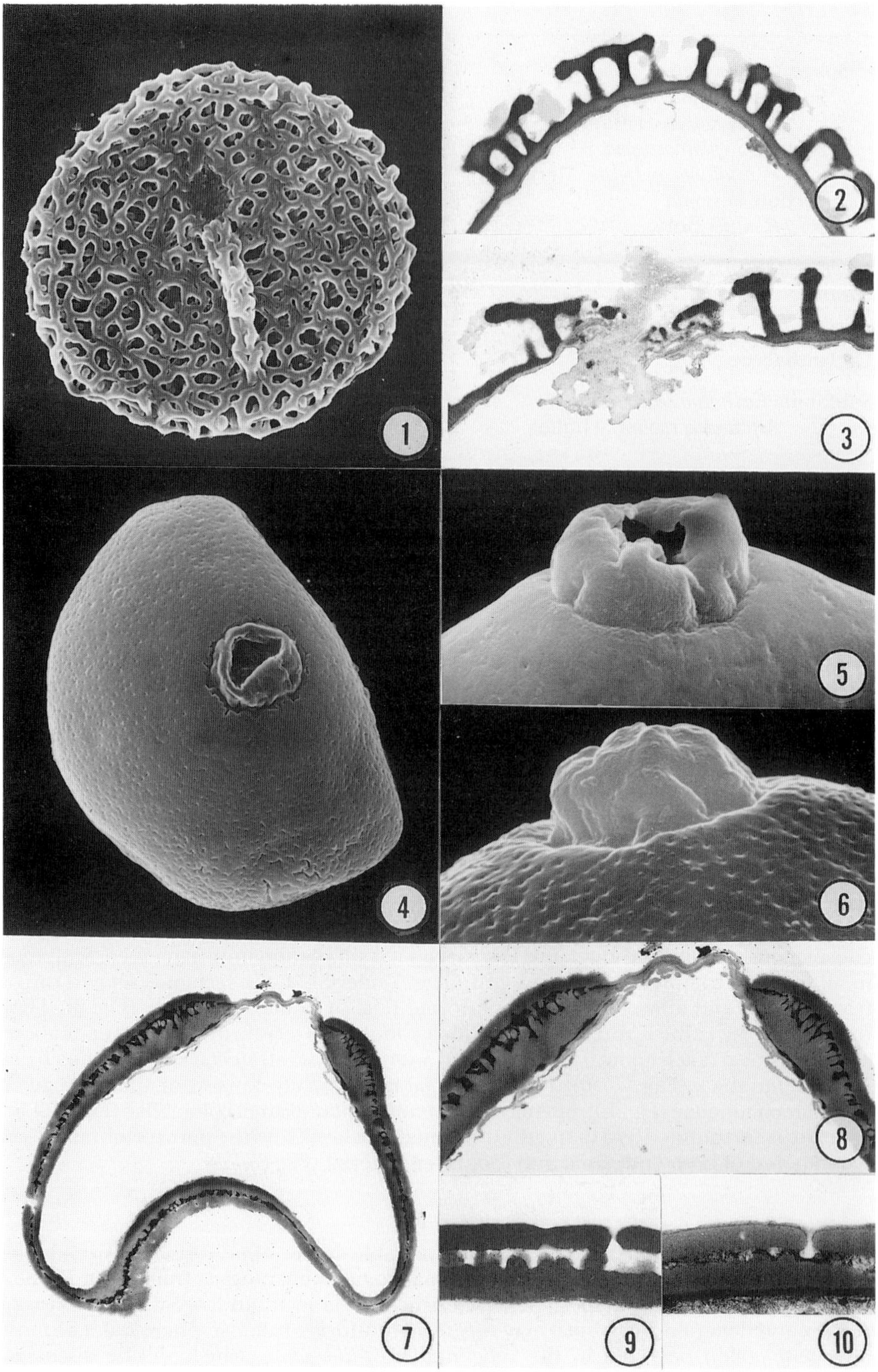
1
2
3
4
5
6
7
8
9
10

ranging from 1.34–1.65. Aspect ratio in the other four subgenera is about 1.2, ranging between 1.07 and 1.29. The aperture consists of a simple pore (never annulate), usually located at the narrow end and toward the flattened face of the grain (Figs. 11 and 14). Sculpture is most commonly echinate, with the spines characteristically set into a slight depression in the exine (Fig. 11). The surface of the exine between spines may be almost smooth to distinctly granular. The exine is frequently thickened beneath spines, a feature which is visible in both LM and SEM, where a verrucate pattern is present on the inside face of the exine (Fig. 12). Sculpture is rarely verrucate, as in *P. papuanus* (Fig. 13), or more commonly scabrate or psilate, as in *P. thwaitesii* (Fig. 14). Psilate pollen appears to be confined to the subgenera *Lophostigma*, *Rykia* and *Acrostigma*, supporting observations by Stone and Merican (unpublished manuscript). TEM reveals that the exine is one-layered and homogeneous in six of the seven subgenera (Figs. 15–19). The exine is essentially imperforate, but occasionally penetrated by small channels (Fig. 15). The aperture is marked by an abrupt thinning of the ektexine and a thickening of the endexine to form a membrane across the pore (Figs. 16–17). Spines vary in shape, but characteristically possess a broad, bulbous base that narrows abruptly to a rounded or capitate tip (Fig. 18). However, sometimes they have a more narrowly pyramidal shape (Fig. 19) or are constricted near the base (Fig. 17). A thickened exine beneath spines and a narrow "moat" surrounding the spine are also visible with TEM (Figs. 18–19). Endexine is usually not recognizable away from the aperture, but in a few specimens it appears as one or several very thin, discontinuous layers (Fig. 15).

Members of the subgenus *Martellidendron* differ in several respects from the above species. The shape is reniform to ovoid, with an aspect ratio of about 1.6 (Fig. 20). The sculpture is microfoveolate or microfossulate with a smooth tectum (Fig. 21). The pore is annulate, and varies in position along the long axis of the grain, as noted by Huynh (1981). Most strikingly, TEM reveals that the exine is three-layered, with distinct columellae, a foot layer and a perforate tectum of roughly equal thickness (Figs. 22–23). Both the foot layer and infratectal layer are thickened around the aperture to form the annulus, while the pore membrane itself consists primarily of endexine (Fig. 22).

In all three genera of Pandanaceae, most grains have a characteristic, ovoid shape with both a narrow and broad end along the long axis of the grain, and a convex and flattened face. Consequently, pollen grains usually display only one axis of symmetry, or if the aperture is displaced to one side, no axis of symmetry. Furthermore, the position of the aperture relative to the long axis of the grain varies among genera. In most species of *Pandanus*, the pore usually occurs at the narrow end of the grain, except in the subgenus *Martellidendron*, where its position varies along the long axis of the grain. In *Sararanga* and *Freycinetia*, the pore position is variable along the convex face of the grain. If the more convex face of a grain represents the distal face and the more flattened face or faces are proximal (Huynh, 1976), the aperture position in *Pandanus* appears to be generally equatorial, whereas in *Freycinetia* and *Sararanga* it is distal or more often subdistal (for example, Fig. 4). If true, *Pandanus* would violate the general rule that monosulcate or derived apertures are distal (Walker and Doyle, 1975). Confirmation of this hypothesis requires observation of microspores in tetrad.

D. *Pandaniidites*

The genus *Pandaniidites* was described by Elsik (1968) for monoporate, subspheroidal, echinate pollen with an annulate pore and imperforate tectum. The type species, *P. texus*,

Figs. 1–10. *Sararanga sinuosa* (Figs. 1–3) and *Freycinetia* (Figs. 4–10). **1.** SEM, whole grain, × 3000. **2.** TEM, detail of exine structure, × 6500. **3.** TEM, detail of aperture, × 7500. **4.** *Freycinetia wilderi*, SEM, whole grain, × 2500. **5.** *Freycinetia arborea*, SEM, detail of aperture and exine sculpture, × 6000. **6.** *Freycinetia ponapensis*, SEM, detail of aperture and exine sculpture, × 6000. **7.** *Freycinetia reineckei*, TEM, whole grain, × 3000. **8.** *Freycinetia reineckei*, TEM, same grain, detail of aperture, × 6000. **9.** *Freycinetia formosana*, TEM, detail of exine structure, × 15,000. 10. *Freycinetia arborea*, TEM, detail of exine structure, × 15,000.

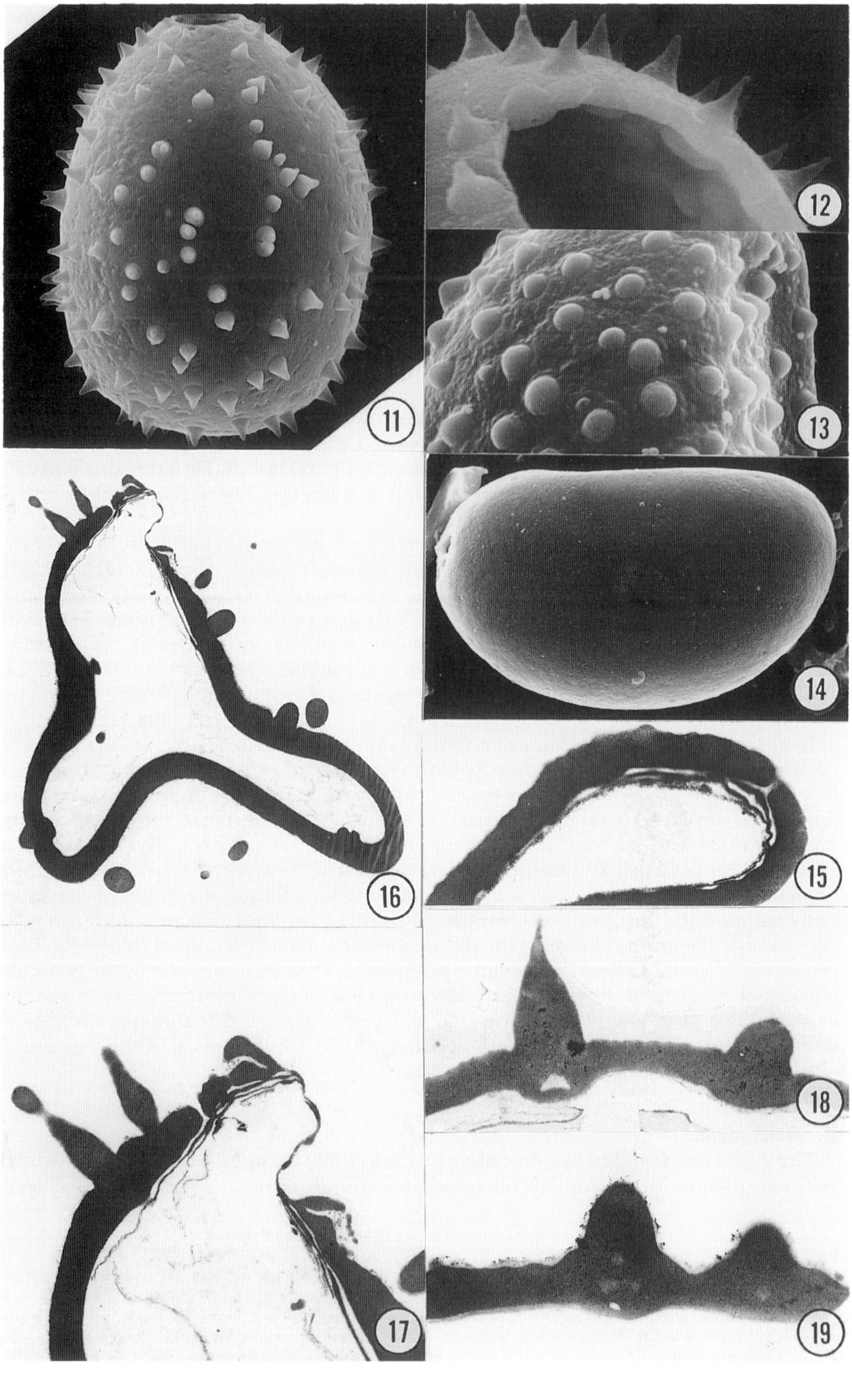
11
12
13
14
15
16
17
18
19

was described from late Paleocene lignites of Texas, USA. *P. radicus* Leffingwell, from the Maastrichtian Lance Formation of Wyoming, USA, is differentiated from the type on the basis of an exine that is less than 0.5 μm (Leffingwell, 1970). *P. typicus* (Norton) Farabee and Canright, described from the Late Cretaceous-early Paleocene of Montana, is reported to have a variably developed annulus (Norton and Hall, 1969), although in our observations of specimens from the type locality the annulus is usually well-developed (see, for example, Fig. 26). Specimens of *P. typicus* from the type locality appear to be significantly larger than specimens attributed to that species elsewhere, as well as the other species, and have a slightly thicker exine. Otherwise, all three species appear very similar in size, shape, aperture position, and spine shape, and it appears likely that they are conspecific. Nevertheless we prefer to provisionally retain the three separate species until they have been subjected to more detailed ultrastructural and morphometric analysis. The following observations apply to *Pandaniidites radicus* and *P. typicus* from their respective type localities; further study is necessary to determine whether these observations apply to the type species and to similar pollen described elsewhere.

Both species of *Pandaniidites* are medium sized: *P. radicus* ranges from 21.8–31.1 × 15.0–24.9 μm (Leffingwell, 1970), while *P. typicus* ranges from 29–44 × 25–40 μm (mean is 34.6 × 30.4 μm). Shape is quite variable, ranging from nearly spheroidal to ovoid, with one end slightly narrower than the other. Aspect ratio varies from 1.0 to 1.3, with an average of 1.15. Some of the variability in shape is undoubtedly due to compression and distortion from fossilization, but some undoubtedly reflects the original shape of the grains. Sculpture is conspicuously echinate, with tall baculate spines rising from bulbous bases, approximately 1–4 μm tall and 1–3 μm wide at base (Figs. 24–25, see also TEM photos in Figs. 27–30). The position of the pore varies, but it is generally located along the midpoint of the long axis of the grain, and only rarely at the end of the grain. The pore is usually conspicuous, round or slightly ellipsoidal and surrounded by an annulus about 2–3 μm wide (Figs. 25–26).

TEM reveals that both species of *Pandaniidites* have a three-layered exine with an essentially imperforate tectum and foot layer of approximately equal thickness, and a very thin infratectal layer consisting of irregular granules to short columellae (Figs. 27–28). *P. typicus* appears to have a slightly thicker exine, with tectum and foot layer about 0.25–0.35 μm thick and infractectal layer about 0.1 μm thick, while *P. radicus* has tectum and foot layer about 0.2–0.25 μm thick and infratectal layer about 0.05 μm thick. Otherwise ultrastructural details appear very similar in both species. Tectum, and sometimes foot layer, are conspicuously thickened beneath spines (Figs. 28, 30). The broad bases of spines are conspicuous in section (Figs. 28–29). Shape of spines varies from napiform or turbinate with a distinct constriction near the base (Figs. 25, 28) to baculate (Figs. 25, 29) or conate (Fig. 30). Spine shape appears to be considerably more variable than that observed within species of *Pandanus*, although quantification is required to confirm the observation. Small channels occasionally penetrate the tectum, but rarely the foot layer (Fig. 29). No sections of the aperture were obtained, but from LM the annulus appears to consist of a thickened foot layer and a slightly thickened infratectal layer. Endexine was not recognized in any of the specimens sectioned, confirming observations at the LM level that exine of *Pandaniidites* is composed entirely of ektexine (Leffingwell et al., 1970).

E. *Lemna*

Pollen of *Lemna* is medium-sized, averaging about 20 μm in diameter. Shape is nearly spheroidal to slightly ovoid, with an aspect ratio of about 1.1. Sculpture is echinate, with

FIGS. **11–19**. *Pandanus*. **11.** *Pandanus furcatus*, SEM, whole grain, × 3500. **12.** *Pandanus tectorius*, SEM, broken grain showing verrucate inner wall of exine and details of sculpture, × 7000. **13.** *Pandanus papuanus*, SEM, detail of exine sculpture, × 7000. **14.** *Pandanus thwaitesii*, SEM, whole grain, × 2000. **15.** *Pandanus thwaitesii*, TEM, detail of exine structure, × 10,000. **16.** *Pandanus leram*, TEM, whole grain, × 3500. **17.** *Pandanus leram*, TEM, same grain, detail of aperture, × 10,000. **18.** *Pandanus antaresensis*, TEM, detail of exine structure, × 7500. **19.** *Pandanus labyrinthicus*, TEM, detail of exine structure, × 15,000.

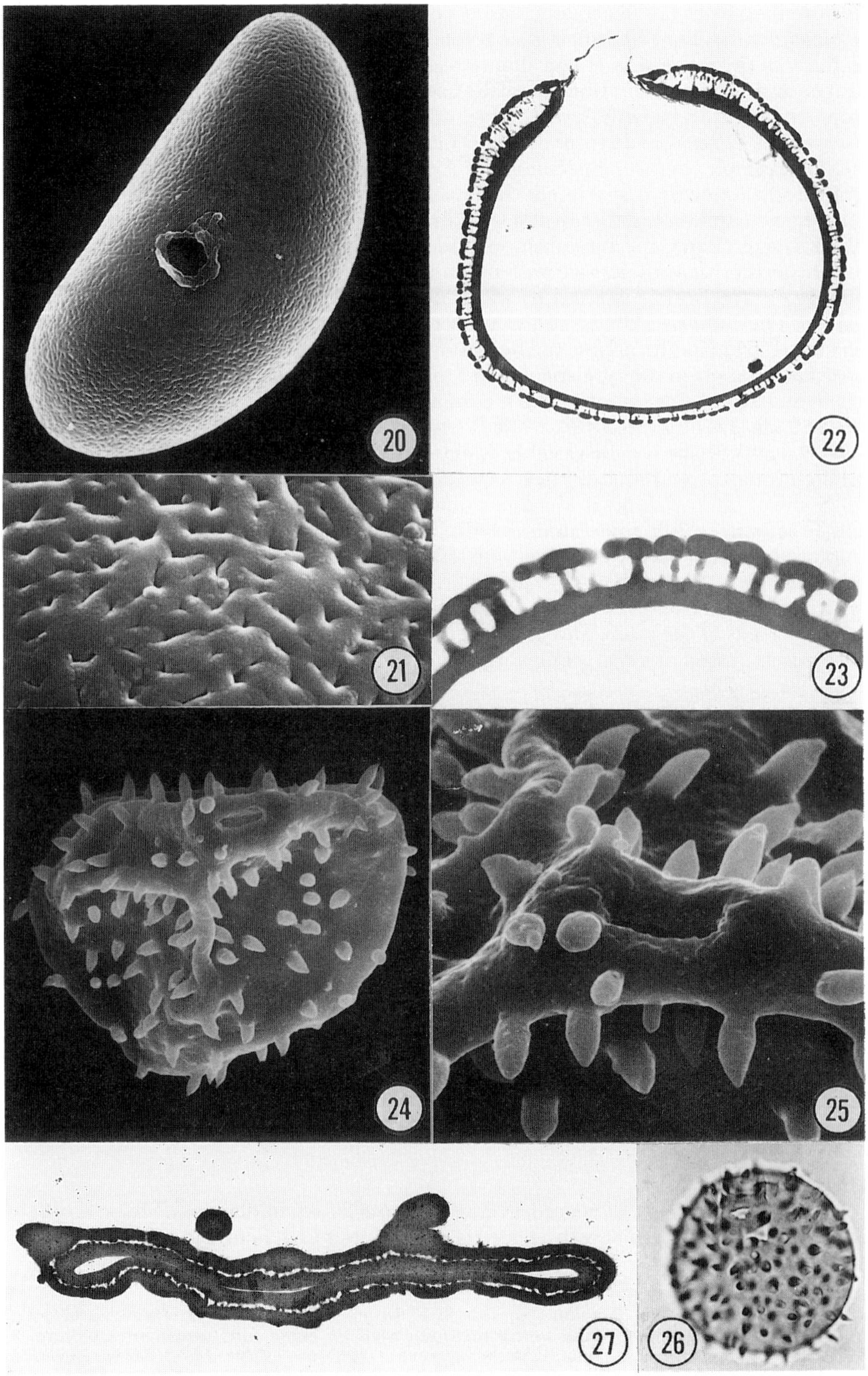
20
22
21
23
24
25
27
26

tall, narrow spines, ranging from 1.5 to 4 μm tall (Fig. 31), often constricted at the base (Fig. 32). The exine may be covered with coarse granules between spines (Fig. 31), and is perforate (Fig. 32). The porate aperture is small, approximately 2 μm in diameter, round, simple and conspicuous (Fig. 31). TEM reveals a very thin tectum and foot layer, with narrow, widely separated columellae (Fig. 33). Spines are hollow in the center, and exine is not thickened beneath them.

Discussion

A. Affinities of *Pandaniidites*

Monoporate pollen with coarse echinate sculpture is unusual; significantly, it occurs primarily among scattered taxa within the Arecidae. *Pandaniidites* resembles few extant pollen types other than *Pandanus*, and most of these differ in too many details to be considered potentially related. Scattered genera among the palms are echinate and monoulcerate or brevisulcate, but their apertures are generally more irregular than in *Pandaniidites* and lack an annulus (Sowunmi, 1972; Thanikaimoni, 1970a, 1970b; Ferguson et al., 1987). Furthermore, there are ultrastructural differences in echinate, monoporate palms: in *Korthalsia*, *Mauritia* and *Mauritiella*, the exine is one-layered (Ferguson, 1986), and in *Ravenea* and *Louvelia* the tectum is perforate, with tall, narrow columellae (Ferguson et al., 1987). Spine morphology differs in these genera as well. Pollen of *Lemna* more closely resembles *Pandaniidites* in possessing a very small, round pore with an imperforate tectum and robust spines. However, *Lemna* possesses a very different ultrastructure, consisting of distinct columellae and a very thin tectum and foot layer. Furthermore, the spines lack the bulbous base and exinal thickenings beneath, and are hollow rather than solid. In addition, *Lemna* is more spheroidal in shape than *Pandaniidites*, and it lacks an annulus.

Ultrastructural details of *Pandaniidites* strongly support its attribution to Pandanaceae. However, it is not referable to modern *Pandanus*; rather, it combines characters of both *Pandanus* and *Freycinetia*. Details of spine morphology, such as the bulbous bases and exinal thickenings beneath spines, as well as aspect ratio, recall echinate species of *Pandanus*. It should be noted that spines in *Pandaniidites* differ in some respects from those of *Pandanus*; for example, they are not recessed into the exine as in most echinate species of *Pandanus*, and they appear to display greater variability in shape. *Pandaniidites* more closely resembles *Freycinetia* in the possession of a three-layered exine with a thin, granular infratectal layer and an annulus, as well as in the position of the aperture generally along the midpoint of the convex face. In most characters, therefore, *Pandaniidites* most closely resembles an echinate species of *Freycinetia*; however, extant species of *Freycinetia* appear to be uniformly psilate or scabrate. Although two species of *Freycinetia* have been reported to be echinate (Stone, 1974; Stone and Merican, unpublished manuscript), specimens attributed to these two species that we have examined were psilate. The discrepancy between our observations and those of Stone and his students may be due to misattribution of the species name to the staminate specimens actually examined. Further study of positively identified specimens will be necessary to resolve the discrepancy, but we suspect that these two species are psilate, just as in other species of *Freycinetia* examined.

FIGS. **20–27**. *Pandanus* (Figs. 20–23) and *Pandaniidites* (Figs. 24–27). **20.** *Pandanus hermaphroditus*, SEM, whole grain, × 2000. **21.** *Pandanus hermaphroditus*, SEM, same grain, detail of exine sculpture, × 15,000. **22.** *Pandanus hermaphroditus*, TEM, whole grain, × 4000. **23.** *Pandanus hermaphroditus*, TEM, detail of exine structure, × 15,000. **24.** *Pandaniidites radicus*, SEM, whole grain, × 3000. **25.** *Pandaniidites radicus*, SEM, same grain, detail of aperture and sculpture, × 6500. **26.** *Pandaniidites typicus*, LM, specimen from near base of the Hell Creek Formation, Montana, USA, × 2000. **27.** *Pandaniidites typicus*, TEM, whole grain, × 5000.

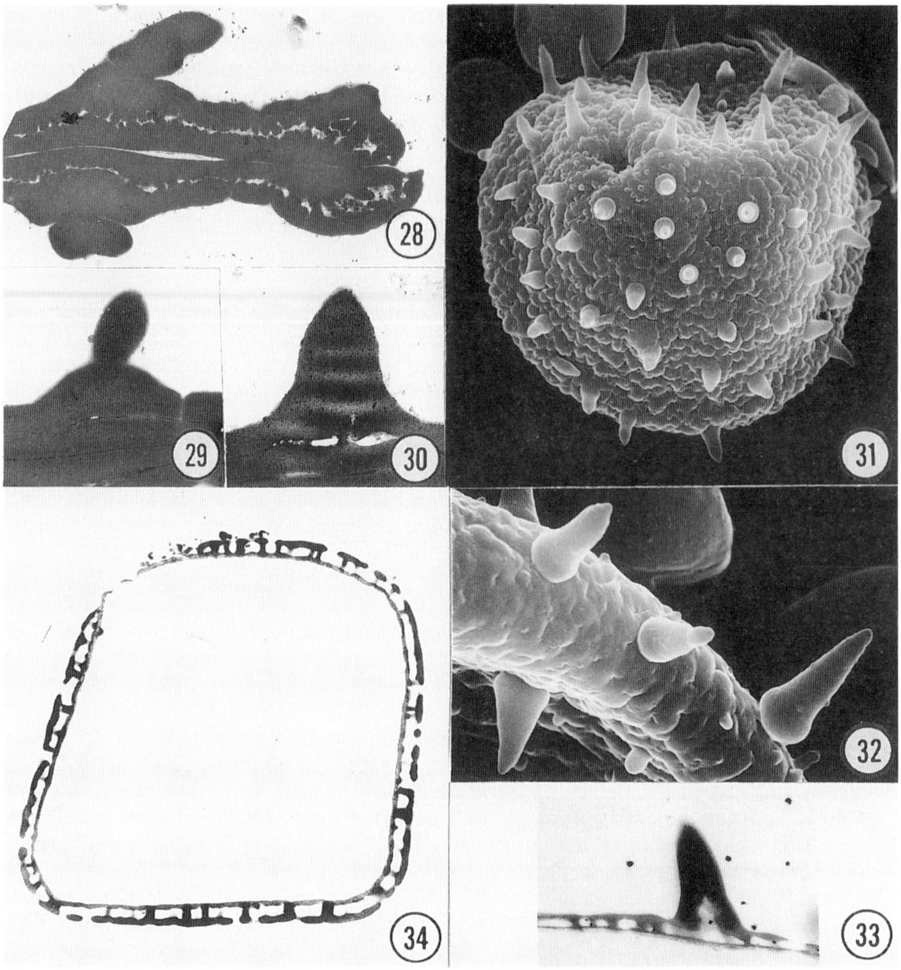

Figs. 28–34. *Pandaniidites* (Figs. 28–30), *Lemna* (Figs. 31–33) and *Asplundia* (Fig. 34). **28.** *Pandaniidites typicus*, TEM, detail of exine, × 8000. **29.** *Pandaniidites radicus*, TEM, detail of spine, × 12,500. **30.** *Pandaniidites radicus*, TEM, detail of spine, × 12,500. **31.** *Lemna gibba*, SEM, whole grain, × 2500. **32.** *Lemna gibba*, SEM, detail of exine sculpture, × 7500. **33.** *Lemna aequinoctialis*, TEM, detail of exine structure, × 15,000. **34.** *Asplundia rhodea*, TEM, whole grain, × 5000.

B. Phylogenetic relationships of *Pandaniidites* and *Pandanaceae*

Having established that *Pandaniidites* most likely represents *Pandanaceae*, how does it fit into the phylogeny of the family, and what does it contribute to our understanding of origin and evolution within the family? Answers to those questions require a consideration of relationships among the modern genera of *Pandanaceae* and between the family and possible outgroups. In the following discussion, we will first consider alternative hypotheses of relationships within the Pandanaceae, then discuss how pollen morphology sheds light on these questions, and finally consider how *Pandaniidites* may be interpreted within a phylogenetic framework, and its implications for the origin, evolution and paleoecology of Pandanaceae.

Relationships among genera of Pandanaceae – There are two views of relationships among genera of Pandanaceae: in the first view, *Sararanga* is most likely primitive because of its arborescent habit, paniculate inflorescence and unspecialized stomatal apparatus (Warburg, 1900; North and Willis, 1971; Stone, 1972); in the second view, *Freycinetia* is considered most primitive, due to its numerous ovules (Takhtajan, 1969). In a recent expansion of the latter view, Cox (1990, 1991) has argued that vertebrate pollination (and hence *Freycinetia*) is primitive within the Pandanaceae, and that anemophily arose in *Pandanus* as an escape from dependence on pollinators, which in turn served as an exaptation (Gould and Vrba, 1982) facilitating colonization of islands and disturbed habitats. Central to his argument is that wind pollination, once evolved, is irreversible due to selective pressures for different functions in male and female flowers (Cox, 1990).

In support of this thesis, Cox (1990) presents a preliminary cladistic analysis of *Freycinetia*, *Sararanga* and *Pandanus*, plus three genera from different outgroups (Arecaceae, Araceae and Cyclanthaceae). The results place *Freycinetia* at the base of Pandanaceae. The cladogram is based on a small sample of characters, as the author himself emphasizes, and does not include potentially informative characters such as pollen, staminate floral morphology, leaf anatomy and architecture, and cuticular morphology, among many others. Furthermore, the two genera chosen to represent palms and aroids as outgroups are advanced within their families (Grayum, 1990; Uhl and Dransfield, 1987), and characters chosen are in some cases questionable. Consequently, we do not accept that the data presented are sufficient to support the conclusion that *Freycinetia* is primitive within the family. We do not discount this hypothesis but a more complete analysis is required to support it.

Furthermore, although pollination biology of *Pandanus* is poorly known, there is indirect evidence both that biotic pollination occurs in the genus, and that wind pollination is by no means irreversible. Pollination biology in the primitive *Pandanus* subgenus *Martellidendron* has not been studied but may be primarily anemophilous, based on the presence of long, pendant stamens that produce abundant, dry pollen in two species. However, more advanced sections show abundant evidence of biotic pollination. For example, although Cox (1990) shows *P. tectorius* to be wind pollinated, the staminate inflorescences of the closely related species *P. odoratissimus* produce a fragrant oil which likely functions as an insect attractant (Dahlgren et al., 1985). Many species of *Pandanus* possess brightly colored inflorescence bracts (Dahlgren et al., 1985). Furthermore, Stone and Merican (unpublished manuscript) note a correlation between psilate pollen and species of low stature inhabiting forest understory, a habit and habitat that together virtually preclude anemophily. Stone (1990b) suggests that beetles or thrips act as pollinators "in many species", and "possible chiropterophily" is reported in one section of *Pandanus* (Cox, 1990).

Evidence of both biotic and abiotic pollination in closely related species of *Pandanus* suggests that pollination in this genus is quite labile. In palms, which have somewhat comparable floral structure and function, adaptations for beetle and wind pollination are quite similar: both syndromes involve small, numerous flowers clustered on an axis with abundant pollen production (Henderson, 1986). In *Pandanus*, the shift between abiotic and biotic pollination plausibly involves minor genetic changes, such as those that affect type and distribution of pollenkitt to promote or reduce pollen clumping and production or inhibition of fragrant essential oils as attractants. In fact, many or most species of *Pandanus* may be amphiphilous, with a generalized, open pollination syndrome that opportunistically utilizes both biotic and abiotic pollen vectors (Wodehouse, 1945).

Pollen evidence on outgroup and ingroup relationships – Uncertainty about relationships among the three genera of Pandanaceae is exacerbated by the isolated position of the family and consequent uncertainty over the appropriate outgroup. Cyclanthaceae (Harling, 1958; Thorne, 1992; Dahlgren et al., 1985), Arecaceae (Cronquist, 1981),

Typhales (Takhtajan, 1987) and Cyperales (Dahlgren et al., 1985) have all been proposed as possible outgroups. Pollen morphology supports coryphoid palms and Cyclanthaceae as most likely outgroups, as well as the primitive position of *Pandanus* subgenus *Martellidendron*. Pollen of Coryphoideae, the most primitive subfamily within the Arecaceae, is generally monosulcate, boat-shaped, columellate and tectate-perforate (Dransfield et al., 1990). Pollen of Cyclanthaceae is generally similar in ultrastructure (Fig. 34; Zavada, 1983), although aperture condition is often ulcoidate or sulcoidate (Harling, 1958). Within Pandanaceae, pollen of *Pandanus* subgenus *Martellidendron* most closely resembles this type of pollen. *Freycinetia* is also similar, but the columellae are generally much reduced, and the tectum in many species is essentially imperforate.

An additional consideration is that tectate-perforate, columellate pollen is primitive within monocots (Donoghue and Doyle, 1989; Doyle and Hotton, 1991). If this is correct, pollen morphology in *Martellidendron* supports its primitive position but does not support specific relationship to palms and Cyclanthaceae; rather, all three retain the primitive monocot ultrastructure.

The pollen of *Sararanga* potentially supports an outgroup relationship with Typhales in possessing an unornamented, excentric pore and a relatively coarse, irregularly reticulate sculpture. However, monads of *Typha angustifolia* and *Sparganium* possess shorter columellae, prominently peaked muri and apparently lack endexine (Nilsson et al., 1977).

If palms or Cyclanthaceae are considered outgroups, pollen morphology of *Sararanga* is most primitive, with its columellate structure, distal pore, and absence of spines and annulus. It is advanced only in possession of a semitectate exine. *Freycinetia* and *Pandanus* subgenus *Martellidendron* both have primitive ultrastructure but possess an annulus. *Freycinetia* appears somewhat more advanced in possessing reduced columellae and primarily tectate-imperforate exine, but is more primitive in possessing a generally distal pore, relative to *Martellidendron*. *Pandanus* species, excluding subgenus *Martellidendron*, are clearly advanced in their one-layered exine, presumed equatorial pore position and possession of spines, in which case absence of an annulus represents a loss. Absence of spines in psilate species of *Pandanus* also presumably represents one or more losses.

Phylogenetic and evolutionary significance of Pandaniidites – The combination of characters in *Pandaniidites* suggests that it could be allied with either *Pandanus* or *Freycinetia*. In a preliminary cladistic analysis of pollen characters by one of us (CLH), *Pandaniidites* and *Freycinetia* consistently form a clade, suggesting that *Pandaniidites* is a sister taxon or basal member of *Freycinetia*. Alternatively, *Pandaniidites* could represent a sister taxon or an extinct clade within *Pandanus* that is united with more advanced *Pandanus* by possession of spines, but that retains a more primitive ultrastructure. Here the annulus is lost in more advanced extant members of *Pandanus*. In any case, *Pandanus* subgenus *Martellidendron* and *Sararanga* consistently occupy a basal position, while *Pandaniidites*, *Freycinetia* and *Pandanus* are more advanced. This suggests that *Pandanus* as currently circumscribed is paraphyletic, and supports the view that *Pandanus* subgenus *Martellidendron* is better treated as a separate genus (B.C. Stone, pers. comm., 1993). Pollen evidence alone is insufficient to choose among competing hypotheses due to the high degree of conflict among characters, which emphasizes the need for cladistic analysis based on a complete character set.

The phylogenetic importance of *Pandaniidites* lies in its establishing a minimum age for the origin of the Pandanaceae. Pollen ascribed to *Pandaniidites* or *Pandanus* has been reported no earlier than Maastrichtian (70 mya) (Muller, 1981; Jarzen, 1983). If, as suggested above, this type of pollen is relatively advanced, origin of the family may have occurred significantly earlier. The earliest arecid record is of palms in the early Senonian (90 mya) (Daghlian, 1981); radiation of arecids would have taken place prior to this, perhaps in the early Late Cretaceous, or even earlier. This view is strengthened by the many primitive aspects of pandanaceous floral morphology, such as absence of a style throughout the family, occurrence of apocarpy in *Sararanga* and *Pandanus*, and incompletely sealed carpels in *Pandanus* (Stone, 1972, 1974, 1990b), which suggests that the ancestor of the Pandanaceae possessed quite primitive floral morphology. Monoporate pollen similar to that of subgenus *Martellidendron* or *Sararanga* may be present but remains unrecognized in pre-Maastrichtian sediments.

If Pandanaceae indeed have their origin significantly earlier than Maastrichtian, *Freycinetia* (or at least vertebrate pollination) is less likely to have been primitive. This is not for lack of vertebrate pollinators: although bats do not radiate until the late Paleocene (Carroll, 1988), marsupials are significant pollinators in modern Australian ecosystems (Turner, 1982), and frugivorous (i.e. seed-eating) marsupials and multituberculates were potential pollinators throughout the Cretaceous (Collinson and Hooker, 1991). However, fossil flowers from the Early and early Late Cretaceous were small and provided only or primarily pollen as a reward (Crepet et al., 1991), and perhaps more importantly, fleshy fruits and floral nectaries did not appear until the Santonian-Campanian (80 mya) (Friis and Crepet, 1987). Given that sugars, in the form of nectar or fleshy floral appendages, rather than pollen are the primary rewards in modern zoophilous pollination syndromes, and that vertebrate pollination most likely arose as a shift from frugivory to floral attractants (Faegri and van der Pijl, 1978), we consider it unlikely that vertebrate pollination evolved before the end of the Cretaceous.

Paleoecological implications – The attribution of *Pandaniidites* to Pandanaceae has interesting implications for paleoclimate as well, especially with respect to the Cretaceous-Tertiary boundary. Interest in the pattern of extinction and survival across the boundary has been stimulated by the hypothesis that terrestrial and marine extinctions at the end of the Cretaceous were due to the impact of a large asteroid (Alvarez et al., 1980). Selective extinctions of broad leaved evergreens and a latitudinal gradient of increasing extinction southward support the hypothesis of a brief period of intense cold as a result of solar attenuation from impact-generated atmospheric dust (Wolfe and Upchurch, 1986). The interval is believed to have been brief because of the continued presence of thermophilic taxa such as crocodiles and palms above the boundary (Wolfe and Upchurch, 1986, 1987; Nichols et al., 1986). *Pandaniidites* is another, apparently thermophilic taxon that survives into the Paleocene in western North America (Nichols et al., 1986). It is of particular interest because in central Montana, it is unusually abundant in sediments immediately overlying the iridium-enriched clay thought to represent the impact fallout layer (Hotton, 1988). This overlying layer is unusually depauperate and dominated by species (usually ferns) thought to represent the earliest stages of recovery from catastrophic disturbance (Tschudy et al., 1984; Wolfe and Upchurch, 1986). Furthermore, pollen attributed to Nymphaeaceae but almost certainly representing *Pandaniidites* (Nichols, 1992) has been reported from an impact layer in Wyoming (Wolfe, 1991). The presence of an apparently thermophilic taxon immediately following the impact is difficult to reconcile with intense cold as a "killing mechanism" in mass extinction. Alternatively, the interpretation of fern-dominated floras overlying the impact layer as representing events on the order of months or years following the impact may be incorrect. Its abundance just above the boundary may be evidence of disturbance, if modern habits may be extrapolated to the fossil form: *Sararanga* and species of *Pandanus* are frequent colonizers of disturbed habitats and particularly of burned-over areas (Stone, 1968; Clunie, 1991).

Conclusions

This analysis underscores the value of ultrastructural evidence in assessing affinities of fossil pollen. Transmission electron microscopy reveals details not otherwise visible that allow a more accurate assessment of characters in *Pandaniidites*. The combination of detailed spine morphology and granular-columellar infratectal layer in *Pandaniidites* places it within Pandanaceae, but not within any living species of *Pandanus*. Many questions about relationships among taxa within the family and with other families remain unanswered. Their resolution will allow us to better interpret the fossil evidence. The following lines of investigation appear to be among the most promising.

1. A cladistic analysis of Pandanaceae and potential outgroups using a complete set of morphological (and molecular) characters is urgently needed to provide an independent data set against which to test competing hypotheses of relationships.
2. A better understanding of pollination biology in *Pandanus*, especially the subgenus *Martellidendron*, would elucidate the distribution and lability of wind and biotic pollination, and test the hypothesis of amphiphily within the genus.
3. Ultrastructural investigation of fossil pollen ascribed to *Pandaniidites* or *Pandanus* from the Paleogene and Neogene should clarify the pattern of evolution in this group, and support or refute the hypothesis of relationship between the Late Cretaceous and later forms.
4. Investigation of fossil leaves attributed to Pandanaceae (Dorf, 1942; S. Wing, pers. comm., 1992), and comparison with modern leaves, if positive, would provide corroborative evidence of Pandanaceae in the Late Cretaceous and Paleogene.

Acknowledgements

The authors gratefully acknowledge the Laboratoire de Phanérogamie, Paris, the U.S. National Herbarium, University of California, Berkeley, Herbarium, the Royal Botanic Gardens at Kew, the Papua New Guinea National Herbarium, the Canadian Museum of Nature and P.A. Cox for providing pollen for this study. We are indebted to B.C. Stone for his generosity in making unpublished data available to us and for valuable insights into the taxonomy and ecology of Pandanaceae, and to D.M. Jarzen for his loan of prepared slides of *Pandanus* and *Freycinetia*. We thank S.L. Wing for helpful discussion, and L.J. Gillespie, D.J. Nichols and the editors, whose comments substantially improved the manuscript. This project could not have been accomplished without the technical skills of M. Bell, N. Hodgkin, P.J. Viola and S. Braden on the scanning electron microscope and W.J. Chissoe on the transmission electron microscope. To all of them we are most grateful. We also thank Unocal Corporation for permission to publish the results contributed by H.A. Leffingwell.

References

Alvarez, L.W., Alvarez, W., Asaro, F. and Michel, H.V. (1980). Extraterrestrial cause for the Cretaceous-Tertiary extinction. *Science* **208**: 1095–1108.

Carroll, R.L. (1988). Vertebrate Paleontology and Evolution. 698 pp. W.H. Freeman, New York.

Clunie, N.M.U. (1978). The vegetation. In: J.S. Womersley (editor). Handbooks of the Flora of Papua New Guinea, Vol. 1, pp. 1–11, Melbourne University Press, Carlton, Victoria.

Collinson, M.E. and Hooker, J.J. (1991). Fossil evidence of interactions between plants and plant-eating mammals. *Philos. Trans., Ser. B* **333**: 197–208.

Cox, P.A. (1982). Vertebrate pollination and the maintenance of dioecism in *Freycinetia*. *Amer. Naturalist* **120**: 65–80.

Cox, P.A. (1984). Chiropterophily and ornithophily in *Freycinetia* (Pandanaceae) in Samoa. *Pl. Syst. Evol.* **144**: 277–290.

Cox, P.A. (1990). Pollination and the evolution of breeding systems in Pandanaceae. *Ann. Missouri Bot. Gard.* **77**: 816–840.
Cox, P.A. (1991). Abiotic pollination: an evolutionary escape for animal-pollinated angiosperms. *Philos. Trans., Ser. B* **333**: 217–224.
Crepet, W.L., Friis, E.M. and Nixon, K.C. (1991). Fossil evidence for the evolution of biotic pollination. *Philos. Trans., Ser. B* **333**: 187–195.
Cronquist, A. (1981). An Integrated System of Classification of Flowering Plants. 1262 pp. Columbia University Press, New York.
Daghlian, C. (1981). A review of the fossil record of monocotyledons. *Bot. Rev.* (*Lancaster*) **47**: 517–555.
Dahlgren, R.M.T., Clifford, H.T and Yeo, P.F. (1985). The Families of the Monocotyledons. 520 pp. Springer-Verlag, Berlin.
Donoghue, M.J. and Doyle, J.A. (1989). Phylogenetic analysis of angiosperms and the relationships of Hamamelidae. In: P.R. Crane and S. Blackmore (editors). Evolution, Systematics and Fossil History of the Hamamelidae, vol. 1, pp. 17–45, Clarendon Press, Oxford.
Dorf, E. (1942). Upper Cretaceous floras of the Rocky Mountain region: II. Flora of the Lance Formation at its type locality, Niobrara County, Wyoming. *Publ. Carnegie Inst. Wash.* **508**: 79–168.
Doyle, J.A. and Hotton, C.L. (1991). Diversification of early angiosperm pollen in a cladistic context. In: S. Blackmore and S.H. Barnes (editors). Pollen and Spores: Patterns of Diversification, pp. 169–195, Clarendon Press, Oxford.
Dransfield, J., Ferguson, I.K. and Uhl, N.W. (1990). The coryphoid palms: patterns of variation and evolution. *Ann. Missouri Bot. Gard.* **77**: 802–815.
Elsik, W.C. (1968). Palynology of a Paleocene Rockdale lignite, Milam County, Texas. I. Morphology and taxonomy. *Pollen & Spores* **10**: 263–314.
Erdtman, G. (1952). Pollen Morphology and Plant Taxonomy — Angiosperms. 553 pp. Almqvist and Wiksell, Stockholm.
Faegri, K. and van der Pijl, L. (1978). Principles of Pollination Ecology. 244 pp. Pergamon Press, Oxford.
Farley, M. (1989). Palynological facies fossils in nonmarine environments in the Paleogene of the Bighorn Basin. *Palaios* **4**: 565–573.
Farley, M. (1990). Vegetation distribution across the early Eocene depositional landscape from palynological analysis. *Palaeogeogr. Palaeoclimatol. Palaeoecol.* **79**: 11–27.
Ferguson, I.K. (1986). Observations on the variation in pollen morphology of Palmae and its significance. *Canad. J. Bot.* **64**: 3079–3090.
Ferguson, I.K., Dransfield, J. and Flawn, J. (1987). A review of the pollen morphology and systematics of the genera *Ravenea* and *Louvelia* (Ceroxyleae: Ceroxyloideae: Palmae). *J. Palynol.* **23–24**: 65–72.
Friis, E.M. and Crepet, W.L. (1987). Time of appearance of floral features. In: E.M. Friis, W.G. Chaloner and P.R. Crane (editors). The Origins of Angiosperms and their Evolutionary Consequences, pp. 145–179, Cambridge University Press, Cambridge.
Gould, S.J. and Vrba, E.S. (1982). Exaptation: a missing term in the science of form. *Paleobiology* **8**: 4–15.
Grayum, M.H. (1990). Evolution and phylogeny of the Araceae. *Ann. Missouri Bot. Gard.* **77**: 628–697.
Harling, G. (1958). Monograph of the Cyclanthaceae. *Acta Horti Berg.* **18**: 1–428.
Henderson, A. (1986). A review of pollination studies in the Palmae. *Bot. Rev.* (*Lancaster*) **52**: 221–259.
Hotton, C.L. (1988). Palynology of the Cretaceous-Tertiary boundary in central Montana, U.S.A., and its implications for extraterrestrial impact. 731 pp. Ph.D. Dissertation, University of California, Davis.

Huynh, K.L. (1976). Arrangement of some monosulcate, disulcate, trisulcate, dicolpate, and tricolpate pollen types in the tetrads, and some aspects of evolution in the angiosperms. In: I.K Ferguson and J. Muller (editors). The Evolutionary Significance of the Exine, pp. 101–124, Academic Press, London.

Huynh, K.L. (1980). La morphologie du pollen de *Pandanus* subg. *Vinsonia* (Pandanaceae) et sa signification taxonomique. *Pollen & Spores* **22**: 173–189.

Huynh, K.L. (1981). *Pandanus kariangensis* (sect. *Martellidendron*), une espèce nouvelle de Madagascar. *Bull. Mus. Hist. Nat., B. Adansonia* **3**: 37–55.

Huynh, K.L. (1982). La fleur mâle de quelques espèces de *Pandanus* subg. *Lophostigma* (Pandanaceae) et sa signification taxonomique, phylogénique et évolutive. *Beitr. Biol. Pflanzen* **57**: 15–83.

Huynh, K.L. (1983). *Pandanus borbonicus* (Pandanaceae), une espèce nouvelle de l'île de la Réunion. *Candollea* **38**: 81–103.

Huynh, K.L. and Stone, B.C. (1981). On a new subsection of *Pandanus* section *Cauliflora* (Pandanaceae) with paniculate staminate inflorescence structure, distinctive leaf anatomy and pollen morphology. *J. Linn. Soc., Bot.* **83**: 213–220.

Jarzen, D.M. (1983). The fossil pollen record of the Pandanaceae. *Gard. Bull. Singapore* **36**: 163–175.

Leffingwell, H.A. (1970). Palynology of the Lance (Late Cretaceous) and Fort Union (Paleocene) Formations of the type Lance area, Wyoming. *Special Pap. Geol. Soc. Amer.* **127**: 1–64.

Leffingwell, H.A. and Hodgkin, N. (1971). Techniques for preparing fossil palynomorphs for study with the scanning and transmission electron microscopes. *Rev. Palaeobot. Palynol.* **11**: 177–199.

Leffingwell, H.A., Larson, D.A. and Valencia, M.J. (1970). A study of the fossil pollen *Wodehouseia spinata*. I. Ultrastructure and comparisons to selected modern taxa. II. Optical microscopic recognition of foot layers in differentially stained fossil pollen and their significance. *Bull. Canad. Petrol. Geol.* **18**: 238–262.

Muller, J. (1981). Fossil pollen records of extant angiosperms. *Bot. Rev.* (*Lancaster*) **47**: 1–142.

Nichols, D.J. (1992). Plants at the K/T boundary. *Nature* **356**: 295.

Nichols, D.J., Jarzen, D.M., Orth, C.J. and Oliver, P.Q. (1986). Palynological and iridium anomalies at Cretaceous-Tertiary boundary, south-central Saskatchewan. *Science* **231**: 714–717.

Nilsson, S., Praglowski, J, and Nilsson, L. (1977). Atlas of airborne pollen grains and spores in Northern Europe. 159 pp. Natur och Kultur, Stockholm.

North, C.A. and Willis, A.J. (1971). Contributions to the anatomy of *Sararanga* (Pandanaceae). *J. Linn. Soc., Bot.* **64**: 411–421.

Norton, N.J. and Hall, J. (1969). Palynology of the Upper Cretaceous and Lower Tertiary in the type locality of the Hell Creek Formation, Montana, USA. *Palaeontographica, Abt. B, Paläophytol.* **125**: 1–64.

Quattrocchio, M.E. and Volkheimer, W. (1990). Paleogene paleoenvironmental trends as reflected by palynological assemblage types, Salta Basin, NW Argentina. *Neues Jahrb. Geol. Paläontol., Abh.* **181**: 377–396.

Skvarla, J.J. and Kelley, A.G. (1968). Rapid preparation of pollen and spore exines for electron microscopy. *Stain Technol.* **43**: 139–144.

Sowunmi, M.A. (1972). Pollen morphology of the Palmae and its bearing on taxonomy. *Rev. Palaeobot. Palynol.* **13**: 1–80.

Stone, B.C. (1968). Materials for a monograph of *Freycinetia* Gaud. IV. Subdivision of the genus, with fifteen new sections. *Blumea* **16**: 361–372.

Stone, B.C. (1972). A reconsideration of the evolutionary status of the family Pandanaceae and its significance in monocotyledon phylogeny. *Quart. Rev. Biol.* **47**: 34–45.

Stone, B.C. (1974). Towards an improved infrageneric classification in *Pandanus* (Pandanaceae). *Bot. Jahrb. Syst.* **94**: 459–540.

Stone, B.C. (1976). On the biogeography of *Pandanus* (Pandanaceae). *Compt. Rend. Séances Soc. Biogéogr.* **458**: 69–90.

Stone, B.C. (1990a). Pandanaceae. In: W.L. Wagner, D.R. Herbst and S.H. Sohmer (editors). Manual of the Flowering Plants of Hawaii, Vol. 2, pp. 1477–1481, Univ. of Hawaii Press, Hawaii.

Stone, B.C. (1990b). New evidence for the reconciliation of floral organisation in Pandanaceae with normal angiosperm patterns. In: P. Baas, K. Kalkman and R. Geesink (editors). The Plant Diversity of Malesia, pp. 33–55, Kluwer Academic Publ., Dordrecht, Boston and London.

Stone, B.C. and Huynh, K.L. (1982). The identity, affinities and staminate floral structure of *Pandanus pendulinus* Martelli (Pandanaceae). *Gard. Bull. Singapore.* **35**: 199–207.

Sweet, A.R. (1986). The Cretaceous-Tertiary boundary in the central Alberta Foothills. II: Miospore and pollen taxonomy. *Canad. J. Earth Sci.* **23**: 1375–1388.

Takhtajan, A.L. (1969). Flowering Plants — Origin and Dispersal. 310 pp. Smithsonian Institution Press, Washington, DC.

Takhtajan, A.L. (1987). Systema Magnoliophytorum. 439 pp. Nauka, Leningrad. (in Russian).

Tarasevich, V.F. (1990). Palynological evidence of the Lemnaceae family in the system of flowering plants. *Bot. Zhurn.* **75**: 959–964.

Thanikaimoni, G. (1970a). Les palmiers: palynologie et systématique. *Trav. Sect. Sci. Techn. Inst. Franc. Pondichéry* **11**: 1–286.

Thanikaimoni, G. (1970b). Pollen morphology, classification and phylogeny of Palmae. *Adansonia* **10**: 347–365.

Thorne, R.F. (1992). Classification and geography of the flowering plants. *Bot. Rev. (Lancaster)* **58**: 225–348.

Tschudy, R.F., Pillmore, C.L., Orth, C.J., Gilmore, J.S. and Knight, J.D. (1984). Disruption of the terrestrial plant ecosystem at the Cretaceous-Tertiary boundary, Western Interior. *Science* **225**: 1030–1032.

Turner, V. (1982). Marsupials as pollinators in Australia. In: S.A. Armstrong, J.M. Powell and A.J. Richards (editors), Pollinators and Evolution, pp. 55–66, Royal Botanic Gardens, Sydney.

Uhl, N.W. and Dransfield, J. (1987). Genera Palmarum. 610 pp. Allen Press, Lawrence, Kansas.

Walker, J.W. and Doyle, J.A. (1975). The bases of angiosperm phylogeny: palynology. *Ann. Missouri Bot. Gard.* **62**: 664–723.

Warburg, O. (1900). Pandanaceae. In: A. Engler, Das Pflanzenreich IV, 9, pp. 1–97, Engelmann, Leipzig.

Ward, J.V., Doyle, J.A., and Hotton, C.L. (1989). Probable granular magnoliid angiosperm pollen from the Early Cretaceous. *Pollen & Spores* **31**: 113–132.

Wodehouse, R.P. (1945). Hayfever Plants. 245 pp. Chronica Botanica, Waltham, Mass.

Wolfe, J.A. (1991). Palaeobotanical evidence for a June "impact winter" at the Cretaceous/Tertiary boundary. *Nature* **352**: 420–424.

Wolfe, J.A. and Upchurch, G.R., Jr. (1986). Vegetation, climatic and floral changes at the Cretaceous-Tertiary boundary. *Nature* **324**: 148–152.

Wolfe, J.A. and Upchurch, G.R., Jr. (1987). Leaf assemblages across the Cretaceous-Tertiary boundary in the Raton Basin, New Mexico and Colorado. *Proc. Natl. Acad. Sci., U.S.A.* **84**: 5096–5100.

Zavada, M.S. (1983). Comparative morphology of monocot pollen and evolutionary trends of apertures and wall structures. *Bot. Rev. (Lancaster)* **49**: 331–379.

Milne, L.A. (1994). Relationship between *Propylipollis annularis* (Tertiary dispersed pollen) and extant *Xylomelum* (Proteaceae). In: M.H. Kurmann and J.A. Doyle (editors). Ultrastructure of fossil spores and pollen, pp. 193–213, Royal Botanic Gardens, Kew.

12. RELATIONSHIP BETWEEN *PROPYLIPOLLIS ANNULARIS* (TERTIARY DISPERSED POLLEN) AND EXTANT *XYLOMELUM* (PROTEACEAE)

LYNNE A. MILNE

Department of Earth Sciences, The University of Queensland, Brisbane 4072, Australia

Abstract

An amplified diagnosis of the long-ranging Australian Tertiary proteaceous-like pollen, *Propylipollis annularis* (Cookson) Martin & Harris, 1974, confines this species to specimens conforming to the type. The tenet that *P. annularis* is comparable to pollen of *Xylomelum occidentale* R.Br., an extant proteacean, is investigated. Pollen grains of *Xylomelum* form two distinct morphological groups that reflect the disjunct geographic distribution of its species. Morphological clines broadly connect the morphological and geographic end members of the genus. Apertural features of immature *X. occidentale* pollen suggest that the mature aperture configuration of this species is due to separation of the nexine in addition to nexinal solution. *Propylipollis annularis* resembles pollen of *X. occidentale* in aperture development and form, but differs in gross wall structure. High-resolution electron microscopy of the tectum surface features of the fossil and modern species reveals a morphological similarity. *Propylipollis* species with comparable apertures to *P. annularis* are morphologically akin to pollen of *Xylomelum* species other than *X. occidentale*. *Propylipollis annularis* is closely related to pollen of *Xylomelum*, but not affiliated with any particular species of the genus. The relatively wide geographic and morphological segregation of extant *Xylomelum* species suggests that they may be relicts of a previously more diverse genus, reflected in the fossil record by *P. annularis* and related dispersed pollen.

Introduction

Cookson's (1950) comparative light microscope study of pollen of the predominantly Southern Hemisphere family Proteaceae, and of Tertiary proteaceous-like pollen, remains a major reference on this subject. A combined light and electron microscopy study of several of Cookson's (1950) fossil and modern proteaceous affiliates forms part of a comprehensive study of the palynostratigraphy and phytogeography of Eocene sediments from the Bremer and western Eucla Basins of Western Australia (Fig. 1).

Cookson (1950) likened the fossil proteaceous-like pollen *Propylipollis annularis* (Cookson) Martin & Harris, 1974, to pollen of extant *Xylomelum* J.E. Smith, in particular to that of *Xylomelum occidentale* R.Br., stating that 'so close is the resemblance between the fossil grains and pollen of *X. occidentale* ... that a relationship between the two parent plants seems probable.' *Propylipollis annularis* and similar fossil pollen often aligned with *P. annularis* are the commonest proteaceous pollen in Australian Tertiary palynomorph assemblages. Ranging from the early Palaeocene (Harris, 1965) through the Miocene (Stover and Partridge, 1973), *P. annularis* is known from all southern Australian Tertiary Basins (Fig. 1). Although most illustrated specimens are closely similar to the type (Cookson, 1950, Pl. 1, fig. 15), the species is commonly used as a repository for similar forms with the same distinctive grevilleoid aperture but different size and wall structure. Biostratigraphically, the long-ranging *P. annularis* is useful only as a general Tertiary indicator; however, the form variants may be more stratigraphically restricted, as demonstrated by Macphail and Truswell (1989).

Following Cookson's (1950) investigation, Tertiary occurrences of *P. annularis* have been regarded as indicative of the presence of *Xylomelum* in the palaeovegetation. *Xylomelum* belongs to subfamily Grevilleoideae (tribe Helicieae) of the Proteaceae. Commonly known as the woody pear after its distinctive fruiting body, *Xylomelum* constitutes medium-sized shrubs to small trees in sclerophyll communities that grow in well-drained heath or woodland (Wrigley and Fagg, 1989). The five known living species are all endemic to Australia.

This study was initiated to examine Cookson's (1950) tenet that the fossil *P. annularis* is closely comparable to pollen of *X. occidentale*. In order to accomplish this: (a) morphological features resolved by electron microscopy have been included in an amplified diagnosis of *P. annularis*, restricting the species to specimens conforming with the type; and (b) the morphology of *Xylomelum* pollen, in particular that of *X. occidentale* and *X. angustifolium* Kipp., has been investigated to assess morphological trends within the genus and their relation to the fossil taxon.

Material and Methods

A. Collection and extraction

Three preparations of both *Xylomelum occidentale* and *X. angustifolium* pollen were examined. One of each species was collected, on separate occasions by G. Playford of The University of Queensland, and D.M. Jarzen of the Canadian Museum of Nature, from living plants growing in Kings Park Botanical Gardens, Western Australia; pollen for the other preparations was taken from sheets in the Brisbane Herbarium (BRI). Pollen of *X. cunninghamianum* D. Foreman, *X. pyriforme* (Gaertn.) Knight, and *X. scottianum* (F. Muell.) F. Muell. were obtained from sheets in BRI. Specimens from *X. salicinum* (R.Br., ex Meissner) A. Cunn. ex Benth., now synonymised with *X. cunninghamianum*, were also taken from BRI sheets to verify that pollen of the former is identical with that of the latter. Pollen from mature anthers of BRI specimens, and complete inflorescences collected by G. Playford, was acetolysed using the standard techniques of Erdtman (1960) and mounted for light microscopy (LM) in Elvacite. Pollen collected by D.M. Jarzen was acetolysed, and mounted for LM in glycerine jelly. Measurements were taken from The University of Queensland preparations (Appendix 1).

Fossil pollen was recovered from clays and lignites of Eocene age, supplied by Centenary International Mining Pty Ltd, from the Balladonia-15 and Zanthus-6 boreholes in the western Eucla Basin, Western Australia (Fig. 1). Organic material was extracted by the processing techniques outlined by Phipps and Playford (1984), involving treatment of the samples with HCl, HF, KOH, and HNO_3, followed by heavy-liquid separation with $ZnBr_2$. Extraneous organic and inorganic matter was removed by 'short-time' or differential centrifugation (Gray, 1965; Barss and Williams, 1973) and by sieving with a 140 μm mesh. Slides were prepared for LM by mounting a portion of the residue in glycerine jelly. Photomicrographs were taken with an Olympus PM35 10AD photographic attachment mounted on an Olympus BH-2 microscope (Appendix 2).

B. Scanning electron microscopy

As the fossil pollen species studied are easily recognisable in a dry state, all fossil and some modern pollen grains were picked with a micromanipulator from a smear of ethanol-suspended residue, and secured on 12 mm glass coverslips with a chloroform-based glue; some modern species were directly smeared onto the glass coverslip. Coverslips were transferred to an aluminium stub, coated with gold in a Denton sputter

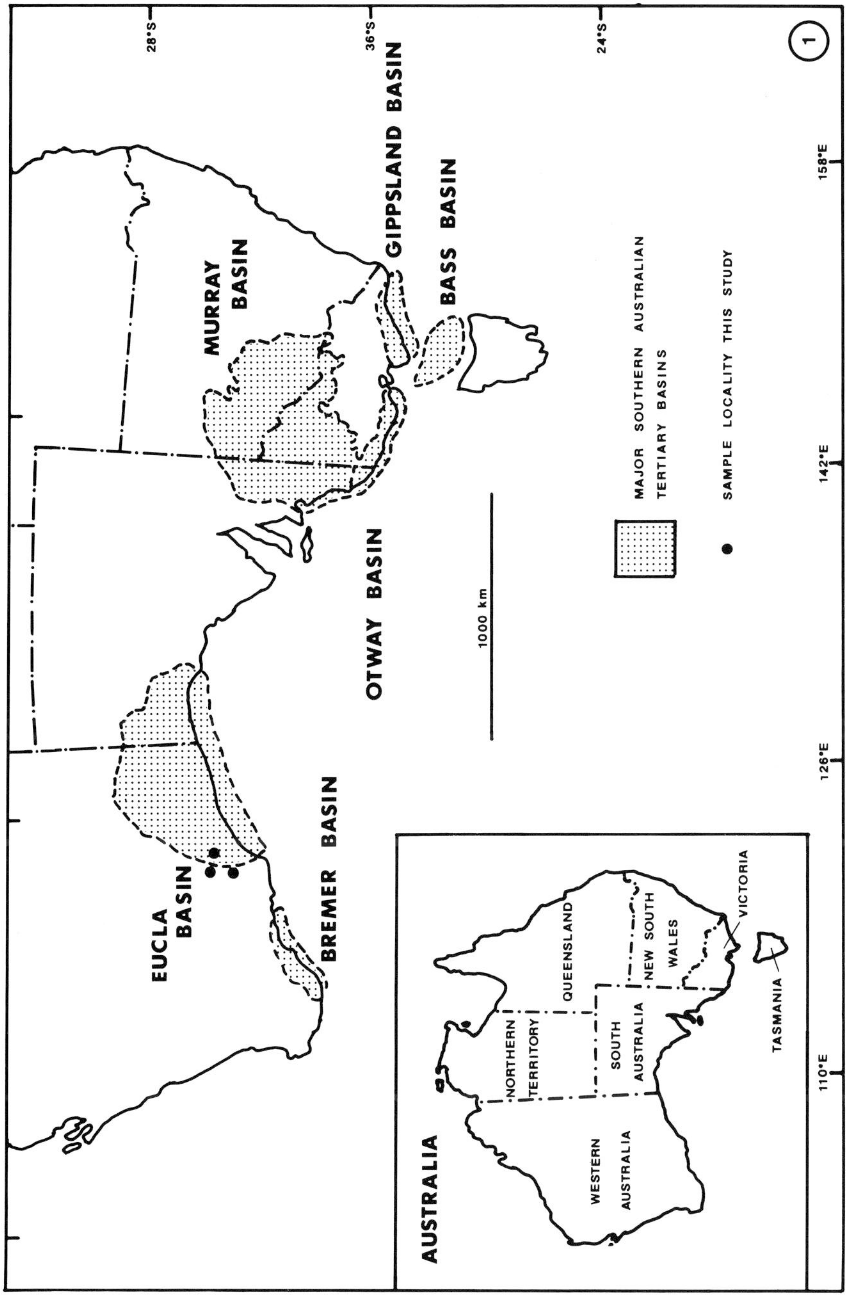

FIG. 1. Locality map showing Eucla Basin boreholes and major southern Australian Tertiary basins (adapted from Harris, 1985).

coater, and imaged on a Philips 505 at variable KV (15–25) and spot sizes (10–20). High-resolution photomicrographs of *X. occidentale* and *X. angustifolium* were taken on a Jeol 6400. When SEM photography was completed, coverslips were removed from stubs and permanently mounted on glass slides with Elvacite for light microscopic verification of specimens.

C. Transmission electron microscopy

Pollen was prepared for TEM using the general preparation procedures of Skvarla (1966), and surface-embedded (Milne and McCaul, in prep.) in medium Spurr's resin. Ultrathin sections cut on an LKB Ultratome III were collected on 200 μm copper grids and post-stained with uranyl acetate and lead citrate. The fossil pollen was little altered by diagenetic processes, and hence was very light, necessitating pre-staining with ruthenium red in glutaraldehyde. Sections were photographed on a Hitachi H-800 at 75 KV.

D. Terminology

Terminology used generally follows Erdtman (1952), Faegri (1956), and Faegri and Iversen (1989). For light microscope observations and comparative remarks, the pollen wall terms sexine and nexine are used *sensu* Erdtman (1952), to be consistent with, and easily related to, the descriptions of other fossil proteaceous pollen; nexine of the fossil pollen is equivalent to footlayer. The wall layer terminology used for electron microscope observations is that of Faegri (1956), which divides the exine into an outer ektexine and an inner endexine, and further divides the ektexine into tectum, infrastructural layer, and footlayer. The infrastructural layer has been variously termed the columellate layer, interstitium, or infratectum. Infratectum (APLF, 1975) is herein preferred as the term to describe the exine layer beneath the tectum; columellae, for example, are infratectal elements (Erdtman, 1952). The relationship of the two wall structure terminologies to each other is illustrated by, amongst others, Larson et al. (1962) and Zavada (1984); the latter discusses pollen wall homologies in detail.

In the original description, Cookson (1950) used the term 'collar' to describe the gross morphology of the aperture of *P. annularis*; it was also used by Martin (1982) with reference to the aperture of his *Propylipollis* sp. (aff. *Xylomelum*). The term 'collar' is retained in this study as a general descriptive term. Consistent with the terminology used by Dudgeon (1983) and Milne (1988) for *Propylipollis*, crassimarginate *sensu* Erdtman (1952) signifies a marked thickening of apertural nexine.

Propylipollis annularis

A. Systematics

The amplified systematic description based on LM observations and the listed synonymies of *P. annularis* hereunder follow the format commonly used to describe and document fossil proteaceous-like pollen. Listed referrals to *P. annularis* and fossil pollen likened to it include only illustrated records. Question marks denote that the photomicrograph or illustration, unaccompanied by descriptive data, shows insufficient sculptural and apertural details for unequivocal identification.

Genus *Propylipollis* Martin & Harris, 1974
1953 in part, *Proteacidites* Cookson ex Couper, p. 42.

Type species: *Propylipollis reticuloscabratus* (Harris) Martin & Harris, 1974; by original designation.

On the basis of aperture morphology, Martin and Harris (1974) subdivided the fossil proteaceous genus *Proteacidites* Cookson ex Couper, 1953, into three form genera:

Proteacidites, *Propylipollis*, and *Cranwellipollis*. This subdivision was not wholly embraced by Australasian palynologists, many preferring to retain all fossil proteaceous species in *Proteacidites* (e.g., Kemp, 1976; Pocknall and Mildenhall, 1984), or retaining in *Proteacidites* those transferred from *Proteacidites* to *Propylipollis* by Martin and Harris (1974), but describing new *Propylipollis* species themselves (Nott and Owen, 1992). As noted by Pocknall and Mildenhall (1984), Jansonius and Hills (1976) did not regard *Propylipollis* as being adequately circumscribed. Problematic are species that do not clearly belong to any of the three genera as diagnosed, and species that display intraspecific variations diagnostic of more than one genus (Milne, 1988). Unfortunately, *Propylipollis* is the repository for many of these and includes pollen with a complex range of aperture morphologies, the common features being some degree of thickening of the nexine at the aperture, and/or some disruption of the apertural nexine. Despite such negative aspects, *P. annularis* is herein retained in *Propylipollis*. In the event of future ultrastructural studies and revision of fossil proteaceous-like pollen systematics, *P. annularis* will likely remain in *Propylipollis* as it conforms with the intention of the original diagnosis in having a propylis (i.e., 'collar').

Propylipollis annularis (Cookson) Martin & Harris, 1974 (Figs. 2–17)

1950	*Proteacidites annularis* Cookson, p. 170, 171, Pl. 1, fig. 15.
?1953	*Proteacidites annularis* Cookson; Couper, p. 42, Pl. 5, fig. 52
1953	*Proteacidites annularis* Cookson; Cookson, p. 466, Pl. 1, fig. 19.
?1960	*Proteacidites annularis* Cookson; Couper, p. 49, Pl. 5, figs. 11, 12.
?1965	*Proteacidites annularis* Cookson; Harris, p. 92, Pl. 28, figs. 12,13.
1972	*Proteacidites* sp. aff. *P. annularis* Cookson; Hekel, p. 16, Pl. 5, fig. 18.
1973	*Proteacidites annularis* Cookson; Stover and Evans, Pl. 2, fig. 13.
1974	*Propylipollis annularis* (Cookson) Martin & Harris, p. 110, Fig. 2C.
?1975	*Proteacidites annularis* Cookson; Haskell and Wilson, Pl. 4, figs. 2, 3.
1975	*Proteacidites annularis* Cookson; Hos, Fig. 41.
?1975	*Proteacidites* sp. cf. *P. annularis* Cookson; Hos, Fig. 54.
1982	*Propylipollis annularis* (Cookson) Martin & Harris; Stover and Partridge, p. 78–79, Pl. 2, fig. 15.
non 1982	*Propylipollis* sp. (aff. *Xylomelum*); Martin, p. 80, Figs. 10, 11.
non 1983	*Propylipollis* sp. cf. *P. annularis* (Cookson) Martin & Harris; Dudgeon, p. 345, Fig. 6A–C.
?1984	*Proteacidites annularis* Cookson; Pocknall and Mildenhall, p. 43, Pl. 23, fig. 1.
1992	*Propylipollis annularis* (Cookson) Martin & Harris; Alley and Broadbridge, Figs. 9A, B.

Amplified diagnosis — Light microscope observations (Figs. 2–6): Pollen grains free, triporate, angulaperturate, isopolar, oblate. Amb triangular, with straight to concave sides. Equatorial diameter (21–)28(–37) μm, determined from 25 specimens. Apertures circular, 4–9 μm in diameter, strongly crassimarginate; encircled by a band of nexine 3–5 μm thick, forming a nexinous 'collar' extending 3–6 μm from the aperture margin. An annulate thinning at the base of the 'collar' is evident in sectional view as an abrupt indentation in the nexine. Nexine disrupted along 'collar', appearing irregularly striate. Non-apertural exine 1.5–3 μm thick, stratified. Sexine tectate-perforate, columellate, 0.5–1 μm thick, thinner than nexine; surface finely reticulate at high focus, perforate at low focus.

Electron microscope observations — *Sculpture* (Figs. 7–13): Tectate-perforate, rugulate. Rugulae 0.5–0.7 μm wide, 0.3–0.5 μm high, 1–3 μm long, and 0.1–0.5 μm apart. Perforations in tectum between rugulae 0.1–0.2 μm in diameter, regularly c. 0.5 μm apart. Tectum may strip to reveal grana and columellae.

Non-apertural structure (Fig. 14): Tectum perforate, upper surface undulate, lower surface more or less regular. Infratectum well-defined, granulate, simpli- and pluri-

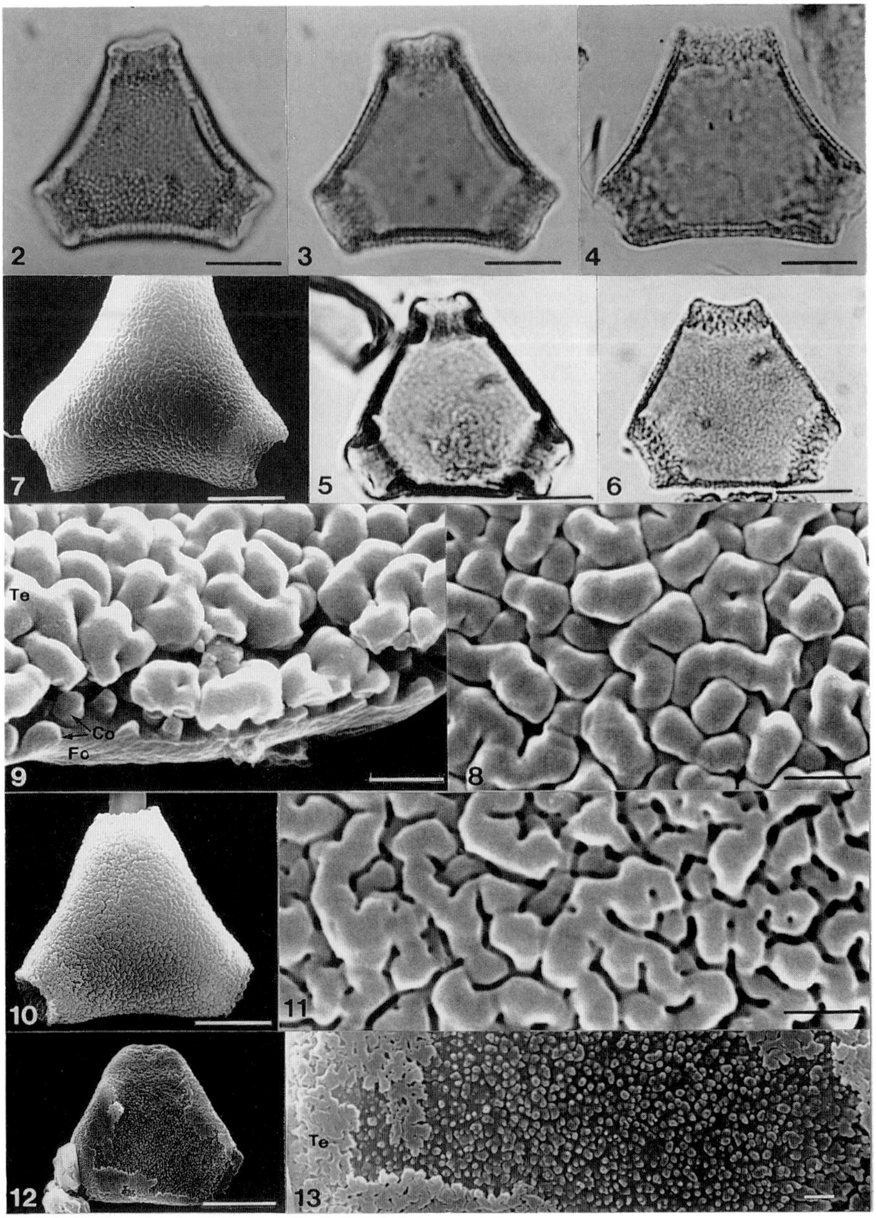

Size bars in figs. 2–23, 25–55. Light micrographs (LM): scale bars = 10 μm. Scanning electron micrographs (SEM): whole grains and part grains, scale bars = 10 μm; sculptural detail and wall fractures, scale bars = 1 μm. Transmission electron micrographs (TEM): all scale bars = 1 μm.

Figs. 2–13. *Propylipollis annularis* (Cookson) Martin & Harris, 1974. **2–3**. Same specimen, high and sectional foci respectively (LM). **4–6**. Different grains showing variable aperture form (LM). **4**. Sectional focus. **5–6**. Median focus. **7–8**. Same specimen (SEM). **7**. Part grain. **8**. Close-up of rugulate sculpture. **9**. Wall fracture, sculptural detail (Te), columellae (Co) and footlayer (Fo) (SEM). **10–11**. Same specimen (SEM). **10**. Whole grain. **11**. Sculptural detail, widely spaced rugulae and perforate tectum. **12–13**. Same specimen (SEM). **12**. Whole grain with partially stripped tectum. **13**. Close-up of tectum (Te) and stripped surface revealing columellae.

columellate; single columellae broad-based, tapering towards tectum; composite columella-like structures more common, with two or more fine branches emanating from a broad irregular base. Footlayer homogeneous, 0.7–2 μm thick, upper surface undulate, lower surface with fine irregular pits.

Apertural structure (Figs. 15–17): Tectum and infratectum thin towards aperture margin, the tectum often extending beyond footlayer and incurving about it. Footlayer constricts sharply at base of apertural region before thickening abruptly to form the apertural 'collar', and tapers close to aperture margin. Outer half of thickened footlayer predominantly homogeneous, with rare fine channels; lower footlayer split and channelled.

Distribution: *Propylipollis annularis* is known from the early Palaeocene through Miocene in Australia (Harris, 1965; Stover and Partridge, 1973), and Palaeocene to early Miocene in New Zealand (Pocknall and Mildenhall, 1984). It has been reported from major southern Australian Tertiary basins (Fig. 1) and associated palaeodrainage systems, isolated freshwater sediments in Australia (e.g., Truswell and Owen, 1988), and DSDP sites between Australia and New Zealand (Haskell and Wilson, 1975).

Affinity: Cookson (1950) compared *P. annularis* to pollen of *Xylomelum* and *Lambertia* J.E. Smith, concluding that it was most closely related to pollen of *X. occidentale.* This supposed relationship is investigated in a following section.

B. Remarks

Wall structure and sculpture of *P. annularis* are consistent among specimens examined, but gross form varies due to preservation or degree of maturation of individual grains. Some granulate forms are preservationally derived, the tectum stripping to reveal columellae and grana (Figs. 12, 13). Grain maturity is reflected by the apertures, which apparently continue to widen after dispersal, resulting in some specimens with compact, distinct apertures (Figs. 3, 5); others with wider, less distinct apertures (Fig. 6); and occasionally specimens display both apertural conditions (Fig. 4). It is possible that this variability has led to a relatively *sensu lato* interpretation of *P. annularis.* Available evidence suggests that *Proteacidites granulatus* Cookson, 1953, synonymised by Stover and Partridge (1973) with *P. annularis,* is, as Cookson (1953) reported, a separate species. The Eucla Basin material, from which this study derives in part, contains a range of fossil pollen not unlike *P. annularis* in aperture form, but exhibiting different wall structures (Figs. 18–23). This is similarly illustrated by *P.* sp. cf. *P. annularis* Dudgeon, 1983, and *Propylipollis* sp. (aff. *Xylomelum)* in Martin (1982), herein considered synonymous, but discrete from *P. annularis.* As noted by Dudgeon (1983) and Martin (1982), this form has essentially the same aperture as *P. annularis,* but sculpturally it is granulate-microverrucate and hence distinct from the reticulo-rugulate *P. annularis.*

Xylomelum

A. Systematics

Xylomelum pollen morphological features reflect the disjunct geographic distribution of species (Fig. 24). Thus, pollen grains of the three eastern Australian species, *X. cunninghamianum* (Fig. 25), *X. pyriforme* (Fig. 26), and *X. scottianum* (Fig. 27), are morphologically very similar, whereas pollen grains of the Western Australian species, *X. angustifolium* (Figs 31, 32) and *X. occidentale* (Figs. 39–41) are markedly different from eastern Australian forms, but bear a distinct resemblance to each other.

Pollen grains of eastern Australian species have subtriangular ambs with convex sides and similar equatorial diameters: viz. *X. cunninghamianum,* (41–) 44(–48) μm, *X. pyriforme* (42–)46(–48) μm, *X. scottianum,* (36–)41(–43) μm. They are tectate and microbaculate;

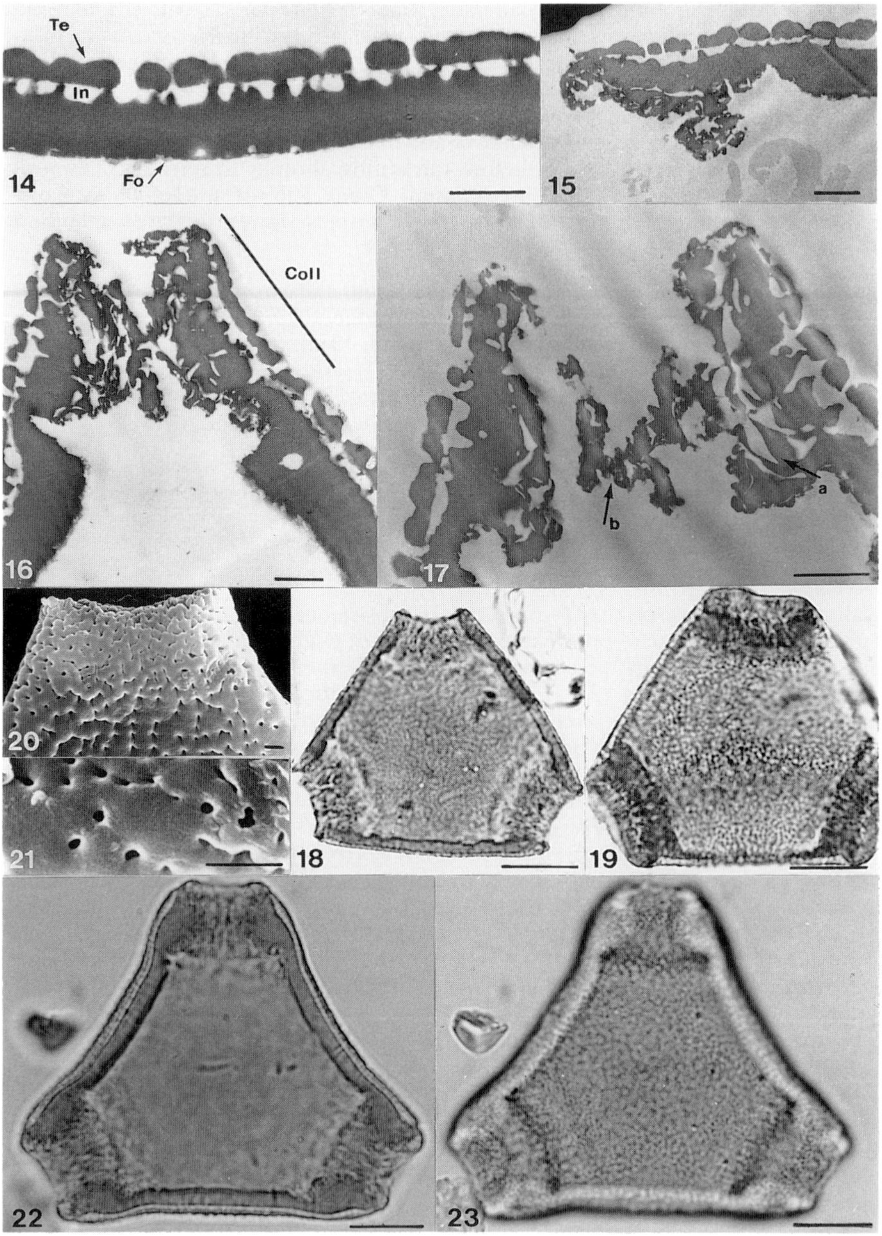

Figs. 14–23. *Propylipollis*. **14–17.** *Propylipollis annularis* (Cookson) Martin & Harris, 1974 (TEM). **14.** Equatorial wall section, tectum (Te), infratectum (In), footlayer (Fo). **15.** Section of one side of aperture. **16.** Full aperture section, collar delineated (Coll). **17.** Close-up of aperture section, arrows indicate (a) angular separation of footlayer, (b) footlayer, i.e. nexinal striae of light micrographs. **18–23.** *Propylipollis* spp. **18–19.** Different grains (LM). **18.** Median focus. **19.** High focus. **20–21.** Same specimen (SEM). **20.** Sculptural detail near aperture. **21.** Close up of equatorial sculpture. **22–23.** Same specimen (LM). **22.** Sectional focus. **23.** High focus on scabrate surface.

bacula are 0.5–1 μm high and 0.2–0.3 μm wide at their base (Fig. 28). The infratectum is almost indiscernible in SEM section; however, where the tectum has broken away, the broad-based infratectal elements (pluricolumellae and grana) can be seen (Figs. 29, 30). Although TEM sections of this group are unavailable, LM clearly shows the aperture structures (Figs. 25–27). *Xylomelum cunninghamianum* pollen has the most distinctive grevilleoid-type aperture, with marked nexine thickening, an indentation in the nexine at the base of the apertural 'collar', and disrupted nexine forming internal striations. In *X. pyriforme*, pollen apertural features are reduced: the nexine is not as thick, the nexinal indentation is barely perceptible, and there is a lesser development of internal striations along the 'collar' than in pollen of *X. cunninghamianum*. The apertures of *X. scottianum* pollen appear to have no indentation at the base of the nexinal thickening, the nexine thickens only slightly; and internal disruption of the apertural nexine is usually absent. In fact, *X. scottianum* has no apertural 'collar'.

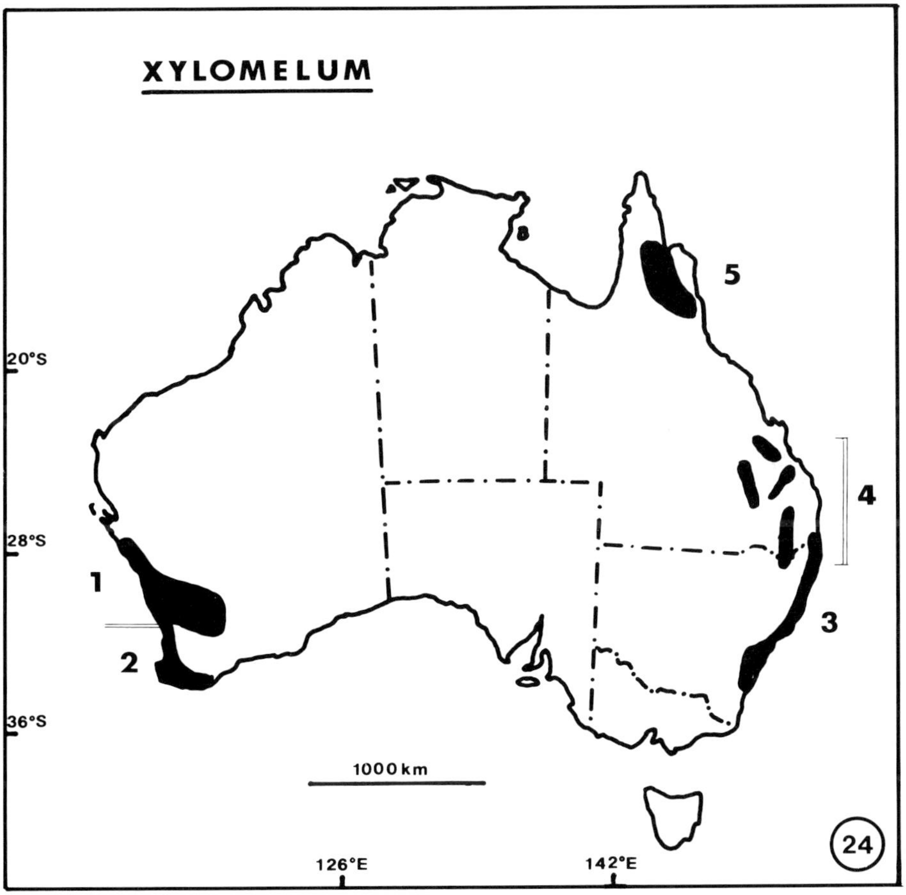

FIG. 24. Distribution of extant *Xylomelum*. 1. *X. angustifolium* Kipp. 2. *X. occidentale* R. Br. 3. *X. pyriforme* (Gaertn.) Knight. 4. *X. cunninghamianum* D. Foreman. 5. *X. scottianum* (F. Muell.) F. Muell. (Compiled from Wrigley and Fagg, 1989).

Pollen of *Xylomelum angustifolium* R.Br. (Figs. 31–38)

1952 *Xylomelum angustifolium;* Cookson & Erdtman in Erdtman, p. 368, Fig. 214A.

Light microscope observations (Figs. 31, 32) — Largest of the *Xylomelum* pollen, *X. angustifolium* grains have an equatorial diameter of (53–)60(–73) μm, determined from 20 specimens, and a subtriangular amb with straight to concave sides. Apertures circular, concave in plan view, (10–)13(–17) μm wide; crassimarginate, encircled by a band of irregularly alveolate nexine, 4–6 μm thick. Non-apertural exine 1.5–3 μm thick; well-stratified, sexine slightly thinner than nexine. Sexine tectate, columellate; surface indistinctly granulo-scabrate. A prominent, often labyrinthine, negative reticulum is formed by basal fusion of sculptural elements.

Electron microscope observations — *Sculpture* (Figs. 33–35): Tectum surface scabrate-pseudoperforate, with irregular protuberances 0.3–1.0 μm in diameter; some coalescing to form elements 1–2 μm long, others connected by low ridges. Pseudoperforations between protuberant elements 0.03–0.07 μm in diameter.

Non-apertural structure (Fig. 36): Tectum apparently complete, 0.5–1 μm thick; upper surface undulate, lower surface irregular. Infratectum 0.5–0.7 μm; columellate. Columellae irregular; bases broader than or equal to height. Footlayer homogeneous, 1–2 μm thick; lower surface undulate and finely pitted. Footlayer thinning slightly at base of apertural area.

Apertural structure (Figs. 37, 38): Tectum thickness constant with that of non-apertural tectum. Footlayer and infratectum expanded to form an irregularly alveolate-like structure, 4–6 μm thick. Endexine finely lamellate.

Remarks: Electron microscopy of *X. angustifolium* proved difficult. All grains prepared from the complete inflorescence were coated with a fine microgranulate layer, the tectum being visible only where it had split (Fig. 35). Many grains of the other preparation were partially covered with a similar coating, or littered with Ubisch bodies; the latter are also common in *X. occidentale* preparations. The coating is not a preparation artefact, but is perhaps a combination of incompletely acetolysed pollenkitt and tapetal material. Memon (1984) also encountered surfaces of uncertain origin on grains of extant Proteaceae.

The surface of *X. angustifolium* is similar to that of a large fossil proteaceous-like pollen (Figs. 20–23) often grouped with *P. annularis* because of its similar aperture. Pollen grains of *X. angustifolium* with straight-sided ambs resemble *Proteacidites granulatus* Cookson, 1953, and could easily be mistaken for large grains of *X. occidentale* if the irregularly alveolate nexinal thickening about the aperture, clearly seen in LM, was not considered. Fossil pollen with a similar nexinal thickening is known from the Eucla Basin material. These grains are similar to *P.* sp. aff. *P. granulatus* (Hekel, 1972).

Pollen of *Xylomelum occidentale* R.Br. (Figs. 39–53)

1952 *Xylomelum occidentale*; Cookson & Erdtman in Erdtman, p. 368, Fig. 214 B.

Light microscope observations (Figs. 39–41): Similar in shape, but slightly larger than *P. annularis, X. occidentale* pollen is (35–)40(–47) μm in equatorial diameter, determined from 20 mature grains. Apertures circular, (5–)7.5(–10) μm wide; strongly crassimarginate, encircled by a band of disrupted nexine 2–4 μm thick. A similarly disrupted area occurs below an annulate thinning and/or separation of the nexine at the base of the aperture 'collar'. Non-apertural exine 1.5–2.5 μm thick; stratified, nexine at least twice as thick as sexine. Sexine tectate, indistinctly columellate; surface scabrate, with small, low protuberances, 1–2 μm apart, which are more prominent on slightly immature grains. Mature grains also display a fine negative reticulum.

Electron microscope observations — *Sculpture* (Figs. 42–44): Tectum surface scabrate-pseudoperforate, undulate with low, broad-based protuberances, 0.5–0.8 μm in diameter and 0.5–2 μm apart; pseudoperforations between protuberant elements delineate finer scabrae 0.2 μm in diameter.

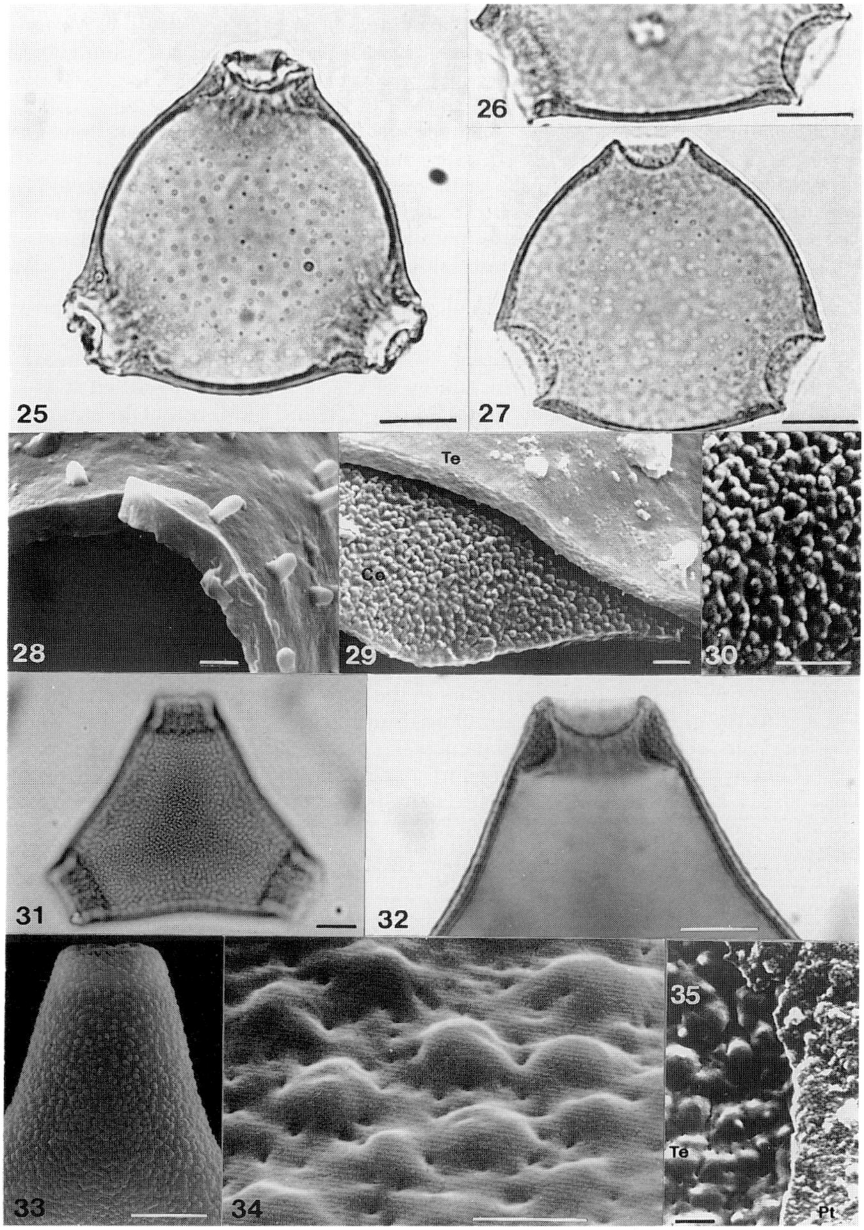

Figs. 25–35. Pollen of extant *Xylomelum* J.E. Smith. **25.** *X. cunninghamianum* D. Foreman, sectional focus (LM). **26.** *X. pyriforme* (Gaertn.) Knight, sectional focus (LM). **27.** *X. scottianum* (F. Muell.) F. Muell., sectional focus (LM). **28.** *X. pyriforme*, fractured grain, tectum surface with bacula (SEM). **29–30.** *X. cunninghamianum*, same specimen (SEM). **29.** Wall detail, tectum (Te) partially removed revealing infratectal grana and columellae (In). **30.** Higher magnification of infratectal elements. **31–35.** *X. angustifolium* Kipp. **31.** Median-high focus on prominent negative reticulum (LM). **32.** Part grain, sectional focus (LM). **33.** Part grain showing scabrate tectum surface (SEM). **34.** High resolution of scabrate pseudoperforate tectum surface (SEM). **35.** Surface detail of grain coated with ?incompletely acetolysed pollenkitt and tapetal material (Pt), split to reveal tectum (Te) surface (SEM).

Non-apertural structure (Figs. 45–46): Tectum apparently complete, 0.5–0.7 μm thick. Infratectum very fine, 0.1–0.2 μm thick; almost indiscernible in SEM section; granulate-columellate, width of columellae greater than or equal to their height. Footlayer homogeneous, 1–2 μm thick, thinning to half this value at base of apertural 'collar'.

Apertural structure (Figs. 45, 47, 48): Tectum 0.5 μm thick, extending beyond the footlayer at aperture opening. Infratectum slightly thicker than non-apertural infratectum; footlayer 1 μm thick at base of apertural area, thickening abruptly to 2.5 μm, then thinning at the aperture margin. Upper footlayer homogeneous; lower footlayer split and channelled, predominantly parallel to pollen surface. Fine channels may delineate further separation of nexine. Endexine finely lamellate (Fig. 48).

Remarks: The very fine negative reticulum, as seen in LM, appears to be an expression of the finer scabration.

Cookson and Erdtman (in Erdtman, 1952) illustrated the striate band below the apertural collar as an additional thickening of the nexine, and described it as such. Technically this is correct but close observations (LM and TEM) reveal the striae of this band to be aggregated or disrupted endexine, not disrupted footlayer.

B. Immature pollen of *Xylomelum occidentale* (Figs. 49–53)

One preparation of *X. occidentale* is from a complete inflorescence with many immature anthers. The immature pollen grains are (23–)26(–29) μm in equatorial diameter, with slightly convex triangular ambs. They feature small apertural openings, and low protuberances (1–2 μm apart) on the tectum surface (Figs. 40, 49).

LM examination of the apertures of variably immature grains showed stages of development of the nexine indentation and related annulate thinning, and also disruption of the nexine. Initially, nexine in the apertural area seems to be a solid plug (Fig. 49). As the grain matures, an annulate thinning at the base of the 'collar' appears, from which, in sectional view, perpendicular slits extend on each side; the apertural plug shows faint areas of nexinal thinning and separation; and the aperture margin widens (Figs. 50, 51). Eventually, as the aperture margin further widens, the nexine disruption and separation become more pronounced; the perpendicular slits open up, separating the apertural and non-apertural nexine to form the characteristic annulate thinning expressed as a nexinal indentation at the base of the 'collar'; and the amb interradii become straight to slightly concave (Fig. 52, 53).

Nexine solution is thought to be responsible for the disrupted apertural nexine common in *Propylipollis* species (Dudgeon, 1983; Milne, 1988) and in pollen of some extant Proteaceae (Dettmann and Jarzen, 1990). The apparent developmental sequence of the aperture of *X. occidentale* suggests that, in this species, the nexinal disruption that appears as irregular striae along the 'collar', and the annulate thinning at the base of the 'collar', may in part be explained by growth and/or expansion of the aperture region, resulting in separation of nexine along lines of weakness initiated and perpetuated by solution.

Variation of aperture form and gross morphology within single preparations of *Xylomelum* pollen, in particular those of *X. angustifolium* and *X. cunninghamianum,* suggest that the above developmental features are not confined to *X. occidentale.* In the fossil record, the exceptionally rare *Propylipollis* sp. B. (Milne, 1988; re-illustrated as Figs. 54, 55 herein) may represent immature *P. annularis.*

C. Summary

The two pollen morphological groups of *Xylomelum* are both tectate-pseudoperforate and exhibit varying degrees of nexinal thickening and disruption in the apertural area. They differ, however, in shape and supratectal features. Although pollen morphological characters may reflect taxonomy, the relationship between the morphological and geographic end members, *X. angustifolium* and *X. scottianum,* is unclear unless the genus

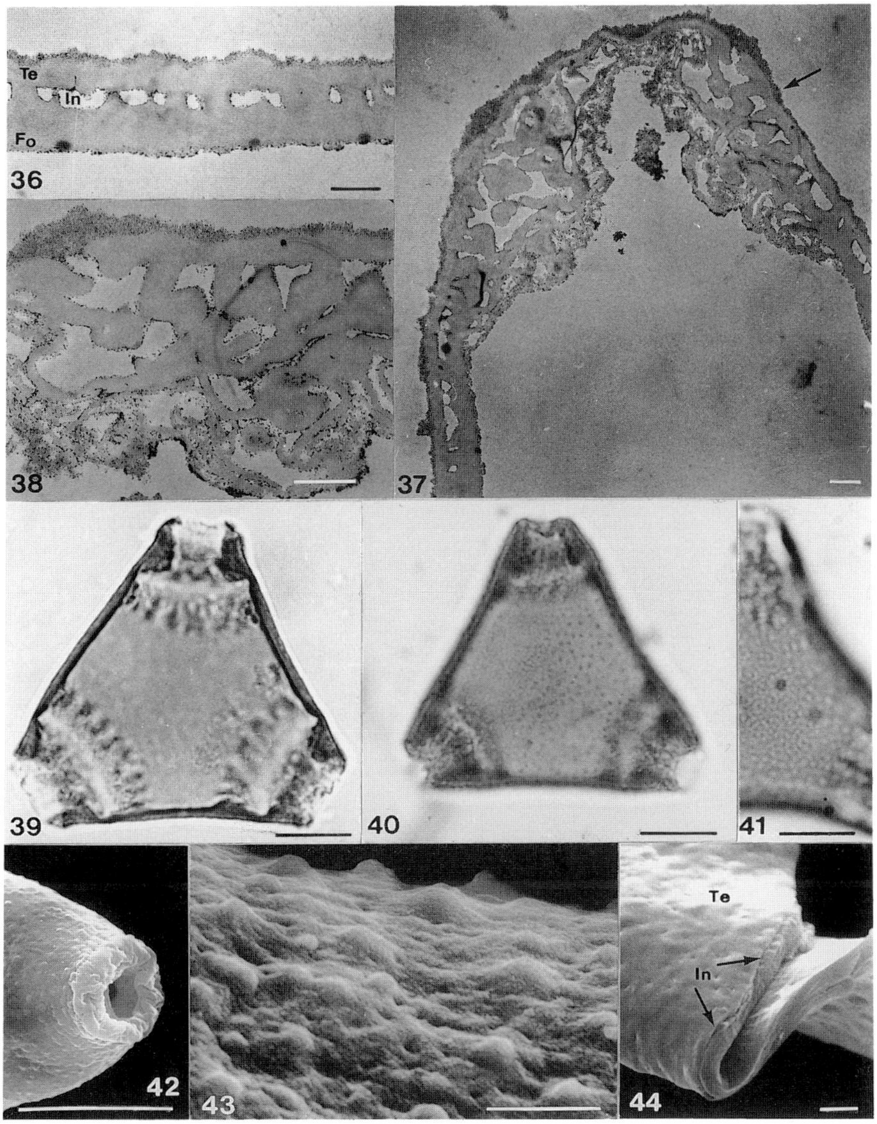

Figs. 36–44. Pollen of extant *Xylomelum*. **36–38.** *X. angustifolium* Kipp. (TEM). **36.** Equatorial wall section, tectum (Te), infratectum (In), footlayer (Fo). **37.** Off-centre apertural section, arrow denotes coating of ?incompletely acetolysed pollenkitt and tapetal material. **38.** Close-up of right side of aperture section. **39–41.** *X. occidentale* R. Br. different grains (LM). **39.** Sectional focus showing aperture form. **40.** Slightly immature grain, focus on larger sculptural elements (low protuberances). **41.** Median-high focus on finely scabrate surface forming negative reticulum. **42–44.** *X. occidentale* R. Br. different grains (SEM). **42.** Part grain, aperture. **43.** High resolution of scabrate-pseudoperforate tectum surface, at an angle to enhance relief. **44.** Fractured specimen showing tectum surface (Te), thin infratectum (In).

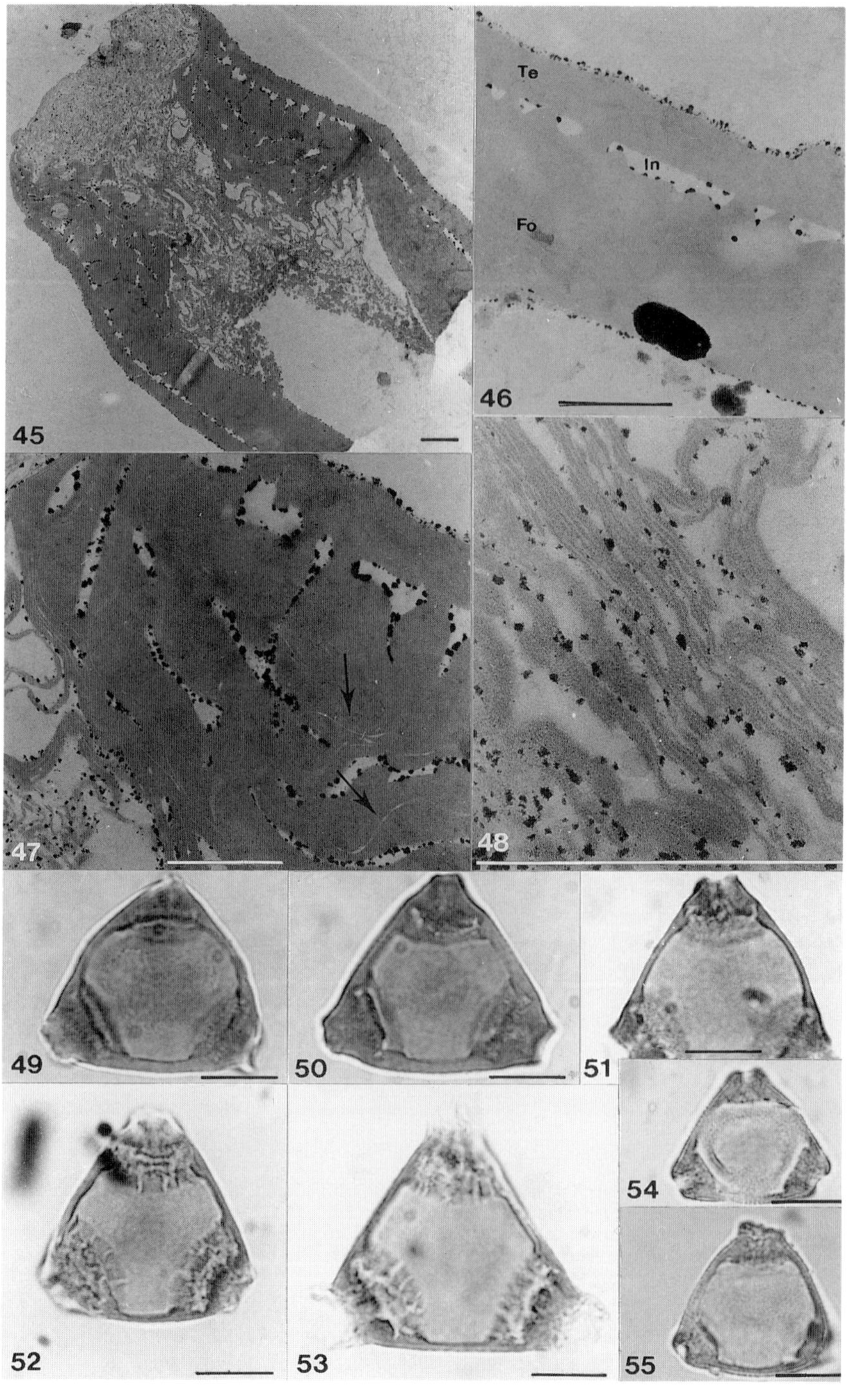
45
46
Te
In
Fo
47
48
49
50
51
52
53
54
55

is viewed overall. With *X. occidentale* as a pivotal species, broadly defined morphological clines can be seen to connect the two extreme species. Equatorial diameter decreases from *X. angustifolium* through *X. occidentale*, *X. cunninghamianum*, and *X. pyriforme*, to *X. scottianum*. Thickness of nexine about the aperture and size of the nexine indentation at the base of the apertural 'collar' decrease from *X. occidentale* to *X. angustifolium*, and from *X. occidentale* through *X. cunninghamianum*, and *X. pyriforme* to *X. scottianum*; tectal elements likewise increase. *Xylomelum occidentale* has morphological features in common with pollen of *X. angustifolium*, and with pollen of eastern Australian *Xylomelum*. It shares a similar amb, apertural structure, and sculptural features with *X. angustifolium*, and has a nondistinct infratectum congruent with those of the eastern Australian species.

Fossil pollen morphologically equivalent to eastern Australian *Xylomelum* pollen is unknown from the Australian Tertiary record, but several likely intermediary forms sharing pollen morphological characters of both the eastern and Western Australian *Xylomelum* pollen do occur. *Propylipollis latrobensis* Harris, 1966, and *Propylipollis* sp. A (Milne, 1988) have distinctive grevilleoid apertures similar to Western Australian *Xylomelum* pollen, and supratectal baculae in common with those from eastern Australia. The ultrastructure of the above fossil pollen has not been investigated, but that of *P. annularis* illustrates similar, although augmented, infratectal elements (broad-based pluricolumellae) to those found in eastern Australian *Xylomelum* pollen.

Discussion

Memon's (1984, 1985) studies of surface features of extant Proteaceae demonstrated that sculptural attributes are not genus, subtribe, or tribe specific, and not necessarily consistent within a genus. It follows that this must also apply to wall structures related to surface features (i.e., tectum, or in some instances, tectum and infratectum). On the other hand, Feuer's (1986, 1989, 1990a, 1990b) TEM studies of Proteaceae, and Dettmann and Jarzen's (1990) illustrated discussion on aperture types in pollen of Proteaceae and fossil proteaceous-like pollen, demonstrate that aperture type can be taxonomically significant within Proteaceae. Although parallelism does occur across subfamilies, closely allied species usually have morphologically similar, if not closely comparable, aperture structure. The importance placed on aperture form in establishing proteaceous affiliations and in erecting proteaceous form genera is illustrated by Cookson's (1950) affiliations of fossil and extant proteaceous pollen and her erection of the proteaceous form genera *Beaupreaidites* and *Banksieaidites*; Martin and Harris's (1974) form genera *Cranwellipollis* and *Propylipollis*, and emendation of *Proteacidites*; and Martin's (1973) emendation of *Beaupreaidites*.

Unless fossil and modern proteaceous species are identical, it is difficult to propose an affiliation without examining the pollen morphological trends within the taxonomic group of the modern species, and trends among fossil types closely comparable to the fossil species. In discussing the possible affiliations of *P. annularis* amongst Recent Proteaceae, Cookson (1950) indicated that apertural features similar to those of the fossil species were most markedly displayed by *Xylomelum* and *Lambertia* (*Lambertia* being another sclerophyllous grevilleoid member of the Proteaceae, but within tribe Macadamieae). She noted, however, that the apertures of *P. annularis* are more 'strongly defined' than those of *Lambertia* pollen, and that *P. annularis* is not sub-isopolar like

FIGS. **45–55**. *Xylomelum* pollen and *Propylipollis*. **45–48**. *Xylomelum occidentale* R. Br. (TEM). **45**. Apertural section. **46**. Wall section, tectum (Te), infratectum (In), footlayer (Fo). **47**. Detail of apertural structure, fine channels (arrowed) probably delineating further 'nexinal separation'. **48**. Lamellate apertural endexine. **49–53**. Different grains of slightly immature *X. occidentale* pollen showing, in sequence, aperture development (LM). **49–51**. Sectional foci. **52–53**. Median foci. **54–55**. *Propylipollis* sp. B (Milne, 1988): different grains, median foci (LM).

Lambertia pollen. LM examination of six of the eleven known species of *Lambertia* has confirmed Cookson's (1950) observations and deductions. The apertures of *Lambertia* pollen have a disrupted nexine along the collar similar to *P. annularis*, but the apertural nexine is not markedly thicker than the non-apertural nexine, and the annulate thinning and indentation in the nexine at the base of the 'collar' are barely discernible in most species. Pollen of some *Lambertia* species does have similar sculpture and non-apertural wall structure to both *Xylomelum* pollen and *P. annularis*; for example, pollen of *Lambertia ericifolia* R.Br. is baculate, like pollen of eastern Australian *Xylomelum* species; and pollen of *Lambertia fairallii* G.J. Keighery is tectate-perforate with similar sculpture to *P. annularis*. Because aperture form is more taxonomically indicative than surface features and related wall structure, *P. annularis* is thought to be more closely allied to *Xylomelum* than *Lambertia*.

Within *Xylomelum* pollen, the aperture of *X. occidentale* is most similar to that of *P. annularis*; it differs in the more gradual thinning of the nexine before the apertural 'collar', and in possessing an additional internally disrupted striate area below the apertural 'collar'. Cookson and Erdtman (in Erdtman, 1952) illustrated this disruption as occurring in the footlayer, and described it as a nexinous thickening. Technically the latter is correct, but TEM of *X. occidentale* shows no disruption of the footlayer below the apertural 'collar'; hence the disruption is in the endexine. As the fossil species lacks endexine, the aperture structures of the modern and fossil species are more similar than Cookson (1950) thought.

Xylomelum occidentale is tectate-pseudoperforate and scabrate, with an indistinct infratectum, whereas *P. annularis* is tectate-perforate and reticulo-rugulate with a relatively substantial infratectum. Although the gross wall structure of the two species is dissimilar, high-resolution SEM of the surface of *X. occidentale* and *X. angustifolium* pollen reveals a morphological similarity between their tectal features and those of *P. annularis*. The tectum surfaces of the two modern species are similar except that the features are more pronounced on *X. angustifolium* than on *X. occidentale*. Additionally, many of the scabrate elements of *X. angustifolium* can be seen to coalesce or are connected by low ridges, expressed in LM as a prominent negative reticulum, which may be complete or labyrinthine, not unlike the sculpture of *P. annularis* as viewed in SEM. Ultrastructurally, the relative size of the non-apertural wall layers of *P. annularis* is most similar to that of *X. angustifolium* pollen, and the infratectal elements resemble those of extant eastern Australian species.

Evidence from the fossil record suggesting affiliation of *P. annularis* with *Xylomelum* comes from *Propylipollis* species with the same distinctive aperture, including those so closely allied to *P. annularis* that they have been considered synonymous with it. The larger *Proteacidites granulatus* Cookson, 1953, *P.* sp. aff. *P. granulatus* (Hekel, 1972), and similar forms from the Eucla Basin (present study), are similar to *X. angustifolium* in overall size and aperture structure. The modern corollary of the close alliance of *P. granulatus* and *P. annularis* may be that specimens of *X. angustifolium* pollen with straight-sided ambs are comparable to similarly shaped specimens of the smaller *X. occidentale* pollen. Of described and/or illustrated *Propylipollis* species with apertures closely comparable with *P. annularis*, *Propylipollis latrobensis* (Harris) Martin & Harris, 1974, and *Propylipollis* sp. A (Milne, 1988) are baculate, as in the eastern Australian *Xylomelum* pollen; and *Propylipollis* sp. (aff. *Xylomelum*) illustrated by Martin (1982), and *Propylipollis* sp. cf. *P. annularis* Dudgeon, 1983, are granulate-microverrucate with an indistinctly stratified exine, and are perhaps even more closely allied to *X. occidentale* than *P. annularis* is. The Eucla Basin material contains several undescribed pollen types with *Xylomelum* morphological characters.

The similarity of the aforementioned fossil types to pollen of *Xylomelum* suggests not only that *P. annularis* is closely related to *Xylomelum*, but also that fossil forms closely resembling *P. annularis* share this affiliation. Further evidence is that the apparent aperture maturation of *X. occidentale*, and the resulting diversity of gross form within the species, is mirrored by *P. annularis*. Additionally, *P.* sp. B (Milne, 1988) is similar to immature pollen of *X. occidentale*. Ontogenetic studies of *Xylomelum* pollen and their parent plants may provide a more affirmative answer, but the available evidence points to *P. annularis* being closely related to *Xylomelum*, in particular to *X. occidentale*.

The two superficially different morphological types of *Xylomelum* pollen, reflecting their wide distribution both geographically and climatically, imply either that taxonomic revision of *Xylomelum* would be appropriate or, more likely, that *Xylomelum* spp. today are relicts of a once more diverse genus. The common occurrence and wide distribution of *P. annularis* and its fossil affiliates suggest that the parent plants belonged to a genus that may have held a position similar to that currently occupied by *Grevillea* R.Br., which contains at least 250 species inhabiting all but extreme deserts, marshes, and high mountains, Australia-wide. If living *Xylomelum* species are relicts of a previously more prominent genus, *P. annularis* and affiliated fossil forms may be representative of its pollen.

Acknowledgements

Tertiary samples were provided by Centenary International Mining Pty Ltd. A University of Queensland Centre for Microscopy and Microanalysis award defrayed the initial costs of scanning electron microscopy. I especially wish to thank Dr G. Playford, for general assistance and for critically reviewing the manuscript; Dr T. McCaul for assisting in the development of a TEM embedding technique, enabling time and cost effective sectioning; and Dr M.E. Dettmann for helpful discussion, the use of her diamond knife, and access to her Proteaceae slide collection containing *Xylomelum* pollen collected by D.M. Jarzen. Also gratefully acknowledged are Mrs W. Armstrong and Mr D. Gowanlock for TEM sectioning; Mr P. Bostock for assistance at the Queensland Herbarium; and Mr M. Butterworth, Mr A. Smith, Dr B. Cribb, Mr R. Webb, Miss D. Millhouse, Mr R. Gould, and Mr P. Gardam for technical assistance. Support from an Australian Postgraduate Research Award is gratefully acknowledged.

References

Alley, N.F. and Broadbridge, L.M. (1992). Middle Eocene palynofloras from the One Tree Hill area, St Vincent Basin, South Australia. *Alcheringa* **16**: 241–267.

APFL (1975). Morphologie pollinique: problèmes de terminologie, taxons-guides et pollen périporés. *Bull. Soc. Bot. France* **122**: 1–272.

Barss, M.S. and Williams, G.L. (1973). Palynology and nannofossil processing procedures. *Canada Geol Surv. Pap.* **73–26**: 1–25.

Cookson, I.C. (1950). Fossil pollen grains of Proteaceous type from Tertiary deposits in Australia. *Austral. J. Sci. Res. Ser. B, Biol. Sci.* **3**: 166–177.

Cookson, I.C. (1953). Difference in microspore composition of some samples from a bore at Comaum, South Australia. *Austral. J. Bot.* **1**: 462–473.

Cookson, I.C. and Erdtman, G. (1952). Proteaceae. In: G. Erdtman (editor). Pollen Morphology and Plant Taxonomy. Angiosperms, pp. 339–369, Almqvist and Wiksell, Stockholm.

Couper, R.A. (1953). Upper Mesozoic and Cainozoic spores and pollen grains from New Zealand. *New Zealand Geol. Surv. Paleontol. Bull.* **22**: 1–77.

Couper, R.A. (1960). New Zealand Mesozoic and Cainozoic plant microfossils. *New Zealand Geol. Surv. Paleontol. Bull.* **32**: 1–87.

Dettmann M.E. and Jarzen, D.M. (1990). The Antarctic/Australian rift valley: Late Cretaceous cradle of northeastern Australasian relicts? *Rev. Palaeobot. Palynol.* **65**: 131–144.

Dudgeon, M.J. (1983). Eocene pollen of probable proteaceous affinity from the Yaamba Basin, central Queensland. *Mem. Assoc. Australas. Palaeontol.* **1**: 339–362.
Erdtman, G. (1952). Pollen morphology and plant taxonomy. Angiosperms. 539 pp. Almqvist and Wiksell, Stockholm.
Erdtman, G. (1960). The acetolysis method. A revised description. *Svensk Bot. Tidskr.* **54**: 561–564.
Faegri, K. (1956). Recent trends in palynology. *Bot. Rev.* **22**: 639–664.
Faegri, K. and Iversen, J. (1989). Textbook of Pollen Analysis. 4th edition. 328 pp. John Wiley, Chichester.
Feuer, S. (1986). Pollen morphology and evolution in the Persoonioideae, Sphalmioideae and Carnarvonioideae (Proteaceae). *Pollen & Spores* **28**: 123–156.
Feuer, S. (1989). Pollen morphology of Embothrieae (Proteaceae): I. Buckinghamiinae, Stenocarpinae and Lomatiinae. *Grana* **28**: 225–242.
Feuer, S. (1990a). Pollen aperture evolution among the subfamilies Persoonioideae, Sphalmioideae, and Carnarvonioideae (Proteaceae). *Amer. J. Bot.* **77**: 783–794.
Feuer, S. (1990b). Pollen morphology of the Embothrieae (Proteaceae): II. Embothriinae (*Embothrium, Oreocallis, Telopea*). *Grana* **29**: 19–36.
Gray, J. (1965). Extraction techniques. In: B. Kummel and D. Raup (editors). Handbook of paleontological techniques, pp. 530–587, W.H. Freeman, San Francisco.
Harris, W.K. (1965). Basal Tertiary microfloras from the Princetown area, Victoria, Australia. *Palaeontographica, Abt. B, Paläophytol.* **115**: 75–106.
Harris, W.K. (1966). *Proteacidites latrobensis*. *Taxon* **15**: 332–333.
Harris, W.K. (1985). Middle to late Eocene depositional cycles and dinoflagellate zones in southern Australia. *Special Publ. S. Austral. Dep. Mines Energy* **5**: 133–144.
Haskell, T.R. and Wilson, G.J. (1975). Palynology of sites 280–284, DSDP Leg 29, off southeastern Australia and western New Zealand. In: J.P. Kennett., R.E. Houtz., *and others* (editors). Initial Reports Deep Sea Drilling Project 29, pp. 723–741, Washington, U.S. Government Printing Office.
Hekel, H. (1972). Pollen and spore assemblages from Queensland Tertiary sediments. *Publs Geol. Surv. Queensland* **355**, *Palaeontol. Pap.* **30**: 1–34.
Hos, D. (1975). Preliminary investigations of the palynology of the upper Eocene Werillup Formation, Western Australia. *J. Roy. Soc. W. Australia* **58**: 1–14.
Kemp, E.M. (1976). Early Tertiary pollen from Napperby, central Australia. *BMR J. Austral. Geol. Geophys* **1**: 109–114.
Jansonius, J. and Hills, L.V. (1976 et seq.). Genera file of fossil spores and pollen. University of Calgary Geology Department, Calgary.
Larson, D.A., Skvarla, J.J. and Lewis, C.W.Jr. (1962). An electron microscope study of exine stratification and fine structure. *Pollen & Spores* **4**: 233–246.
Macphail, M.K. and Truswell, E.M. (1989). Palynostratigraphy of the central west Murray Basin. *BMR J. Austral. Geol. Geophys.* **11**: 301–331.
Martin, A.R.H. (1973). Reappraisal of some palynomorphs of supposed proteaceous affinity. I. The genus *Beaupreaidites* Cookson ex Couper and the species *Proteacidites hakeoides* Couper. *Geol. Soc. Austral. Special Publ.* **4**: 73–78.
Martin, A.R.H. (1982). Proteaceae and the early differentiation of the central Australian flora. In: W.R. Barker and P.J. Greenslade (editors). Evolution of the Flora and Fauna of Arid Australia, pp. 77–83, Peacock Publ., Adelaide.
Martin, A.R.H and Harris, W.K. (1974). Reappraisal of some palynomorphs of supposed proteaceous affinity. *Grana* **14**: 108–113.
Memon, H.R. (1984). Classification of pollen surface patterns of modern Proteaceae using scanning electron microsope. *Pakistan J. Bot.* **16**: 205–229.
Memon, H.R. (1985). Implication of pollen surface pattern for species and generic level classification in Proteaceae. *Pakistan J. Bot.* **17**: 87–106.
Milne, L.A. (1988). Palynology of a late Eocene lignitic sequence from the western margin of the Eucla Basin, Western Australia. *Mem. Assoc. Australas. Palaeontol.* **5**: 285–310.

Nott, J.F. and Owen, J.A.K. (1992). An Oligocene palynoflora from the Shoalhaven catchment N.S.W. and the Tertiary evolution of flora and climate in the southeast Australian highlands. *Palaeogeogr. Palaeoclimatol. Palaeoecol.* **95**: 135–151.
Phipps, D. and Playford, G. (1984). Laboratory techniques for extraction of palynomorphs from sediments. *Queensland Dep. Geol. Pap.* **11**: 1–23.
Pocknall, D.T. and Mildenhall, D.C. (1984). Late Oligocene-early Miocene spores and pollen from Southland, New Zealand. *New Zealand Geol. Surv. Paleontol. Bull.* **51**: 1–66.
Skvarla, J.J. (1966). Techniques of pollen and spore electron microscopy. Part I. Staining, dehydration and embedding. *Oklahoma Geol. Notes* **26**: 179–187.
Stover, L.E. and Evans, P.R. (1973). Upper Cretaceous-Eocene spore-pollen zonation, offshore Gippsland Basin, Australia. *Geol. Soc. Austral. Special Publ.* **4**: 55–72.
Stover, L.E. and Partridge, A.D. (1973). Tertiary and Late Cretaceous spores and pollen from the Gippsland Basin, southeastern Australia. *Proc. Roy. Soc. Victoria* **85**: 237–286.
Stover, L.E. and Partridge, A.D. (1982). Eocene spore-pollen from the Werillup Formation, Western Australia. *Palynology* **6**: 69–95.
Truswell, E.M. and Owen, J.A. (1988). Eocene pollen from Bungonia, New South Wales. *Mem. Assoc. Australas. Palaeontol.* **5**: 259–284.
Wrigley, J.W. and Fagg, M. (1989). Banksias, Waratahs and Grevilleas, and all plants in the Australian Proteaceae family. 584 pp. Harper Collins, London.
Zavada, M.S. (1984). Angiosperm origins and evolution based on dispersed fossil pollen ultrastructure. *Ann. Missouri Bot. Gard.* **71**: 444–463.

APPENDIX 1: Register of *Xylomelum* species. BRI = Queensland Herbarium

Species, authority	**Collector**	**Provenance**	**BRI AQ**
X. angustifolium Kipp.	G. Playford	Australia (WA)	living spec.
	A. Morrison	Australia (WA)	106300
	D. Jarzen	Australia (WA)	living spec.
X. cunninghamianum D. Foreman	n/a	Australia (QLD)	317218
X. occidentale R. Br.	G. Playford	Australia (WA)	living spec.
	A. Morrison	Australia (WA)	106296
	D. Jarzen	Australia (WA)	living spec.
X. pyriforme (Gaertn.) Knight	F.E. Davies	Australia (QLD)	517042
X. scottianum (F. Muell.) F. Muell.	R.L. Specht	Australia (QLD)	102104

APPENDIX 2: Register of illustrated specimens. Coordinates derived from standard 'England Finder' slide; no coordinates are given for extant specimens on strew stubs. Prefix 'Y' denotes slides in the microfossil collection, Department of Earth Sciences, The University of Queensland; uncatalogued residues are also held in this collection. 'UWA' denotes slides from the collection of the Geology Department, University of Western Australia. The one CMN slide is in the collection of Dr M.E. Dettmann, Botany Department, The University of Queensland.

Figure	Prep/slide	Coordinates	Cat. No.
Propylipollis annularis			
2,3	LA31/4	K44/3	Y6846
4	LA35/4	D33/4	Y6847
5	LA35/4	N29/3	Y6848
6	LA34/4	D55/4	Y6849
7,8	LA28/S19	N32/3	Y6850
9	LA28/S19	N32/3	Y6851
10, 11	LA30/S23	J41/4	Y6852
12, 13	LA27/S18	L38/4	Y6853
14–17	LA31	—	—
Propylipollis sp. B			
54	Z6/31	O29/4	UWA92019a
55	Z6/39	U29/0	UWA92024a
Propylipollis spp			
18	LA34/4	J54/3	Y6854
19	LA35/4	F40/2	Y6855
20, 21	LA27/S18	L38/2	Y6856
22,23	LA30/4	M20/0	Y6857
Xylomelum angustifolium			
31	LABRI72/1	R32/0	Y6858
32	LABRI72/1	D39/0	Y6859
33	LABRI72/S38	—	Y6860
34	LABRI72/S38	—	Y6861
35	B1893/S22	M42/3	Y6862
36–38	B1893	—	—
Xylomelum cunninghamianum			
25	LABRI8/1	N46/3	Y6863
29, 30	LABRI8/S22	N41/2	Y6864

Appendix 2 continued

Figure	Prep/slide	Coordinates	Cat. No.
Xylomelum occidentale			
39	B1894/4	F27/4	Y6865
40	LABRI76/1	Q37/4	Y6866
41	CMN14351	M35/1	—
42	B1894/S20	L38/4	Y6867
43	LABRI76/S39	—	Y6868
44	B1894/S20	N43/3	Y6869
45–48	B1894	—	—
49	B1894/4	N28/1	Y6870
50	B1894/4	H37/0	Y6871
51	B1894/4	L37/0	Y6872
52	B1894/4	J34/4	Y6873
53	B1894/4	H34/0	Y6874
Xylomelum pyriforme			
26	LABRI4/1	H46/2	Y6875
28	LABRI4/S22	N40/0	Y6876
Xylomelum scottianum			
27	LABRI6/1	N32/1	Y6877

TAXONOMIC INDEX

Bold type indicates a main text reference; italics refer to figures.